Water, Rhetoric, and Social Justice

Environmental Communication and Nature: Conflict and Ecoculture in the Anthropocene

Series Editor: C. Vail Fletcher, University of Portland

This interdisciplinary book series seeks original proposals that examine environmental communication scholarship. In the Anthropocene era, the period during which human activity has become the dominant influence on climate and the environment, the need for highlighting and recentering nature in our worldviews and policies is urgent, as collapsing ecosystems across the globe struggle to survive. Topics might include climate change, land use conflict, water rights, natural disasters, non-human animals, the culture of nature, ecotourism, wildlife management, human/nature relationships, food studies, sustainability, eco-pedagogy, mediated nature, eco-terrorism, environmental education, ecofeminism, international development, and environmental conflict. Ultimately, scholarship that addresses the general overarching question "how do individuals and societies make sense of and act against/within/out of nature?" is welcomed. This series is open to contributions from authors in environmental communication, environmental studies, media studies, rhetoric, political science, critical geography, critical/cultural studies, and other related fields. We also seek diverse and creative epistemological and methodological framings that might include ethnography, content analysis, narrative and/or rhetorical analysis, participant observation, and community-based participatory research, among others. Successful proposals will be accessible to a multidisciplinary audience.

Recent titles in the series:

Tracing China's Environmental Protest over Wild Public Networks: Becoming Activists, by Elizabeth Brunner

Critical Environmental Communication: How Does Critique Respond to the Urgency of Environmental Crisis? by Murdoch Stephens

Natural Disasters and Risk Communication: Implications of the Cascadia Subduction Zone Megaquake, edited by C. Vail Fletcher and Jennette Lovejoy

Water, Rhetoric, and Social Justice

A Critical Confluence

Edited by Casey R. Schmitt,
Christopher S. Thomas,
and Theresa R. Castor

LEXINGTON BOOKS
Lanham • Boulder • New York • London

Published by Lexington Books
An imprint of The Rowman & Littlefield Publishing Group, Inc.
4501 Forbes Boulevard, Suite 200, Lanham, Maryland 20706
www.rowman.com

6 Tinworth Street, London SE11 5AL, United Kingdom

British Library Cataloguing in Publication Information Available

Library of Congress Cataloging-in-Publication Data

Library of Congress Control Number: 2019954113
ISBN 978-1-7936-0523-8 (pbk. : alk. paper)
ISBN 978-1-7936-0521-4 (cloth : alk. paper)
ISBN 978-1-7936-0522-1 (Electronic)

To our families, for your patience, support, and love.

Contents

Introduction

Stirring the Waters: Justice, Injustice, and the Springs of Rhetorical Response

Casey R. Schmitt, Christopher S. Thomas, and Theresa R. Castor

Without water, not a living thing on this planet could survive. Roughly 60 percent of the human body is composed of water, and more than 70 percent of the planet's surface is covered in it. It sustains forests, fuels agriculture, carries waste and sewage from inhabited spaces, and powers cities, schools, and hospitals with its currents and flow. Water is essential, and, in theory, there is more than enough water on the planet to sustain even present-day populations and demands.

Yet, in recent decades, a series of spectacular, tragic, and highly visible global events have illuminated a cruel paradox for water and life in the Anthropocene. Though plentiful and abundant, water is variously denied and inaccessible to billions. According to the Centers for Disease Control and Prevention (CDC), the World Health Organization, and UNICEF, in the twenty-first century, a time of unrivaled comfort and progress for much of the world, 780 million people do not have access to clean, safe drinking water, and an estimated 2.5 billion people—over 35 percent of the global human population—lack access to improved sanitation. Amid increased demands for water due to population growth and industrial development, climate change is altering sea levels, surface water flow, and precipitation patterns. Increased industrial demand has paired with decreased federal funding and regulation for clean water.

In the United States, portions of the landmark Clean Water Act are being retracted and rewritten as numerous water-related natural and unnatural disasters play out: children lack clean drinking water in Flint, Michigan; Lakota communities that vehemently protested installation of the Dakota Access and Keystone Pipelines now experience oil spills from those very same pipes; and residents of the Carolinas and southeast Texas in the aftermath of recent hurricanes grapple with waters contaminated by sewage and livestock waste. In California's San Joaquin Valley, industrial agricultural runoff has tainted drinking water with nitrates and carcinogens, while in Navajo communities mining practices have laced the water with arsenic and uranium (Food and Water Watch 2017, 3). In Qidong, China, people marched in the streets after waterways were polluted with industrial discharge. In Brazil, standing pools of water contributed to the outbreak of the mutated Zika virus, which led to a spike in prenatal deformities. CDC estimates suggest more than eight hundred thousand children die each year from diarrheal diseases, directly linked to lack of access to clean water and sanitation. In Europe, North America, and Australia, extreme heat and drought conditions have produced the largest and most destructive wildfires on record. In India, temperatures have topped 115 degrees Fahrenheit in the west, while in the east cities like Chennai, home to more than seven million people, have faced unprecedented municipal water shortages. Glacial retreat in the Tibetan Plateau threatens the source feeding the Yangtze, Yellow, Mekong, Ganges, and Indus Rivers, the primary source of fresh water for nearly one-third of the world's population. In the Arabian Peninsula, literal wars are waged over access to water, and in Cape Town, South Africa, bureaucratic stalemates produced the looming specter of "Day Zero," on which the city of 4.3 million's public water system would be completely exhausted.

And all this is happening while access to free-flowing, abundant, clean, and safe water is broadly considered—from the United Nations to religious authorities—to be a basic human, legal, and God-given right.

Water is central to human health, agriculture, industrial production, hygiene, transportation, and recreation, but the competition for access, use, and control of water is intense and reflects both competing values and divergent perspectives among water stakeholders.

This tension, this paradox, between the universal right and competition over access makes water a central case in the emergent exigences of the Anthropocene era. The series of recent global examples demonstrates how biophysical changes, threats, and crises emerge directly from human action—whether industrial, cultural, or political—in ways and locations where they would not have emerged before.

This collection examines the human response to a human-made global environmental exigence, with particular attention to the tension between access and competition. The twenty-first century has demonstrated that as clean water access ebbs across the planet, it is often the marginalized classes, the minority groups, and the poor or otherwise disenfranchised who bear the heaviest of burdens. The collection applies critical communication methods and perspectives to interrogate the pressing perennial yet mind-boggling dilemma faced in environmental studies and policy: that clean water, the very stuff of life, which flows freely from the tap in affluent areas, is also denied to huge populations, materially and fluidly exemplifying the currents of justice, liberty, and equity.

Our project takes its thematic inspiration from the oft-quoted appeal to "let justice roll down like waters and righteousness like a mighty stream," originating in the Book of Amos and deployed frequently as a rallying cry for social justice and equity. The collection applies themes of social justice and rhetorical demonstration to the literal question of water itself. *Water, rhetoric, and justice, while bound metaphorically in public imagination are, indeed, bound literally in the Anthropocene.*

The chapters in this collection examine how individuals and communities have responded to present-day water crises as matters of social justice, through oratory, mass demonstration, deliberation, testimony, and other rhetorical appeals. Social justice, in these chapters, refers to human-established and -maintained systems of equity, reciprocity, circularity, and balance. *Water justice*—the socially bound and maintained systems ensuring access to water for drinking, cleaning, bathing, and sanitation—is part of social justice, and while it directly governs physical states and changes in oceans, rivers, lakes, and groundwater, it is the product of human regulation, activism, and education that resists pollution and overconsumption. *Water injustice* is a breach in the balance, resulting in threats, costs, and repercussions thrust upon the most vulnerable—including the poor, the young, and those less able or unable to advocate for themselves (extending, at times, to animals, plants, and physical features affected by human actions and choices).

It is water injustice, just as other forms of social *injustice*, that has inspired springs of rhetorical response. In systems of injustice and imbalance, the physical constraints are often insurmountable through individual actions, but through persistence, coalition, and collaborative symbolic action, these physical constraints may indeed be questioned, challenged, and even overturned. Cases of water activism and advocacy demonstrate that water justice, as a social justice issue, is also inherently a site of rhetorical strategy, engagement, and innovation.

This collection fills a gap in the existing literature of water justice and environmental communication by focusing attention on *discourses, activism, and other rhetorics campaigning for water justice*. While current circumstances have drawn attention to the themes of water, rhetoric, and social justice separately at conferences and in single scholarly articles, this collection compiles the varied conversations to identify and analyze a larger trend. Its chapters focus on activism, advocacy, and other public discourse cases, in

North America and around the globe, to argue that water, rhetoric, and social justice are inherently interrelated in the Anthropocene.

The chapters here analyze rhetorical responses to water injustice to provide understanding for what does and does not work in water justice rhetoric as activists, policy-makers, and educators prepare for future looming water challenges, including increased needs for water infrastructure due to population growth, climate change, and shifting technological demand.

As part of the Lexington Environmental Communication and Nature: Conflict and Ecoculture in the Anthropocene series, this book's themes engage the global interaction of human rhetorics/constructs and biophysical realities/necessities. The question of water justice illuminates anthropocentric actions and impacts, while directly engaging the series key concepts of water rights, climate change, natural disasters, and nonhuman animals, and the series's key question: *How do individuals and societies make sense of and act against/within/out of nature?*

ENVIRONMENTAL JUSTICE

Water justice is an extension and subset of ecological and environmental justice, terms with multiple and sometimes competing definitions. Low and Gleeson (1998) define *ecological justice*, at its core, as that concerning the "fair distribution of environments among all the inhabitants of the planet." Walker (2012) applies a more anthropocentric lens, arguing *environmental justice* is "about the intertwining of environment and social difference—how for some people and some social groups the environment is an intrinsic part of living a 'good life' of prosperity, health and well-being, while for others the environment is a source of threat and risk, and access to resources such as energy, water and greenspace is limited or curtailed" (1).

Walker focuses on framing and claim-making. He distinguishes *environmental justice from environmental inequality* by distinguishing between what *is* (inequality, "a *descriptive* term, describing a condition of difference or unevenness of something") and what *ought* to be (justice) (12–13). Distinguishing *environmental justice* from Low and Gleeson's *ecological justice*, Walker emphasizes "distributive justice; blame and responsibility focused on industry, corporate actors, and institutionalized practices; interconnected scales of analysis; and parallel grassroots and government versions" (20–23). Building on Schlosberg (2004; 2007), he includes in his definition justice with respect to *distribution* (i.e., "who gets what in the environment") (42), *procedure* (i.e., how things are done) (47), and social *recognition* (i.e., simply being acknowledged) (50). Wrapped up in explorations and audits of environmental justice are "the distribution of vulnerability to harm amongst different demographic and social groups, and how this varies over and beyond the spatial patterning of who lives in hazardous places" and "the ways in which people across different intersecting social and demographic categories can be subject to multiple and interacting forms of inequality" (129).

Bullard (2005) describes environmental justice not as a single concept but as a "framework" that "attempts to uncover the underlying assumptions that may contribute to and produce unequal protection," concerned with "the ethical and political questions of 'who gets what, why, and how much'" (25). It is grounded in the principle that "all individuals have a right to be protected from environmental degradation" (25) and it "shifts the burden of proof to polluters and dischargers who do harm, who discriminate, or who do not give equal protection to racial and ethnic minorities" (28). It is defined by the questions it raises and addresses: "What groups are most affected by a specific environmental problem? Why are they affected? Who caused the problem? What can be done to remedy it? How can communities be justly compensated and reparations be paid to individuals harmed by industry and government actions? How could the problem have been prevented?" (29). He writes:

> At its core, environmental justice is about incorporating environmental issues into the broader intellectual and institutional framework of human rights and democratic accountability. The term necessarily encompasses the widest possible definition of what is considered "environmental" and is unmistakably anthropocentric in its orientation—placing people, rather than flora and fauna, at the center of a complex web of social, economic, political, and environmental relationships. Most important, it concerns itself primarily with the environmental *in*justices of these relationships and the ways and means of rectifying these wrongs and avoiding them in the future. (258)

Within this broad framework, there is vast variance and diversity, with focuses on different regions and groups, different theoretical underpinnings, and concentrations on different modes of justice, but Bullard argues that the "lack of coherence" in environmental justice discourses is not necessarily problematic, as it ensures that more perspectives and groups are included in the broader campaign (259). Agyeman and Evans (2004), likewise, argue that environmental justice, as a term and concept, has provided a "vocabulary of political opportunity" and "an important way of bringing attention to previously neglected or overlooked patterns of inequality which can matter deeply to people's health, well-being and quality of life" (Walker 2012, 1).

WATER JUSTICE

Water justice applies environmental justice principles to water, treating water as a social and political issue (Bakker 2012) understood and managed differently by different cultures in different locations but equally necessary to life and welfare across populations (Barnes and Alatout 2012). In the Anthropocene, water justice grapples with both the extreme shifts in clean water availability brought on by climate change (droughts, flooding, massive storms, etc.) and increasing neoliberal and industrial privatization of water reserves and resources, whereby poor communities are set in competition with wealthy multinational corporations over the finite resource (Sultana 2018, 485). Simply put, with a finite amount of fresh water available, some get more and others get less, and geopolitical conflict ensues. Water-rich nations block water access to water-poor nations. Those with water constraints avoid bathing and washing, and lack of sanitation impacts health. Citing Garrett Hardin (1968), who warned of the human tendency to exhaust resources when each person or group seeks its own share without attending to the future needs of others, author Barbara Kingsolver (2010) calls water the "ultimate commons" (49): once seemingly abundant and now, legitimately, inconceivably, at risk of depletion.

While 70 percent of the planet's surface is indeed covered in water, only about 2.5 percent of that cover is fresh water (Bullard 2005, 282). Moreover, much of that fresh water is polluted, and a "lack of clear water and adequate water-sanitation conditions contributes each year to approximately 2 billion diarrhea infections and 4 million deaths" (282). Biological inequity extends into broader threats to health and dignity. In developing countries and the Global South, absence of water infrastructure means people (mostly women) must spend much of their day travelling to fetch water from far away, leaving less time for work, leisure, education, or infrastructural development (Bakker 2010; Sultana 2018). "Where clean water is scarcest, fetching it is almost always women's work" (Rosenberg 2010, 102), and lack of water equity perpetuates gender inequality, tantamount, in some cases, to emotional and physical water slavery (Rosenberg 2010; Sultana 2018). In developed countries of the Global North, aging infrastructure and mounting water bills restrict access to clean, healthy water, especially impacting poor communities and communities of color, shown to pay higher average household water bills (Food and Water Watch 2017). Those who cannot pay water bills are threatened with water shutoffs, and these shutoffs lead to "collateral consequences" (2), including loss of child custody and foreclosures, exacerbating existing social inequalities, and thus making water injustice a key focus of efforts against environmental racism (Bullard 2005, 32). Worldwide, nations, institutions, and corporations with financial and political influence have more ready access to water than local and indigenous communities that have relied on meltwater, groundwater, and waterways for their way of life for generations.

In 2010, the United Nations passed the Resolution on the Human Right to Water and Sanitation, which states: "The right to safe and clean drinking water and sanitation is a human right that is essential for the full enjoyment of life and all human rights" (UN, 2010), but little has changed in the years since that moment. The persistent and rapidly increasing problem has prompted many to campaign, with increasing urgency, for humans to consider the political, social, and cultural aspects and ramifications of water (Orlove and Caton 2010; Bijker 2012; Sultana 2013).

Farhana Sultana (2018) has penned, perhaps, the definitive essay on water justice to date. "Water," Sultana writes, "is essentially about power—the power to decide, control, allocate, manage—thereby affecting people's lives. This is intersectionally experienced by gender, class, race, and other axes of social difference, therefore affecting different groups of people in varied ways" (485). "Water justice," she writes, "requires appreciation that there are no easy, simple or singular solutions to the global water crisis, and that water problems cannot be resolved through technical solutions alone but require broader recognition that they are inherently ecological, political and social issues simultaneously" (487).

Efforts toward promoting water justice include antiprivatization movements, democratizing water governance, addressing equity discrepancies, and simply spreading information. Sultana reminds readers, water justice "is not just about physical quantities of water but also about quality, accessibility, affordability and reliability" (484), and it involves concerted attention to climate change disparities and crises, neoliberal attempts to privatize and control water sources, and the links between access to safe water and access to education, health, and employment across gender and race divides.

THE ROLE OF RHETORIC IN ACHIEVING WATER JUSTICE

This collection is about that work: the campaigning for water justice through education, deliberation, advocacy, and agitation. It is about the creative rhetorical means of "stirring the waters" and pressing the conversation. Existing book collections on water justice focus largely on definitions, policy, politics, and scientific or technical perspectives (Shiva 2002; Soll 2013). Boelens, Perrault, and Vos's 2018 collection, *Water Justice*, engages similar global themes of sovereignty, access, damming, and sanitation, but its unifying focus is on governance and policy. When the chapters do engage communication studies, the focus is more upon organizational communication, public relations, and risk communication than on social movements, activism, or rhetoric. At this early stage in scholarly engagement with water justice (Neal, Lukasiewicz, and Syme 2014; Zwarteveen and Boelens 2014), the conversation is still being framed and established, though the interest in and urgency of water justice as a subject of public concentration is readily documented in a series of recent books targeting popular readerships alongside scholarly audiences (Barlow 2007; Salzman 2012; Schiffler 2015; Schwartz 2016; Simon 1998). There are recent books on indigenous water rights advocacy (Hoover 2017), and, especially, on the water crisis in Flint (Clark 2018; Hanna-Attisha 2018; Pauli 2019), evidencing the wealth of material and interest in recent crises; yet, here, too, the central analytical focus is on governance, policy, and economics. Currently, there are no other books specifically addressing the discourses, activism, and rhetorics campaigning for water justice in the communication and rhetoric fields, despite the rise in highly publicized and highly visible water crisis cases. Individual essays, chapters, and conference papers on the topics of rhetoric and water justice do exist (see, e.g., Krajewski, Schumacher, and Dalrymple 2017, and Hopke 2012), but no single collection gathers case examples in one place for an extended and comparative discussion of water justice rhetorics for use in courses, advocacy, or general consideration.

This collection highlights discourse and water justice movements in non-official spheres, from activists, artists, and the grassroots. It's about the droplets and trickles of rhetorical invention that convene and combine as torrents of social force and change the world over. It applies a *rhetoric* perspective, *analyzing how symbolic actions in a civic sphere create, perpetuate, challenge, and occasionally upend social systems of meaning, identity, and power*. In extending the technical, economic, moral, and political conversations on water justice, it applies *special focus on the novel rhetorical concepts and responses not necessarily unique to but especially enacted in water justice situations.*

We follow definitions of environmental communication established by Killingsworth and Palmer (1992), Cantrill and Oravec (1996), Benton and Short (1999), and Pezzullo (2007) to argue that human discourses both shape and are shaped by circumstances in the biophysical environment, and that symbolic, creative, disruptive, and resistive actions play a significant role in defining environmental exigences and constraints as much as industrial action, legal regulation, or governmental policy. Our collection uses this foundation to extend the already thriving discussion of water justice on the technical, economic, legal, and political stages.

We seek to illuminate advocacy and the perspectives of indigenous, impoverished, or otherwise marginalized communities, with a specific focus on water.

This collection includes chapters that engage especially prominent recent cases alongside other and less well-known cases to promote a broader conversation, in search of patterns of water, rhetoric, and social justice that can be applied to discussion of emergent cases in coming years. It includes an intentionally eclectic variety of voices and perspectives, highlighting writers from all points in their academic careers. The collection, therefore, offers a unique perspective on water justice while consolidating and synthesizing varied case examples, in the interest of inspiring focused discussion going forward.

Rhetoric scholar Nancy Fraser (2008) argues that justice of any kind is rooted in a "parity of participation" in social life and that "overcoming injustice means dismantling institutionalized obstacles that prevent some people from participating on a par with others, as full partners in social interaction" (16). Roosvall and Tegelberg (2018), building from Fraser, suggest that there are three primary remedies to such obstacles—redistribution of means, recognition of identity, and representation (37)—and that all of these can be championed, forwarded, and achieved through public discourse and communication.

Since, as Agyeman and Evans (2004) propose, "environmental justice" has provided a "vocabulary of political opportunity," attention to "water justice" has likewise opened channels for new movement and engagement. In the current century, women, indigenous activists, and young people have found and played especially prominent roles in global campaigns for water justice (Bullard 2005, 24). Novel rhetorical strategies, including artwork, public demonstration, theater, and digital activism, outside historically sanctioned spheres of decision-making, have used affect and experience to reconfigure public perceptions (Hall et al. 2009). Environmental justice and mainstream environmental movements have joined forces and flowed together (Endres et al. 2009). Ecoactivists and those campaigning for water justice have become savvy of media representations and adapted appeals to wrest control of the narratives and definitions surrounding cases of water injustice (Newlands 2018). Writers like Cox (2009), Fishman (2012), and Sultana (2018) have all rallied their readers in the current moment, acknowledging the challenges ahead in the struggle for water justice but also emphasizing that if concerted efforts are taken now, substantial change can indeed occur. Sultana (2018) stresses that resistance and justice have "raised awareness of the broader issues linked to water, highlighting both local crises but also global connections, and argued for a right to water and sanitation in the face of dispossession, exclusion and inequity," especially crediting the "cumulative effort of activists, scholars, practitioners, community members and others" (487). She lauds how "creative ways that challenge average citizens to learn more, to get involved, and to cultivate opportunities for friends and family to learn and be involved are increasing through the efforts of various non-profits, community groups, and activists" (490). And she advocates for education, investigation, advocacy, and democracy with a water justice rallying cry: "Advancing an ethic of care towards water is the first step for revolutionary changes in our relationship to water and to promote water democracy. This is why understanding and appreciating water justice is important for transformations that are equitable and inclusive. Fighting for water justice means creating a more equitable and just world" (490–91).

As editors of this collection, we could not agree more, and we have compiled the following chapters to highlight and celebrate the varied forms of water justice rhetorics advancing this ethic in the current day.

PREVIEW OF CHAPTERS AND CONTENT

The chapters of this collection are organized into four thematic sections, based on prominent themes of water, rhetoric, and social justice facing the world at the current time. These section themes are:

1. Pipelines, Dams, and Water Sovereignty
2. Access and Water Wars
3. Crisis and Disaster
4. Posthuman and More-Than-Human Justice

Each of these areas correlates with a *significant threat to* and *corresponding goal for* water justice:

- Constriction/Obstruction/Privatization vs. Free-Flowing Waterways
- Lack of Access/Water Shortage vs. Abundant Access
- Contamination and Disaster vs. Clean Water and Public Safety
- Threats to More-Than-Human Life vs. Justice for Plants, Animals, and Water Itself

Each section is preceded by a brief introduction to its unifying theme, with summaries of the chapters it contains. Each individual chapter situates its analysis through a specific case study and rhetorical concept or term to guide future writing and thought on environmental communication and water justice rhetorics.

Section I, "Pipelines, Dams, and Water Sovereignty," examines local and indigenous rhetorical resistance to installation of industrial dams and pipelines that constrict and obstruct free-flowing waterways. Its chapters explore rhetorical constructions of local moral frameworks, false dichotomy, spectacle demonstrations, and alternatives to settler colonial logics. Its cases include rhetorics that stir the waters in the Standing Rock Sioux Reservation, Chile, the Pacific Northwest, and New England.

Section II, "Access and Water Wars," examines discourses and actions that create and respond to human-made water shortage and access crises. Its chapters explore rhetorical constructions of "public" and "community," greenwashing as a political justification, vernacular attempts to "preach beyond the choir," and hydro-anthropomorphism. Its cases include rhetorics that stir the waters in the Great Lakes; Kansas; Detroit, Michigan; and the Global South.

Section III, "Crisis and Disaster," examines community and grassroots responses to public health and safety crises, with special attention to how poor and otherwise impoverished communities respond to water injustice. Its chapters explore rhetorical constructions of dwelling in a crisis location, mobilizing marginalized identities as authority, creative online tactics, and boundaries. Its cases include rhetorics that stir the waters in Louisiana; Flint, Michigan; China; and the Salish Sea.

Section IV, "Posthuman and More-Than-Human Justice," extends the conversation beyond the human, to examine advocacy for animals, plants, and material environments affected by human actions and discourse. Its chapters explore rhetorical constructions of satire, grief, post-humanism, and ecospheric perspectives. Its cases include rhetorics that stir the waters in the Animas River, the eastern Australian coast, India, and the central Pacific Ocean.

The collection ends with a reflection by Dr. Aimee Kendall Roundtree, synthesizing the chapters' core themes and proposing new directions for future research and discussion on the rhetorics of water justice.

There are, of course, highly relevant examples of water, rhetoric, and social justice not included in the chapters due to space constraints and the sake of cohesion. These examples include, for instance, current mass media depictions of water justice, frames and claims surrounding the commercial fracking industry, unnatural disasters like superstorm hurricanes and droughts that exacerbate social inequity through uneven distributions of safe and healthy water, and environmental crises related to water but not explicitly "water issues," like wildfires and methane release in the Arctic, among others. We aim to acknowledge these other areas for study in the coming chapters, if only in passing, but also to encourage future work, writing, advocacy, and attention on them through the frames presented in this work.

Water is fluid, ever-moving and ever-shifting through different states, and central to sustainable life on this planet. The same attributes apply to rhetoric, and to justice. In moments of water injustice, the three elements come together and stir the water, turn the tide, break stagnation, and roll down like a mighty stream.

We thank the contributors to this collection and the educators, activists, scientists, policy-makers, communities, individuals, and other rhetors that make their chapters possible. And we thank you, the reader, for engaging water justice and keeping your attention ever downstream.

REFERENCES

Agyeman, Julian, and Bob Evans. 2004. "'Just Sustainability': The Emerging Discourse of Environmental Justice in Britain?" *The Geographical Journal* 170, no. 2: 155–64.

Bakker, Karen. 2012. "Water: Political, Biopolitical, Material." *Social Studies of Science* 42, no. 4: 616–23.

Barlow, Maude. 2007. *Blue Covenant: The Global Water Crisis and the Coming Battle for the Right to Water*. New York: The New Press.

Barnes, Jessica, and Samer Alatout. 2012. "Water Worlds: Introduction to the Special Issue of *Social Studies of Science*." *Social Studies of Science* 42, no. 4: 483–88.

Benton, Lisa M., and John R. Short. 1999. *Environmental Discourse and Practice.* Oxford: Blackwell.

Bijker, Wiebe. 2012. "Do We Live in Water Cultures? A Methodological Commentary." *Social Studies of Science* 42, no. 4: 624–27.

Boelens, Rutgerd, Tom Perreault, and Jeroen Vos. 2018. *Water Justice*. Cambridge: Cambridge University Press.

Bullard, Robert D. 2005. *The Quest for Environmental Justice: Human Rights and the Politics of Pollution.* San Francisco: Sierra Club Books.

Cantrill, James G., and Christine L. Oravec. 1996. *The Symbolic Earth: Discourse and Our Creation of the Environment.* Lexington: University Press of Kentucky.

Clark, Anna. 2018. *The Poisoned City: Flint's Water and the American Urban Tragedy*. New York: Metropolitan Books.

Cox, Robert. 2009. "SIU and Beyond: Challenges of Scale and the Strategic." In *Social Movement to Address Climate Change: Local Steps for Global Action*, edited by Danielle Endres, Leah Sprain, and Tarla Rai Peterson, 393–422. Amherst, NY: Cambria Press.

Endres, Danielle, Tracylee Clarke, Autumn Garrison, and Tarla Rai Peterson. 2009. "Toward Just Climate-Change Coalitions: Challenges and Possibilities in the Step It Up 2007 Campaign." In *Social Movement to Address Climate Change: Local Steps for Global Action*, edited by Danielle Endres, Leah Sprain, and Tarla Rai Peterson, 179–209. Amherst, NY: Cambria Press.

Fishman, Charles. 2012. *The Big Thirst: The Secret Life and Turbulent Future of Water*. New York: Free Press.

Food and Water Watch. (2017). *Water Injustice: Economic and Racial Disparities in Access to Safe and Clean Water in the United States*. Washington, DC: Food and Water Watch.

Fraser, Nancy. 2008. *Scales of Justice: Reimagining Political Space in a Globalizing World.* New York: Columbia University Press.

Hall, Damon M., Leigh A. Bernacchi, Tema O Milstein, and Tarla Rai Peterson. 2009. "Calling All Artists: Moving Climate Change from My Space to My Place." In *Social Movement to Address Climate Change: Local Steps for Global Action*, edited by Danielle Endres, Leah Sprain, and Tarla Rai Peterson, 53–79. Amherst, NY: Cambria Press.

Hanna-Attisha, Mona. 2018. *What the Eyes Don't See: A Story of Crisis, Resistance, and Hope in an American City*. New York: One World.

Hardin, Garrett. 1968. "The Tragedy of the Commons." *Science* 162, no. 3859: 1243–48.

Hoover, Elizabeth. 2017. *The River Is in Us: Fighting Toxics in a Mohawk Community*. Minneapolis: University of Minnesota Press.

Hopke, Jill E. 2012. "Water Gives Life: Framing an Environmental Justice Movement in the Mainstream and Alternative Salvadoran Press." *Environmental Communication* 6, no. 3: 365–82.

Killingsworth, M. Jimmie, and Jackie Palmer. 1992. *Ecospeak: Rhetoric and Environmental Politics in America*. Carbondale: Southern Illinois University Press.

Kingsolver, Barbara. 2010. "Water Is Life." *National Geographic* 217, no. 4: 36–49.

Krajewski, Joanna M. T., Amy C. Schumacher, and Kajsa E. Dalrymple. 2017. "Just Turn on the Faucet: A Content Analysis of PSAs about the Global Water Crisis on YouTube." *Environmental Communication* 13, no. 2: 255–75.

Low, Nicholas, and Brendan Gleeson. 1998. *Justice, Society and Nature: An Exploration of Political Ecology*. London: Routledge.

Neal, Marian J., A. Lukasiewicz, and G. J. Syme. 2014. "Why Justice Matters in Water Governance: Some Ideas for a 'Water Justice Framework.'" *Water Policy* 16, no. 2: 1–18.
Newlands, Maxine. 2018. *Environmental Activism and the Media*. New York: Peter Lang.
Orlove, Ben, and Steven C. Caton. 2010. "Water Sustainability: Anthropological Approaches and Prospects." *Annual Review of Anthropology* 39, no. 1: 401–15.
Pauli, Benjamin J. 2019. *Flint Fights Back: Environmental Justice and Democracy in the Flint Water Crisis*. Cambridge, MA: MIT Press.
Pezzullo, Phaedra C. 2007. *Toxic Tourism: Rhetorics of Pollution, Travel, and Environmental Justice*. Tuscaloosa: University of Alabama Press.
Rosenberg, Tina. 2010. "The Burden of Thirst." *National Geographic* 217, no. 4: 96–111.
Roosvall, Anna, and Matthew Tegelberg. 2018. *Media and Transnational Climate Justice: Indigenous Activism and Climate Politics*. New York: Peter Lang.
Salzman, James. 2012. *Drinking Water: A History*. New York: Overlook Duckworth.
Schiffler, Manual. 2015. *Water, Politics and Money: A Reality Check on Privatization*. Cham, Switzerland: Springer.
Schlosberg, David. 2004. "Reconceiving Environmental Justice: Global Movements and Political Theories." *Environmental Politics* 13, no. 3: 517–40.
Schlosberg, David. 2007. *Defining Environmental Justice: Theories, Movements, and Nature*. Oxford: Oxford University Press.
Schwartz, Judith D. 2016. *Water in Plain Sight: Hope for a Thirsty World*. New York: St. Martin's Press.
Shiva, Vandana. 2002. *Water Wars: Privatization, Pollution, and Profit*. Cambridge, MA: South End Press.
Simon, Paul. 1998. *Tapped Out: The Coming World Crisis in Water and What We Can Do about It*. New York: Welcome Rain.
Soll, David. 2013. *Empire of Water: An Environmental and Political History of the New York City Water Supply*. Ithaca, NY: Cornell University Press.
Sultana, Farhana. 2013. "Water, Technology, and Development: Transformations of Development Technonatures in Changing Waterscapes." *Environment and Planning: Society and Space* 31, no. 2: 337–53.
Sultana, Farhana. 2018. "Water Justice: Why It Matters and How to Achieve It." *Water International* 43, no. 3: 483–93.
Walker, Gordon. 2012. *Environmental Justice: Concepts, Evidence and Politics*. London: Routledge.
Zwarteveen, Margreet, and Rutgerd Boelens. 2014. "Defining, Researching and Struggling for Water Justice: Some Conceptual Building Blocks for Research and Action." *Water International* 39, no. 2: 143–58.

Section I

Pipelines, Dams, and Water Sovereignty

Struggles for water sovereignty—local community control over waterways and water bodies, how they are used, what human products enter into them, and whether and how they are blocked or rerouted—are struggles for distributive justice, procedural justice, and justice of social recognition. While water sovereignty has been central to water justice for millennia, the global conversations surrounding it are just beginning to take form.

The 2016 demonstrations against installation of the Dakota Access Pipeline through the Standing Rock Sioux Reservation, for instance, garnered widespread attention in the United States, awakening audiences to rhetorical campaigns for water sovereignty all over the world (Moore 2019). In Brazil, indigenous communities of the Amazon, including Kayapó, Arara, Juruna, Araweté, Xikrin, Asurini, and Parakanã peoples, have publicly marched and campaigned against installation of the Belo Monte mega-dam on the Xingu River (Survival International, n.d.). The dam, soon to be completed, is set to become the fourth largest dam on Earth, though its installation will flood some regions, parch others, and impact fish stocks crucial to thousands of local communities. Kayapó Chief Raoni, among others, has used image rhetorics to gain global attention to the cause, leading marches through city streets and attending government hearings in traditional Kayapó dress, disrupting status quo expectations of bureaucratic deliberation. In Laos and Thailand, farmers have spoken out against China's damming of the upper Mekong (Lancang) River, wresting control of a waterway that crosses international boundaries. In the Philippines, when silt and toxic runoff from mines flowed into the Boac and Makulapnit rivers, the Cordillera People's Alliance organized a People's Regional Mining Conference, bringing indigenous and grassroots groups together to ultimately halt mining activities (Bullard 2005, 173). And in Canada, indigenous activists have led the resistance to the installation of Keystone XL and other pipelines disregarding treaty terms and local input on crucial water decisions.

Bullard (2005) calls neoliberal globalization and modernization a kind of "developmental genocide" for "those who stand in the way of the economic exploitation of valuable resources" (169). The "basic element of this process," he writes, "is the devaluation of the victims, which makes them seem inferior or worthless" (169), and he details how indigenous and local communities are most frequently its targets. Colonialism and postcolonial neoliberal policies, promoting multinational corporations and privatization of resources, wrested control of water and waterways from native peoples. Policies promoting "self-regulating markets, privatization and governance at a distance" have "resulted in the opening of markets in areas that had once been considered sacrosanct—in areas such as water, education, health care, worker safety, social security and environmental pollution" (Mascarenhas 2007, 566).

But here, as in other cases of water injustice, local communities have indeed struck back, using identity and media attention as rhetorical tools. Capitalizing on the authority granted by and constitutive attraction of local identity (Feldpausch-Parker et al. 2009), local communities have "organized themselves and made alliances with a wide variety of groups to publicize their situation and apply pressure on multinational corporations, holding them accountable for their behavior" (Bullard 2005, 168). Through image events and spectacle demonstrations, tribes and local groups can gain the attention of mass media and alter the public narratives of industrial progress (Moore 2019). With transnational coalitions sustained through social media and other digital formats (Roosvall and Tegelberg 2018), water activists can foster "cosmopolitan solidarity" (60) and networks of connective action, enabling the small fish to counter large ones in the struggle for water sovereignty.

This section examines such rhetorical invention in four cases of water sovereignty activism and deliberation, with particular attention to pipelines and dams, and the ways they disrupt and obstruct waterways that would otherwise be free-flowing. Each chapter highlights rhetorical concepts useful to conceptualizing public campaigns for water sovereignty and justice.

In chapter 1, Emilie Falc presents a multivalent and ecofeminist visual rhetorical analysis of photographs, videos, artwork, political cartoons, and other images from the 2016 Standing Rock demonstrations. Falc argues that the framing of discourse in the demonstrations acted both as power-within and as a *moral framework*, and that the communication of three recurrent themes—indigenous peoples as water protectors, water as life, and Earth as mother—amass as a visual argument and moral system for demonstrators and their audiences. The repetition of the images creates a compelling visual argument as a master trope of nonviolent strength.

In chapter 2, Gabi Mocatta focuses on the case of Chile's failed HidroAysén megaproject to examine both how hydroelectricity can be constructed in public discourse as low-carbon energy and how such discourse may displace other arguments on the unsustainability of large dams. The chapter uses longitudinal analysis to trace the linkage of discourse on HidroAysén and climate change in key Chilean print and online media across the whole trajectory of the megaproject, from 2006 to 2017. It demonstrates how dam building and climate change mitigation were initially discursively aligned as "competing risks" and how the resulting *false dichotomy*, first used as justification for building the dams, became an opportunity for anti-dams social movements in public discourse.

In chapter 3, Casey R. Schmitt applies rhetorical criticism to the material and image rhetorics of the "Free the Snake Flotilla" demonstration in Washington state, an annual mass kayak, canoe, and paddleboard float along the Snake River, calling for the removal of "deadbeat dams." The chapter challenges arguments that leisure-structured spectacle gatherings function as persuasive/disruptive culture-jams and instead argue that *leisure spectacle demonstrations* serve primarily as a space of constitutive identity performance and connective community work.

In chapter 4, Emma Lundberg, Caroline Gottschalk Druschke, and Alicia Lehrer challenge popular appeals for the removal of aging dams as a solution to the *settler colonial logics* of dam construction. The authors argue that dam removals in the United States instead may be just as likely to reinscribe and reinforce dominant settler colonial logics of domination and control, acting as a palliative but not a remedy to settler colonial river management practices. The management discourse and logics that drive dam removals in the northeastern United States, they demonstrate, seem to repeat logics of efficiency and disconnection, and seem easiest to accomplish in economically privileged and politically savvy communities. To offer an alternative, the authors build from work on the Woonasquatucket River Watershed Council in Providence, Rhode Island, a group that proposes alternative modes for dam removal and fish passage projects in ways that *amplify* the voices of marginalized community residents and encourage the co-creation of trans-species relationships. They suggest that a collaborative model of river restoration might usefully undermine—rather than reinforce—settler colonial logics and achieve pragmatic goals for rivers and the communities they support.

REFERENCES

Bullard, Robert D. 2005. *The Quest for Environmental Justice: Human Rights and the Politics of Pollution.* San Francisco: Sierra Club Books.

Feldpausch-Parker, Andrea M., Damon Hall, Christi Horton, Jodi Minion, Anna Munoz, Israel Parker, and Tarla Rai Peterson. 2009. "Step It Up in the Lone Star State: How Identity and Myth May Impact a Movement." In *Social Movement to Address Climate Change: Local Steps for Global Action*, edited by Danielle Endres, Leah Sprain, and Tarla Rai Peterson, 23–51. Amherst, NY: Cambria Press.

Mascarenhas, Michael. 2007. "Where the Waters Divide: First Nations, Tainted Water and Environmental Justice in Canada." *Local Environment* 12, no. 6: 565–577.

Moore, Ellen. 2019. *Journalism, Politics, and the Dakota Access Pipeline: Standing Rock and the Framing of Injustice*. New York: Routledge.

Roosvall, Anna, and Matthew Tegelberg. 2018. *Media and Transnational Climate Justice: Indigenous Activism and Climate Politics*. New York: Peter Lang.

Survival International. n.d. "Belo Monte Dam." https://www.survivalinternational.org/about/belo-monte-dam.

Chapter One

Water Is Life

Shared Destinies

Emilie Falc

Xiuhtezcatl Martinez (Aztec) (2017), youth director for Earth Guardians, recalls how the International Indigenous Youth Council formed at Standing Rock in the spring of 2016. It began, Martinez says, with the Sacred Stone Camp, a youth prayer camp, along the Missouri River. The camp sought to bring together tribal youths in the community to heal from an epidemic of suicides, domestic and sexual violence, missing and murdered indigenous girls and women, and drug and alcohol abuse. It also sought to stop the Energy Transfer Partners' sponsored project from drilling under the Missouri and laying the Dakota Access Pipeline (DAPL) upriver from eighteen million people who rely on the region's aquifer as a water source. He explains that, after several weeks, the youths organized their own five-hundred-plus mile relay run to the Omaha, Nebraska, Army Corps of Engineers office to convince an Army Corps representative to meet with them. After that meeting, the young people decided to run another relay from Standing Rock to Washington, DC. However, in the meantime, the Corps gave DAPL a green light. Martinez recalls that when the youth returned from spreading their message across the country, they thus called for "water protectors" to join them at Sacred Stone Camp.

In October 2016, the water protectors from Sacred Stone created the Oceti Sakowin camp along Lake Oahe in the treaty territory as a nonviolent, direct action occupation of the land next to the Missouri River and as a place for thousands of people from across the globe to stand in solidarity with Standing Rock (Indigenous Environmental Network 2016). Water protector Winona LaDuke, White Earth Ashinaabe, explained, "We heard the call from LaDonna Allard" and responded and that the gathering included "water protectors from all over the continent" (APTN 2017). Former camp coordinator at Oceti Sakowin, Phyllis Young (Standing Rock Sioux), spoke these words: "[I]t's time for the woman to stand up. We owe it to our ancestors to fight to live and we have answered that challenging spirit that is in each and every one of us. So the linear age of man is over and it's time for the woman to step up. The circular motion must be put in place" (Lafleur Vetter 2016).

Soon, the news program *Democracy Now!* captured the moment when protectors rushed to stop a bulldozer attacking an ancestral burial site and the militarized police force sprayed mace while their attack dogs bit protectors. Martinez (2017) argues that this moment is when the hashtag #WaterIsLife began trending. He describes that conflict escalated on Thanksgiving weekend in November 2016, when police fired water cannons, rubber bullets, and a concussion grenade at protectors, causing hypothermia and injuries. The power of the Standing Rock demonstrations brought international support and media attention to the Standing Rock Sioux and their allies (Moore 2019).

In December 2016, President Obama's administration paused DAPL by seeking an Environmental Impact Assessment; however, when President Trump took office in January 2017, he ordered DAPL to proceed, and it has done so, in spite of law requiring the Army Corps to consider the impacts of an oil spill on fishing rights, hunting rights, and environmental justice (Martinez 2017).

Progressive campaigns often fail to generate enough popular support to draw sustained media coverage and international support. Yet Lakoff (2010) argues that the messages that frame the environmental movement can have a significant impact on swaying public opinion and creating social change. In the pages that follow, my analysis of anti–fossil fuel rhetoric retrieved from social media sources seeks to underscore the effective rhetorical framing of the collective direct-action movement. In this chapter, rhetoric includes slogans of the No Dakota Access Pipeline (No DAPL) movement, naming or labeling practices, and images of women depicted in the No DAPL movement art, banners, posters, and political cartoons. The slogans and labels repeated by organizers and advocates—including "*mni wiconi*," "Water is Life," "protect the sacred," "water protectors," and "Mother Earth"—act as rhetorical frames, critical to the image management and self-determination of the communicators. These phrases coincide with images of women that invoke and address power.

The communication of themes in the images themselves creates a visual argument. The repetition of the images makes this compelling argument a master trope of nonviolent strength in the face of riot police, and the images defy historically dehumanizing images of Native women in popular culture which anthropologist Elizabeth Bird (1999) notes as sexualized or in service to white men.

This chapter identifies the coherent moral framing of the Standing Rock movement rhetoric through the repetition of visuals and art of indigenous women in social media, and its success at mobilizing thousands to participate in the resistance to white supremacist, colonialist, capitalist patriarchy. Within the aims of this collection, it highlights a visual rhetorical campaign for water justice, and interrogates its successes to inform future work.

FRAMING, ECOFEMINISM, AND INDIGENOUS AMERICAN ONTOLOGIES

Rhetorical scholar Randall Lake (1983) explains that within Native American protest rhetoric, "Language is instrumental largely in the sense that it can invoke the supernatural. Discourse may be addressed, but its audience is Power, not other humans. And Power is addressed through ritual, through the proper recitation of songs, prayers, and dance" (136). Additionally, the rhetoric can be interpreted through an ecofeminist understanding of "power-with," "power-within," and "power-toward" (Hunt 2014), as the water protectors seek to articulate their traditional, cultural values as resistance and in direct response to the hegemony of the white supremacist, colonialist, and capitalist patriarchy. Paula Gunn Allen (Laguna Pueblo/Sioux) (1986) writes that "Indian control of the image-making and information-disseminating process is crucial" to resist cultural and spiritual genocide (42). The images recirculate a visual trope of internal strength and courage of nonviolent resistance.

My analysis underscores the importance of creating repetitive visual messaging in the social media context that employs a moral frame that resonates with the audience's values across cultures. Nevertheless, I agree with Cox (2010) who indicates that Lakoff's approach to effective rhetorical and moral framing alone will not protect the waters. Cox's insistence on a campaign's ability to exert leverage to redress an imbalance in the system will apply in this case both with legal means and through divestment.

Lakoff (2010) claims words and slogans without attention to moral framing are not enough to create a successful environmental campaign. Rather, in the case of environmental communication, the messaging needs to counteract the prevailing politically conservative moral system that privileges consumption over conservation with its ideology that "man is above nature in a moral hierarchy, that nature is put there by God purely for human use and exploitation," the neoliberal frame expressed in appeals to "let the market decide," and other economic principles that ignore the fundamental value of life (74; see also, chapter 6). Progressive values that he sees as more aligned with environmental campaigns are "empathy, responsibility (personal and social), and the ethic of excellence (make the world better, starting with yourself)"

(74). Developing messaging that expresses and connects with these values can both move an audience to action and connect with an audience's need for community.

Furthermore, Lakoff points out that the concept of the "environment" is unfortunately a "terribly false frame," meaning a frame that is like a fallacy (77). An environment that exists outside of oneself excludes one from seeing oneself as part of nature. Allen (1986) writes, "The notion that nature is somewhere over there while humanity is over here forces English-speaking people into a position of alienation from the world they live in. Each creature is a part of a living whole" (59–60).

Native American worldviews create a moral system that is more aligned with the progressive values regarding the natural world than the conservative moral system. Allen (1986) also asserts, "American Indians are not merely doomed victims of western imperialism or progress; they are also the carriers of the dream that most activist movements in the Americas claim to be seeking. The major difference between most activist movements and tribal societies is that for millennia American Indians have based their social systems, however diverse, on ritual, spirit-centered, woman-focused worldviews" (2). Further, she states, "Among gynocratic or gynocentric tribal peoples the welfare of the young is paramount, the complementary nature of all life forms is stressed, and the centrality of powerful women to social well-being is unquestioned" (3). She writes:

> Symbols in American Indian systems are not symbolic in the usual sense of the word. The words articulate reality—not "psychological" or imagined reality, not emotive reality captured metaphorically in an attempt to fuse thought and feeling, but that reality where thought and feeling are one, where objective and subjective are one, where speaker and listener are one, where sound and sense are one. (71)

Allen's views align with ecofeminist theorists' views of the importance of recognizing how natural ecosystems are models of the balance of power, integrity, and regeneration.

Ecofeminists seek to revalue women's roles and life-giving processes while emphasizing interconnections as the web of life is viewed as less hierarchical and more interdependent (King 2005). This is not to say that the artifacts under study are ecofeminist, but rather ecofeminists can actively resist seeing nature as separate from oneself in solidarity with the water protectors. Ecofeminists have been critiqued as essentializing the connection between women and nature (Warren 2002). Noel Sturgeon (1997) points out: "The figure of the Native American woman as the 'ultimate ecofeminist' mediates, for white ecofeminists, the conflict between the critique of the patriarchal connection between women and nature and the desire for that very connection" but that it could also reinforce negative stereotypes of primitivism and the "noble savage" in hunting and gathering societies (268). Furthermore, Sturgeon argues that Native American environmental activists are not likely to identify as ecofeminists even while they generally agree with ecofeminist positions, since they frequently ground their ethics in cultural beliefs. Sturgeon sees possibilities for ecofeminists to begin to see nature in urban and constructed spaces and ally environmentalists with feminists and antiracists. King (2005) articulates an ecofeminist future which includes learning "holistic health and alternate ecological technologies, living in communities that explore old and new forms of spirituality which celebrate all life as diverse expressions of nature, considering the ecological consequences of our lifestyles and personal habits, and participating in creative forms of public resistance, including nonviolent civil disobedience" (474). These views can be described as shifting to a "power-toward" stance, which rejects life-denying systems in order to embrace a holistic view and healthy relationships (Hunt 2014). The political expression of our lives through mindful actions brings our whole selves to a communication campaign.

Tom Goldtooth (Diné and Dakota) and Daniel Wildcat (Yuchi, Muscogee) explain Native beliefs in a statement from the 2012 International Indigenous Conference:

> Our Indigenous lifeways are the original "green economies." This is more than an abstract philosophy. Our Mother Earth is the source of life. Water is her lifeblood. The well-being of the natural environment predicts the physical, mental, emotional and spiritual longevity of our Peoples. Mother Earth's health and that of our Indigenous Peoples are intrinsically intertwined. When our homelands are in a state of good health our Peoples are truly healthy. This inseparable relationship must be respected for the sake of our future generations and for the well-being of the Earth herself.

Ed McGaa (Oglala Sioux) (1990), reminds us:

> Now our planet is in great danger. Why not turn to ceremony, at least to get the feeling, the message that our planet must live? She is speaking to us quite strongly already. Let her speak also in ceremony. We can gain a special resolve by communicating within the ceremonies. By listening to nature through nature-based ceremonies, we can be like the Sioux. (11)

This is to say that the framing of discourse in the social movement acts both as power-within (a special resolve) and as a moral framework for how to address the problem in our natural world that we face. McGaa's view is consistent with Lakoff's (2010) advice that "what needs to be done is to activate the progressive frames on the environment (and other issues) and inhibit the conservative frames. This can be done via language (framing the truth effectively) and experience (e.g., providing experiences of the natural world)" (76). We see both at play in the rhetorics of pipeline protest.

IMAGES AND ANALYSIS

The voices of the organizers and advocates belong to those from Sacred Stone Camp and Oceti Sakowin Camp. This chapter approaches the artifacts from a perspective of discourse analysis which contextualizes the images within the broader sign systems and provides examples of the movement rhetoric that express the repetition of the core messaging through words and images, following methods of visuality in social media laid out by Martin Hand (2017). Hand notes that "inter-textuality" is significant in "discourse analysis of visual materials, which focus on the sites of images themselves in relation to a broader discursive framing" (219).

Search terms in Google Images and Google Videos included #NoDAPL, #WaterIsLife, and Standing Rock. Examples were selected through available sampling of popular video blogs about Standing Rock seen by thousands of viewers, and banners and art depicted in those video blogs that were part of the direct actions. While researching, additional examples of political cartoons and photos that inspired the art were found on social media sites like Twitter and Tumblr. According to Hand (2017): "On the one hand, the image is an *evidential document* of something else (communities, identities, events), where on the other hand, the image is more like an inter-textual site of discursive *relationships*. Both approaches may treat images as representational or pictoral forms, meaning that images are taken to represent *something*, regardless of whether this is thought to 'mirror' or 'construct' that object" (220). Thus, the moral frames which I identify here were not the only frames employed by the water protectors and their allies, but were the *prevalent* frames employed that help to express the shift of the moral framing away from topics of land management and rights to a reframing of the survival of people and all living beings, recognizing interconnectedness through solidarity.

Further, "the significance of the 'stream' now outweighs that of the single image" (Hand 2017, 221). Olson's (2009) explication of the persuasive potential of recirculation emphasizes that recirculation of an image through a close but slightly altered rendering can be analyzed with attention to the producers' and public's reuses in their specific contexts. Within this environmental movement, the images recirculate among the supporters and activists of Standing Rock in North Dakota but also draw from earlier images from the Mi'kmaq women's anti-fracking resistance in New Brunswick, Canada, still in circulation during and after the Standing Rock protests. This online "stream" of discourse directs our attention to the protectors of the water and how the public responds.

I identify three primary themes in the pages that follow, united in an indigenous American and ecofeminist moral appeal.

Water Protectors

In the images circulated by and among DAPL protestors, a persistent self-identification as "water protectors" frames protestors' actions as a collective involving other indigenous people in a decolonizing effort to emphasize relationships instead of individual voices in the struggle to support human needs for survival. The theme of a need for protection rings true at a time of fearfulness over damaging toxins, oppressive regimes, and environmental instability of the planet due to global warming. A focus on water allows for representatives of other tribes to participate in the communal element of water and helps to underscore the sharing of the commons.

This frame is also consistent with the prayerful approach to resisting the pipeline. Ladonna Brave Bull Allard (Standing Rock Sioux Tribe), founder of Sacred Stone Camp, for instance, called for action in a Facebook video posted in the spring of 2016, writing, "I will stand to protect the water and the land. I am asking each of you to come stand with us." Jodi Gillette (Hunkpapa and Oglala Lakota), former White House advisor for Native American Affairs, explains, "We don't have any other place to go. These are our only remaining homelands. We have to protect them. Enough is enough. Really awful things happen to different tribes and it is deeply wounding, and when people see this their spirit is called to act" (Divided Films 2016).

The images most heavily circulated on protestors' social media enact a stream of visual "water protector" imagery (see figures 1.1 through 1.6). They depict nonviolent strength as the courage to stand, with images of women in the center and foreground.

The repeated message of protecting the water in these images is consistent with rhetoric of the water protectors. A Sicangu/Oglala Lakota youth, Iyuskin American Horse (2016) writes from Standing Rock:

> We are not protesters. We are protectors. We are peacefully defending our land and our ways of life. We are standing together in prayer, and fighting for what is right. We are making history here. We invite you to stand with us in defiance of the black snake.

The framing of the direct action as protecting the sacred is also expressed as defending the people from the pipeline (also known as the "Black Snake") and the bulldozers run by the unfeeling Energy Transfer Partners (ETP) Corporation, backed by the state government. Dave Archambault (2016), Chairman of the Standing Rock Sioux Tribe, explains that the Black Snake is

Figure 1.1. A large assembly of indigenous people and their allies stand on the highway that runs between the Standing Rock Sioux Reservation and Bismarck, North Dakota, with a banner that reads "Defend the Sacred" and includes a figure of a woman in a shawl. The figure could be Asdzáá Tl'ógí or Juanita Manuelito, a Diné woman who fought for indigenous rights of the Diné people according to Indigenous Action Media (www.IndigenousAction.org). *Source:* Goldtooth, D. (2016). *Defend the Sacred* [Photo]. Retrieved from Twitter.

"made of nothing but Greed. There is nothing good that has ever come from Greed. Greed is pure poison. It blinds and twists thinking. It is what my people have endured and continue to endure." Stopping DAPL would mean stopping the Black Snake through strength and the community working together as women lead the way. The Black Snake, too, then, also plays a role in the visual enactment of the "water protectors" theme (see figures 1.7 and 1.8). Each of these images indicate that the women will continue to resist and hold back the menacing Black Snake, representing the oil pipelines that threaten the water.

The "water protector" theme also built heavily from existing water sovereignty demonstrations, campaigns, and struggles. For instance, prior to the ETP's application submitted for the Dakota Access Pipeline in 2014, Mi'kmaq water protectors of the Elsipogtog First Nation in New Brunswick, Canada, in 2013 employed the moral frame of protecting water in their resistance to fracking, or hydraulic fracturing, that began with a peaceful blockade of the highway and escalated into a conflict involving Royal Cana-

Figure 1.2. Jesus Barraza (Chicano) and Melanie Cervantes's (Chicana) poster design portrays a young indigenous woman in a Stand with Standing Rock T-shirt in the foreground and calls for solidarity with Standing Rock to stop the Dakota Access Pipeline, defend the land, and protect the water. In the background, horses and riders, including an indigenous rider carrying a warrior staff, lead a group of people. *Source:* Barraza, J., and Cervantes, M. (2016). Solidarity with Standing Rock [Poster]. Retrieved from https://justseeds.org/graphic/solidarity-with-standing-rock/. Licensed under Creative Commons Zero on Just-Seeds.org.

Figure 1.3. The political cartoon by Marty Two Bulls (Oglala Lakota) depicts an indigenous woman in a traditional dress with a feathered warrior staff in one hand and her other hand protecting a young child behind her. She is standing up courageously to fierce guard dogs baring their teeth. *Source:* Two Bulls, Marty. (2016). Make a Stand: Water is Life [Political Cartoon]. Retrieved from https://www.herbblock-foundation.org/herblock-prize/prize-winners/marty-two-bulls-sr, © 2016 Marty G. Two Bulls, Sr.

dian Mounted Police (RCMP) anti-riot tactics like shooting rubber bullets and using pepper spray against the Elsipogtog First Nation citizens (Troian 2013). Women stood face-to-face with the RCMP, at times shouting at them, and, then, one woman, Amanda Polchies, knelt before the RCMP and held up an eagle feather. She explained, "I prayed for the women that were in pain, I prayed for my people, I prayed for the RCMP officers. . . . I prayed that everything would just end and nobody would get hurt" (Troian 2013; see figure 1.9).

The resulting iconic photo, taken by Ossie Michelin (Inuk), became the basis for several artistic images for the campaign and gave a visual for the moral frame of the sacredness and prayerfulness of the resistance. Doris Copage, a Mi'kmaq elder explains, "I want to call it 'protect' [. . .] rather than 'protest.' We are here to protect our water, our land. We have a river. It's a beautiful river, we love it and we respect it" (Troian 2013). Through such

Figure 1.4. The poster depicts a woman with arm raised, holding a braid of sweetgrass looking over the lake and an eagle flying high above the water. The text reads "Protect the water. Defend the land." ***Source:*** **Cervantes, Melaine. (2016). Protect the Water [Poster]. Retrieved from https://justseeds.org/graphic/protect-the-water-defend-the-land/. Licensed under Creative Commons Zero on JustSeeds.org.**

Figure 1.5. The screenprint by Issac Murdoch (Anishinaabe) of Thunderbird Woman, a woman in a traditional dress with wings, a feather tucked in her hair pointed skyward, heart symbol on her chest and surrounded by lightning, rain, and four thunderbirds. The thunderbird woman is a strong symbol of the experience of heart, water, and sky. This screen print continues to appear in its grayscale design in social media in images of actions in resistance to the pipelines. *Source:* Murdoch, I. (2016). Thunderbird Woman. Permission for noncommercial use granted by the artists.

framing, the photo depicts Polchies in a position of leadership through defense. The significant moment of resistance is replayed in artfully enhanced images created by Gregg Deal and Nicholas Lampert that honor the significance of the role of women as protectors of the water (see figures 1.10 and 1.11).

The Polchies photo is a key example of how the repetition of an image of a woman showing nonviolent strength persisted in the social movement opposing fossil fuel before, during, and after the Standing Rock protests. Lampert's image merges a Mi'kmaq woman with a Standing Rock slogan, "Water is Life," as a recirculation of nonviolent strength, faith, and spirit. The shift from a woman kneeling to a woman standing in some artistic renderings

Figure 1.6. The screenprint by Christi Belcourt, Thunderbird Baby, shows a pregnant woman standing in a stream with a raised fist who is connected intimately with water and the understanding that water flows through us. The woman and thunderbird in the cloud have heart symbols on their chests. The cloud rains into the stream and the kettle. *Source:* Belcourt, C. (2016). Thunderbird Baby. Permission for noncommercial use granted by the artist.

Figure 1.7. This art of an indigenous woman wearing traditional dress, having long black hair, shows her standing in the water while holding over her head the severed head of the black snake baring its fangs and forked tongue. The image circulated on Twitter. *Source:* Two Bulls, M. (2016). Stop DAPL [Political Cartoon]. Retrieved from https://twitter.com/indiancountry/status/795835390779330560. © Marty Two Bulls, Sr.

and adaptations of the Lambert/Polchies image resonated with the place name, Standing Rock, and reinforced the sense of a women-led movement. Additionally, the point of view of Lampert's image as well as Michelin's photo is that of being an active, frontline water protector, not just an onlooker from a distance. The raised arm in the image signals, potentially, a connection to the sky and spiritual power. The gesture of holding an eagle feather up could resonate in a U.S. context with echoes of the torch-bearing arm of the Statue of Liberty, holding up a light to peace and freedom. Clearly, the repetition in these multiple images of women facing environmental injustice, circulating in social media, invites the viewer to empathize with and draw strength from the iconic symbol of water protectors.

Figure 1.8. Promotional poster of the *First Daughter and the Black Snake* documentary shows Winona LaDuke (Anishinaabe) in a cowboy hat and jeans sitting on a horse in the northern plains below an image of a black snake with oil derricks in the background. LaDuke articulates solidarity with Standing Rock in using similar black snake imagery to fight pipelines in Minnesota as in North Dakota. *Source:* Pickett, K. (2017). First Daughter and the Black Snake [Poster]. Retrieved from https://www.keripickett.com/portfolio-item/first-daughter-and-the-black-snake/. Copyright Pickett Pictures.

Mni Wiconi and Water is Life

A second recurrent frame, in repeated reference to the maxims "Water is Life" and/or "*Mni Wiconi,*" emphasizes how water is essential to survival. It stresses a human interconnectedness to water, as it is needed by everyone and within everyone, and implicitly argues that water needs to be respected and valued.

The phrase "Water is Life" speaks to "the popular Lakotayapi assertion 'Mni Wiconi'—water is life or, more accurately, water is alive" (Dhillon and

Figure 1.9. Amanda Polchies (Elsipogtog) kneels on the road holding up an eagle feather in face of the Royal Canadian Mounted Police in navy blue uniforms and baseball caps. *Source:* Photo credit: Michelin, O. (2013). Amanda Polchies. APTN News. Retrieved from https://www.cbc.ca/news/indigenous/inuk-journalist-ossie-micheline-best-image-museum-1.4174047. © Ossie Michelin. Used with permission of the photographer.

Estes 2016). This framing that decommodifies water resurfaces again and again in video narratives posted online. The *Mni Wiconi* frame was part of the discourse of the Standing Rock Sioux Tribe's listening session with the Army Corps of Engineers in the spring of 2016, currently circulating via YouTube (Unicorn Riot 2016). The theme was expressed in this listening session through song and through repeated arguments by LaDonna Allard and others who expressed, "We cannot live without the water. This is not an Indian issue. It's a people issue. This is an everybody issue. Everybody needs water, so if we don't stand up and fight for water we're [*sic*] lost our humanity."

The *Mni Wiconi* frame is also uniquely connected to women through messages and images in video interviews of the water protectors circulating via Vimeo. In the fall of 2016, Melaine Stoneman (Sicangu Lakota) noted:

> Some people say Mni Wiconi is just water, but for us Lakotas it's more than just water. It's more than just praying for water. It's our way of life. Water of life. And our way of life, you can't find it in the dictionary. (App 2016b)

She then draws an explicit relationship between water and *women*, adding:

Figure 1.10. The poster created by indigenous artist Gregg Deal shows the woman raising an eagle feather to the sky in prayer in the resistance to fracking and drilling. *Source:* Deal, G. (2016). Mi'kmaq Protest [Poster]. Retrieved from http://www.ethicalactionalert.com/2013/10/download-this-anti-fracking-protest.html. Open source.

Figure 1.11. Lampert's screen print depicts a woman holding an eagle feather with an upraised arm and the message is "Water is life." ***Source:*** **Lampert, N. (2016). Water is life. Retrieved from https://justseeds.org/graphic/water-is-life-3/. Licensed under Creative Commons Zero on Justseeds.org.**

> We're here to make a connection, here to make a connection with the water and the sacredness it has here for all of us, for all of us women. Watching our women hit the ground. Watching the women on their knees. Watching them bleeding and rubber bullets to their head is very disheartening to see that they had some kind of intentions to hurt us women. . . . So there is a reason why we must continue on as women nation. So there's a reason why we must stand together, not only in solidarity, not only in prayer, but it is in unity all over the world, not just here at Standing Rock, but globally we must stand together as one. (App 2016b)

The *Mni Wiconi* frame hails women as included in the spiritual life of the community as well as a part of the basis of power in the community. Stoneman insists, "A woman should never be yelled at. [They] targeted our women," yet the response of the water protectors is to nonviolently "stand in prayer" (App 2016b).

In another video circulating on Vimeo, J. C. Begay (Sicangu Lakota) reiterates that "if we don't have water, we don't have life," and explains that one of the songs sung at Standing Rock includes words of healing and to "pray for clean water, and everlasting life with our water. And, again, that praying that our sacred hoop be mended" (App 2016b). Water is imbued with power in its status as sacred in these recirculating online videos.

Lakota historians Mark St. Pierre and Tilda Long Soldier (Oglala Lakota) (1995) confirm these views, thus reinforcing the power of the imagery of water and its significance to indigenous people. St. Pierre and Long Soldier write, "Mni (water) is essential and seen as connecting all life. It is viewed as having sacred powers of healing. Water is used in conjunction with almost all ceremonies as the most precious and basic of gifts" (46). St. Pierre and Long Soldier continue: "In Lakota society, the spiritual and economic powers of women were not only acknowledged but well respected." (48). Such views reinforce respect for women and water in traditional indigenous ways.

In the videos of the Standing Rock activists, the *Mni Wiconi*/"Water is Life" frame is highly visible in the graphic images on demonstrators' banners and posters, many of which were designed by artists from the JustSeeds artist collective. Simpson (2017) notes that the resurgence of artist collectives is "a disruptive and deliberate act of turning away from the colonial state . . . grounded in a coded articulation . . . of Indigenous intelligence as theory and process and as affirmative refusal" (198). The message shifts from "water of life" or *aqua vitae* to "Water is Life" and "*agua es vida*" across multiple digital images available on the Justseeds.org website (see figure 1.11).

Several iconic images depict the message of *Mni Wiconi*. The banner image created by Issac Murdoch (Anishinaabe) and Christi Belcourt (Michif Metis), for instance, depicts Thunderbird Woman with wings, with drops of water, lightning bolts, and the message "We are here to protect. Water is life. Mni Wiconi" (see figure 1.5). Belcourt (2016), in her keynote speech, "The Revolution Has Begun," at the Maamwizing Conference in Sudbury, Ontario, explains, "The waters are viewed as the lifeblood of Mother Earth." The message of *Mni Wiconi* and "Water is Life" is reinforced in an image circulating on INDIANZ.com of Bobbi Jean Three Legs, citizen of the Standing Rock Sioux and organizer of the indigenous youth relay run across the nation. Madonna Antoine Eagle Hawk (Rosebud Sioux) explains in a video circulating on Vimeo: "The prophecy was that all the generations, the indigenous, all tribes would come together to fight the black snake and the youth would be the ones to lead it. And they were the first to run all the way to

Washington, DC, and lead it" (App 2016a). The repetition of the slogans over time reinforces the message as a coherent moral frame.

Mother Earth

Thirdly, a recurrent hailing of the land as "our mother" or "Mother Earth" reinscribes a respectful relationship with the Earth that revalues and honors the natural world and the beings on it. This framing of "Mother Earth" is also a moral frame, which Lakoff (2010) argues is needed to advocate for the environment in the face of fierce opposition. "The Mother Earth is Hunci Waka the grandmother of everything and the water is her blood. And through this blood we live," explains LaDonna Allard (Lakota) in a video circulating on YouTube (Divided Films 2016). The hailing of the Earth as mother was echoed throughout the Standing Rock campaign by many leaders, including Chief Arvol Looking Horse, 19th Generation Keeper of the Sacred Buffalo Calf Pipe Bundle of the Lakota, Dakota, and Nakota Nations.

Along with the theme of water, the Lakota belief in the Mother Earth as sacred and vulnerable to assault was expressed at the listening session in the spring of 2016 with the Army Corps of Engineers in a video circulating via YouTube (Unicorn Riot 2016). In the fall of 2016, Eryn Wise (Jicarilla Apache/Laguna Pueblo), a youth leader in the International Indigenous Youth Council from Sacred Stone Camp, stood at the doors of the Army Corps of Engineers and spoke to Colonel John Henderson in front of the gathered crowd. The following intense questioning was recorded and shared via social media: "Have you ever been raped?" Wise asks. Henderson replies, "No, I have not." Wise continues, "Okay. I have. When they put that pipeline in our Mother, every single day they do it reminds me of what it was like when somebody forcibly put something inside of me that I did not request to be there. And on behalf of the youth that want to wake up with clean water tomorrow and want to have children consensually, we are asking you to not grant that easement. We're asking you to stand up on behalf of the people that you swore to protect" (Becker 2016).

The framing of the raping of Mother Earth is consistent with the protection framing and the call to respect life and American lives.

DISCUSSION

The three major frames of the No DAPL campaign at Standing Rock discussed here—water protectors, water is life, and Mother Earth—are more than just slogans. The frames evoke an indigenous moral system that includes the responsibility to be life-supporters and to have empathy for all who share our ecosystem, invoking power and asserting "power-toward." The frames are effective because they are internalized, inspiring, and wholesome in emphasizing "power-with," a coalition-building power among indigenous people. The repetition of the moral frame visually in images and text, and voiced in videos circulating in social media among the water protectors, also flows beyond reservation borders, state borders, and national borders and streams into the political consciousness. The moral frame of the water protectors aligns with conservationists' perspective on protecting ecosystems to improve habitats and ecofeminists' perspectives on recognizing how respecting and caring for nature means caring for and respecting women. Water protectors' framing aligns with health-conscious consumers who seek to protect themselves and their families from the onslaught of industrial hazards. In addition to the representatives from international tribal groups, Standing Rock drew in numerous clergy and an estimated two thousand veterans as demonstrators and allies (Healy 2016). Water protectors' framing aligns with the values of clergy oriented toward social justice and veterans who took an oath to serve and protect Americans from threats, foreign and domestic.

Protecting the sacred rejects a consumerist narrative of unstoppable progress and affirms the indigenous peoples' return to traditions and indigenous wisdom. The prevalence of this stream of images of strong indigenous women in social media builds empathy, whereas traditional images of indigenous men might evoke stereotypes. Respecting indigenous women as courageous leaders of their own communities exemplifies a vision in which gender equity is a presumed component of a transition to a more sustainable future. The "Water is Life" frame and its call for unity invites us to stand in solidarity and as witnesses to the disrespect of sacred land and water. The frame of *Mni Wiconi* exists in a context of an America facing threats to water on multiple fronts. Solidarity is essential to see this social movement continue to grow and for seeking justice. And the Mother Earth frame expands upon and unites these other two prevalent themes in the image rhetorics of the movement.

Of course, moral frames are not enough on their own to impact the outcomes of the conflict. Cox (2010) argues that in order for an environmental campaign to be successful, it needs to apply leverage to the weakest point in the prevailing power structure which could be in the form of a lawsuit regarding upholding regulations. Still, in the big picture of Standing Rock, the direct actions of the water protectors might have had an impact on the Standing Rock Sioux Tribe's legal efforts by raising awareness and engagement, and thus directing more legal funds toward the tribe's ongoing legal battles with the Army Corps of Engineers (Meyer 2017). Earth Justice lawyers continue to support Standing Rock's legal fight at the time of this writing (by, for instance, filing a "supplemental complaint" on November 1, 2018, in their lawsuit against the Army Corps to stop DAPL). The confrontations with the private security workers also brought to light a larger concern about the increase in state tolerance of the use of violence to suppress freedom of expression and to attempt to manipulate peaceful demonstrations into the appearance of a riot. Also in the bigger picture, the "Water is Life" campaign boosted the awareness of the No DAPL campaign and its online push urging customers to divest from Wells Fargo and other major banks that finance the corporation constructing the pipeline. With the attention from activists from multiple tribal affiliations on the water, a successful online divestment campaign hindered Kinder Morgan's Trans Mountain pipeline expansion project (Gignac 2017). A kind of indigenous internationalism has been described by artist and environmental activist Leanne Betasamosake Simpson (Michi Saagiig Nishnaabeg) (2017), as an international co-resistance. Without the unifying moral frame and the stream of images, such campaigns would have a much harder time developing necessary solidarity and on-the-ground coalitions to pressure banks and to work on municipalities divesting as well.

Moreover, the No DAPL campaign and its stream of images—of women, in particular—could have impacts beyond pipelines. More indigenous women ran for office in America in 2018 (Trahant 2018). Minnesota stands in a historical moment electing as Lieutenant Governor Peggy Flanagan (White Earth Ojibwe), on the Democratic ticket, while Donna Bergstrom (Red Lake Ojibwe), ran on the Republican ticket. These are tangible results of respecting indigenous women's strength, courage, and leadership.

Public support and visibility for the No DAPL campaign is not enough to turn away the bulldozers, security helicopters, and riot-geared security workers; however, the campaign's rootedness in multiple moral frames shows a path to integrity through steadfastness in understanding the nature of life. Sky Bird Black Owl, a water protector, explains, "they talked about that first it would be the children, and then the women, and finally our men would stand up. I firmly think that our men need our women to stand up and be strong and to be reminded they need to make the world safe for us and our daughters, and their mothers, sisters, aunties, [and] the women they don't know" (Kring 2016). The solidarity of working together and collaborating via our natural diversity strengthens the movement toward respecting Mother Earth and begins to redistribute the wealth of our nation's generosity, caring, kindness, and fortitude as these values resonate in the images and voices of indigenous women.

REFERENCES

Allen, Paula Gunn. 1986. *The Sacred Hoop: Recovering the Feminine in American Indian Traditions*. Boston: Beacon Press.

American Horse, Iyuskin. 2016. "'We Are Protectors, Not Protesters': Why I'm Fighting the North Dakota Pipeline." *The Guardian*. August 18, 2016. https://www.theguardian.com/us-news/2016/aug/18/north-dakota-pipeline-activists-bakken-oil-fields.

App, Brittany. 2016a. "Where There Once Was Water: Standing Rock Stories: Donna and Curly." Video, 13:07. https://vimeo.com/194112853.
App, Brittany. 2016b. "Where There Once Was Water: Standing Rock Stories: Melaine and J.C." Video, 14:02. https://vimeo.com/194266511.
APTN. 2017. "Dennis Ward Sits Down with Renowned Activist, Winona LaDuke." *APTN News*. May 16, 2017. https://aptnnews.ca/2017/05/16/winona-laduke/.
Archambault, David. 2016. "Demonstrations a Cry to Stop Desecration of Mother Earth." *Native Sun News*. August 25, 2016. http://www.nativesunnews.today/news/2016-08-24/Voices_of_the_People/Demonstrations_a_cry_to_stop_desecration_of_Mother.html.
Becker, Eddie. 2016. "Army Corps Listens to Standing Rock Pipeline Voices." Filmed November 16, 2016. Video, 7:25. https://www.youtube.com/watch?v=CnqGwheuTUo.
Bird, S. Elizabeth. 1999. "Gendered Construction of the American Indian in Popular Media." *Journal of Communication* 49, no. 3: 61–83.
Cox, J. Robert. 2010. "Beyond Frames: Recovering the Strategic in Climate Communication." *Environmental Communication* 4, no. 1: 122–33.
Dhillon, Jaskiran, and Nick Estes. 2016. "Introduction: Standing Rock, #NoDAPL, and Mni Wiconi." *Hot Spots, Cultural Anthropology*. December 22, 2016. https://culanth.org/fieldsights/1007-introduction-standing-rock-nodapl-and-mni-wiconi.
Divided Films. 2016. "Mni Wiconi: The Stand at Standing Rock." Video 8:26. https://www.youtube.com/watch?v=4FDuqYld8C8.
Gignac, Julien. 2017. "Environmental and Indigenous Groups Demand Banks Avoid Supporting Trans Mountain Expansion." *The Star*. June 8, 2017. https://www.thestar.com/news/canada/2017/06/08/environmental-and-indigenous-groups-demand-banks-avoid-supporting-trans-mountain-expansion.html.
Goldtooth, Tom B. K., and Daniel Wildcat. 2012. "Rights of Mother Earth: Restoring Indigenous Life Ways of Responsibility and Respect." Paper presented at the International Indigenous Conference, Haskell Indian Nations University, Lawrence, Kansas. April 4, 2012. http://www.ienearth.org/rights-of-mother-earth-restoring-indigenous-life-ways-of-responsibility-and-respect/.
Hand, Martin. 2017. "Visuality in Social Media: Researching Images, Circulations and Practices." In *The SAGE Handbook of Social Media Research Methods*, edited by Luke Slone and Anabel Quan-Haase, 219–21. Los Angeles: SAGE.
Healy, Jack. 2016. "As North Dakota Pipeline Is Blocked, Veterans at Standing Rock Cheer." *New York Times*. December 5, 2016. https://www.nytimes.com/2016/12/05/us/veterans-north-dakota-standing-rock.html.
Hunt, Kathleen P. 2014. "It's More than Planting Trees, It's Planting Ideas": Ecofeminist Praxis in the Green Belt Movement. *Southern Communication Journal* 79, no. 3: 235–49.
Indigenous Environmental Network. 2016. "Citing 1851 Treaty, Water Protectors Establish Road Blockade and Expand Frontline #NoDAPL Camp." *Indigenous Rising*. October 23, 2016. https://indigenousrising.org/citing-1851-treaty-water-protectors-establish-road-blockade-and-expand-frontline-nodapl-camp/.
King, Ynestra. 2005. "The Ecology of Feminism and the Feminism of Ecology." In *Feminist Theory: A Reader*, edited by Wendy K. Kolmar and Frances Bartowski, 469–75. Boston: McGraw-Hill.
Kring, Shannon. 2016. "Exclusive Interview: Sky Bird Black Owl, First Woman to Give Birth at the Standing Rock Protests." Video 2:24. https://www.youtube.com/watch?v=7vaKoaG3Nlo.
Lafleur Vetter, Sara. 2016. "Standing Rock Rising." Video 7:27. https://vimeo.com/190829243.
Lake, Randall. 1983. "Enacting Red Power: The Consummating Function in Native American Protest Rhetoric." *Quarterly Journal of Speech* 69, no. 2: 127–42.
Lakoff, George. 2010. "Why It Matters How We Frame the Environment." *Environmental Communication* 4, no. 1: 70–81.
Martinez, Xiuhtezcatl. 2017. *We Rise: The Earth Guardians Guide to Building a Movement That Restores the Planet*. Emmaus, PA: Rodale.

McGaa, Ed. 1990. *Mother Earth Spirituality: Native American Paths to Healing Ourselves and Our World.* New York: Harper.

Meyer, Robinson. 2017. "The Standing Rock Sioux Claim 'Victory and Vindication' in Court." *The Atlantic.* June 14, 2017. https://www.theatlantic.com/science/archive/2017/06/dakota-access-standing-rock-sioux-victory-court/530427/.

Moore, Ellen. 2019. *Journalism, Politics, and the Dakota Access Pipeline: Standing Rock and the Framing of Injustice*. New York: Routledge.

Olson, Lester C. 2009. "Pictorial Representations of British America Resisting Rape: Rhetorical Re-Circulation of a Print Series Portraying the Boston Port Bill of 1774." *Rhetoric and Public Affairs* 12, no. 1: 1–35.

Onaman Collective. 2016. "The Revolution Has Begun—Keynote by Christi Belcourt." Video 33:53. Retrieved from https://youtu.be/XqBXDPzyLm0.

Simpson, Leanne Betasamosake. 2017. *As We Have Always Done: Indigenous Freedom through Radical Resistance*. Minneapolis: University of Minnesota Press.

St. Pierre, Mark, and Tilda Long Soldier. 1995. *Walking in the Sacred Manner: Healers, Dreamers, and Pipe Carriers—Medicine Women of the Plains Indians*. New York: Touchstone Books.

Sturgeon, Noel. 1997. "The Nature of Race: Discourses of Racial Difference in Ecofeminism." In *Ecofeminism: Women, Culture, Nature,* edited by Karen J. Warren and Nisvan Erkal, 260–78. Bloomington: Indiana University Press.

Trahant, Mark. 2018. "Native Candidates Are Rewriting History." *Indian Country Today*. July 27, 2018. https://newsmaven.io/indiancountrytoday/news/native-candidates-are-rewriting-history-mDgpDEAPtk2JNsyEMn9t6Q.

Troian, Martha. 2013. "Mi'kmaq Anti-Fracking Protest Brings Women to the Front Lines to Fight for Water." *Indian Country Today.* November 10, 2013. https://newsmaven.io/indiancountrytoday/archive/mi-kmaq-anti-fracking-protest-brings-women-to-the-front-lines-to-fight-for-water-alK1jV5GE0Sa-gXWtatWwQ/.

Unicorn Riot. 2016. "Standing Rock Spirit Camp Visits Army Corps of Engineers." Video, 6:53. https://vimeo.com/165321281.

Warren, Karen J. 2002. "Response to My Critics." *Ethics and the Environment* 7, no. 2: 39–59.

Chapter Two

When Water Is Energy

Tracing Mediatized Discourses in Chile's Mega-Hydro Debate

Gabi Mocatta

Debates in the water justice literature have most often centered on equitable access to *water*. However, debates around water for hydroelectric energy generation, and socio-environmental justice for those most affected by hydroelectric developments, have also now increasingly become intertwined with the literature on water justice (Hoffman 2017; Islar 2013; Martínez and Castillo 2016; Schneider, Buser and Graefe 2014; Zanotti 2015). This is particularly so in relation to the Global South (Harris, Goldin, and Sneddon 2013), where privatization and marketization of water, together with a new spate of hydroelectric dam building (Ansar et al. 2014; Zarfl et al. 2015) have spurred activism against dams and mediatized conflict surrounding water justice. Due to this renewed growth in hydroelectric dam initiatives, especially in the current context of energy transition, understanding discourses around hydroelectricity and water justice is now particularly timely.

This chapter offers insights into a discursive dimension for understanding struggles for water justice. It examines what happens in water justice debates when water represents energy, by tracing the mediatized enactment of conflict (Hutchins and Lester 2015) over the never-realized HidroAysén hydroelectric megaproject, planned for Patagonia, Chile. Using a detailed case study of conflict surrounding this project, this chapter focuses on the ways that anti-dams activism and pro-hydroelectric development discourses around water and energy emerged and interacted across the trajectory of the debate over the dams. It examines discourse surrounding national interest, climate change, and energy transition in Chile that emerged as part of the lengthy dams debate, and it also investigates how activist discourses ultimately sought to connect "water-based forms of injustice" (Zwarteveen and Boelens 2014, 148) with wider questions of social inequality in Chile. The chapter shows how, by connecting place-based water injustices in the far southern reaches of Chile with intergenerational social inequality and incomplete democratization on a national level, anti-dams activism made HidroAysén representative of much more than a conflict around water and environment. This chapter argues that it was ultimately this ability to discursively connect seemingly disparate issues that ensured the HidroAysén case ended with a win for socio-environmental justice in Chile.

The research here contributes to a growing field of studies of discourse in relation to hydroelectric dam construction in diverse national contexts, including the work of Ozen (2014), Wooden and de la Croix (2015), Borgias and Braun (2017), and Tur et al. (2018). This chapter builds on such analyses by making a detailed study of pro- and anti-dams discourse in media over an extended time period, and by connecting hydroelectric dam building with wider questions of water and socio-environmental inequality, (in)justice, and the mediatized enactment of environmental conflict.

WATER JUSTICE, INEQUALITY, AND THE DISCURSIVE CONSTRUCTION OF HYDROPOWER

Zwarteveen and Boelens (2014) have conceptualized water as simultaneously natural and social, so that conflicts over water are inherently problems of justice. Boelens, Perreault, and Vos's (2018) assessment of the evolving field of water's political ecology highlights the power and politics inherent in water use and management. These authors describe "hydrosocial networks" (2) in which unequal power structures govern decision-making over water development procedures and practices. Such power structures, underpinned by global neoliberalism, ensure that water governance has escaped the control of the nation state, and now increasingly rests with multinational financial institutions, not publicly elected governments or local water user groups (Zwarteveen 2015; Bakker 2010; Swyngedouw 2005).

Importantly, the power relations inherent in water's political ecology are constructed and maintained *discursively*. Foucault conceptualized discourse as "a group of statements which provide a language for talking about—a way of representing the knowledge about—a particular topic at a particular historical moment" (Hall 1997, 44). Discursive practices create "a delimitation of the field of objects" and define "a legitimate perspective for the agent of knowledge" thereby determining "norms for the elaboration of concepts and theories" (Foucault 1977, 199). Discursive practices are therefore inherently linked with power, so that certain discourses serve certain interests and interest groups better than others. Discursive practices around water management include technical discourses on the efficiency and productivity of industrial and large-scale water users, as opposed to the "backwardness" of small-scale water users (Boelens and Vos 2012). They also encompass, for example, discourses on the national interest, so that those who relinquish local water resources, or who are moved for large dams, act selflessly for the good of their fellow citizens.

Discourses around water in relation to hydroelectric dams, specifically, have also habitually constructed large dams as being sources of "progress" and "necessary for development" (Tur et al. 2018, 132)—this despite overwhelming research evidence that shows they manifest social injustice and underscore inequality (Berkun 2010; Carruthers and Rodriguez 2009; Hoffman 2002, Khagram 2004; McCully 1996; Siciliano and Urban 2017) inflicting social trauma and "slow violence" (Blake and Barney 2018) on both environment and people. In the face of climate change in particular, hydroelectric dams have also been discursively constructed as sources of "clean energy" (Bratman 2015), as "low carbon" (Lutsey and Sperling 2008), and as essential to a new "green economy." Such discourses used by dam developers, governments, international funding bodies, and transnational corporations have therefore characterized hydroelectric dams as "climate friendly," despite ample research which shows that dams can be significant greenhouse gas emitters (Fearnside 1997, 2004, 2015; Maeck et al. 2013; Wehrli 2011). Anti-dams social movements are also discursively constructed in conflicts surrounding dams. Those who oppose dams have often been characterized negatively (Bratman 2015; Da Silva and Rothman 2011), as being "anti-development," as "criminals" and "bandits," "terrorists" (Ozen 2014), and even "demonic forces" (Elizondo 2009).

To fully comprehend water justice struggles and outcomes, it is therefore crucial to understand both the *inequalities* that underpin such struggles, and the *discourses* that support them. Water justice, and its inverse, are inherently linked with inequality: those who suffer the most from water injustices are those who suffer most from inequality. Likewise, water justice is profoundly connected with discourse. Discourses in water management serve to support particular ways of organizing decision making, and to legitimize particular forms of authority over other forms, privileging some groups over others (Zwarteveen and Boelens 2014). In conflicts over water, the definitions that prevail are often those of the powerful. Power and politics therefore pervade not only the ways that water is managed and allocated, but also the ways of thinking and talking about water. One of the key arenas for constructing discourses and pursuing struggles in conflicts over water is the media. I therefore now turn to a brief exploration of mediatization of environmental conflicts, before discussing in detail the chapter's primary case study.

MEDIATIZATION OF ENVIRONMENTAL CONFLICT

The role of media as a site for struggles for issue definition and to achieve symbolic power has been analyzed and conceptualized by several scholars (Anderson 1997; Castells 2007; Hansen 2010; Lester 2010). Gamson and Wolfsfeld (1993) have described social movements and media as mutually dependent "interacting systems," and Wolfsfeld (1997) has conceived of the media as an *arena* in which conflict is played out, and where groups and their issues or causes "compete for media attention as a means to achieve political influence" (2). In Cottle's (2006) concept of "mediatized conflict," more than simply reporting conflicts, the media enact or perform them. Media thereby become key *actors* in such conflicts, changing the course of mediatized debates.

A key theme underpinning the role of media in mediatized conflict is also access to *visibility*. For Thompson (1995, 2005, 2011, 2013) achieving visibility through the media "is to gain a kind of presence or recognition in the public space" (2005, 49). Ultimately, an actor's (or network of actors') degree of visibility has a strong correlation with their political power. Creating visibility can also be a case of bringing to light information that industry or government would rather keep *invisible*. While in the age of deep mediatization (Couldry and Hepp 2018) social movements are more visible than ever, and are able to rely on "movement media" and "self-mediation" (Cammaerts 2015) to articulate their messages in order to reach the public at large, "the key channel was and is getting access to and coverage by the established media" (Rucht 2013, 262). This is a site which Rucht (2013) and others (Hutchins and Lester 2011) have noted that environmental movements still struggle to access.

Building on Castells's (2007, 2009) work on the Network Society, which identifies power in the control of flows of information, and using Arsenault and Castells's (2008) conceptualization of "switching points," which describes the sites of interaction between media, individuals, and civic institutions through digital and mobile communication, Hutchins and Lester (2015) have theorized the mediatized enactment of conflicts related to the environment. These scholars identify the mutually constitutive interactions at the sites where activism campaigns, journalism practices, politics and decision-making, and industry intersect. They conceptualize mediatized environmental conflict as indicative of power dynamics in a globalized world, with the groups and decision-makers who come to exercise greatest influence being those most able to control mediated visibility and invisibility, to construct discourses, frame meaning, and determine the content, direction, and *interconnection* of flows of information. Using the lens of mediatized environmental conflict, I now turn to the case study, to identify how power dynamics and discourses interacted with inequality, water-based (in)justice, and social movement struggle in the case of Chile's HidroAysén.

THE CASE OF HIDROAYSÉN, ELITE CLOSURE, AND ENVIRONMENTAL CONCERN

Established in 2006, HidroAysén was Chile's largest ever planned infrastructure project—a hydroelectric megaproject of five dams and a 2,300-kilometer-long power line. The US$8 billion project was a transnational joint venture between Chilean utility Colbún (owned by the elite Chilean Matte family) and the Spanish company Endesa (subsequently acquired by Italian energy conglomerate Enel). Water rights to many of Chile's rivers had been privatized in 1981 under the Pinochet regime (Latta 2010), and Endesa/Enel owned nonconsumptive water rights on the rivers Baker and Pascua in the remote southern Aysén region of Chilean Patagonia. The project intended to generate power from glacier and snow melt in these fast-flowing rivers and feed this power into the country's arid central region, home to 90 percent of Chile's population and much of its industry. An interconnection between Chile's central and northern grids by 2017 was also expected to make HidroAysén's energy available for mining in the country's far north.

The five HidroAysén dams would have flooded sections of two wild, previously undammed rivers, destroying ancient temperate forests, habitat of the endangered huemul deer. The dams would have also compromised fresh water habitats, and altered ecosystems in the saltwater fjords at the mouth of the rivers by trapping silt and river-borne nutrients in dams. They would have flooded part of a national park and, together with the transmission line, the project would have impacted six other national parks and eleven national reserves. The transmission line would have necessitated the clearing of some twenty-three thousand hectares of native vegetation. Social impacts included the opening up of a sparsely peopled region, characterized by small villages with a uniquely regional identity, to five thousand workers for over twelve years. An incipient eco-tourism industry in Aysén would also have been negatively affected. Importantly, of the 2,750 MW generating capacity of HidroAysén, none was to stay in the region, with all power exported to the north, despite the region being socio-economically disadvantaged. Chile's Ministry of Social Development reported in 2015 that Aysén had higher levels of multidimensional poverty (20.9 percent) than the country average (16.9 percent), and also higher levels of inequality than the Chilean average (CASEN 2015). Though the developing company promised jobs and progress in the region as a result of the dam project, previous Chilean and international experience has shown that large dams commonly increase poverty and underscore inequality, particularly in the developing world, where their stated mission is usually to do the opposite (Khagram 2004; McCully 1996).

HidroAysén was planned and being debated at a time when the issue of inequality in Chile was a pressing one, but postmaterialist values (Inglehart 1997; Inglehart et al. 2000) including environmental concern, were becoming more widespread (Franzen and Vogl 2013). Both these factors fed into discourse around the project. To turn first to inequality, Chile has experienced a remarkable trajectory of economic growth over the last forty years. Its 2017 GDP per capita of $24,248 (World Bank 2018), adjusted for purchasing power parity, puts it into the realm of a "developed economy." However, income distribution is highly unequal. In the middle of the HidroAysén timeframe, in 2014, Chile's GINI coefficient (a measure of a country's income distribution inequality where 0 represents perfect equality and 1 perfect inequality) was 0.5. This was the highest figure for an OECD country and substantially higher than the OECD average of 0.31 (OECD 2014, 3). Much research has also identified the persistence of intergenerational economic elites in Chile (Núñez and Miranda, 2011; Solimano, 2014; Solimano and Pollack, 2006; Torche, 2005). A section of 0.1 percent of the population, made up by some 4,460 families (CASEN 2013)—including the part-owners of HidroAysén—constitutes this elite in Chile. Such elites are often referred to as *los dueños de Chile*—"the owners of Chile" (Carmona 2002), and they have historically shaped and most importantly, *maintained* the structure of the country's economy to their own advantage. The maintenance of this status quo, sometimes termed "elite closure" (Núñez and Miranda 2011; Torche 2005), is achieved through elites' ability to influence politics (Joignant et al. 2015) and economic policy—especially in relation to income taxation—thereby "blocking any serious attempt to shift income distribution" (Solimano 2011, 10). At the time HidroAysén was being planned, "elite closure" and decades of neoliberal economic policy had therefore been effective in entrenching intergenerational disadvantage, and thwarting social mobility.

Despite this, research into inequality in age cohorts shows that for *some* Chileans the social and economic divide may have been diminishing during the HidroAysén timeframe, the GINI index for those born from the end of the 1960s to the late 1970s being eight to twelve points below the index for the nation as a whole (Sapelli 2011). As Inglehart et al. (1997, 2000) tells us, societies that are focused primarily on the attainment of material well-being have different values than postmaterialist societies, where participation in civil society, defense of the natural environment, and the pursuit of individual goals take precedence over more immediate material concerns. In the HidroAysén timeframe, therefore, as inequality among specific age cohorts began to reduce, frustration at social unfairness and elite closure began to extend to Chileans' thinking on the environment. Environmental concern had been increasing since the early 1990s, often in response to persistent environmental damage or new threats. Research into environmental attitudes in Chile, particularly among young people, indicates high levels of environmental awareness and concern, and a strong willingness to engage in pro-environmental behaviors (Cordano et al. 2011). At the time the HidroAysén project was being planned, therefore, frustration at intergenerational social inequality and aspiration to overcome it, coupled with new postmaterialist environmental concern, were key social trends in Chile.

"CLEAN, RENEWABLE, AND CHILEAN": DISCOURSES IN THE HIDROAYSÉN DEBATE

Understanding discourse in a mediatized conflict over water that stretched for eleven years, and in which mainstream media, the networks of "mass self-communication" (Castells 2009), "movement media" (Cammaerts 2015), and corporate strategic communications output are all important sites of rhetoric, is no small task. The research presented below is based on data that included thirty-eight semistructured interviews with key actors in the debate; a corpus of 2,385 print and online media articles from three key Chilean media outlets (*El Mercurio*, *La Tercera*, and *El Mostrador*); a wide range of movement-produced media; HidroAysén's corporate communications material; and, finally, social media output from both HidroAysén and *Patagonia sin Represas* and from key commentators on the project. In the sections that follow, I provide an overview of key features in the discourse surrounding the dams, and show how various discourses built on each other over time.

A "Nation Building Project" versus Destruction of the "National Image"

In October 2006, HidroAysén's then CEO, Hernán Salazar, made the first public presentation on the project in the town of Coyhaique, Aysén. The conservative newspaper *El Mercurio* described him as "nervous" as he presented to a "more than difficult public" (*El Mercurio* 07/10/2006). During Salazar's session, the newspaper reported, a group of activists entered the venue wearing T-shirts bearing the words "Aysén: Without Dams" and silently stood up on chairs to signal their rejection of the project.

This was the first salvo in the mediatized HidroAysén debate. The social movement campaign *Patagonia sin Represas*, or Patagonia without Dams (run by the Committee for the Defence of Patagonia, or CDP, a collective of local and international activist groups) had been established in response to HidroAysén and other dam projects in early 2006. Though it relied on local expertise and volunteering, the campaign was largely funded by foreign grant-making institutions and individuals. These included the American conservationist Douglas Tompkins, who owned large tracts of land in Chilean Patagonia, which were being restored as biodiverse habitat, with the aim of creating national parks to be donated to the Chilean state.

As Mitzi Urtubia, journalist and communications consultant for NGO Ecosistemas, a CDP member, recalled of the start of the *Patagonia sin Represas* campaign:

> Our campaign was highly proactive. It began even before the project began its own campaign. . . . Because of this, over time, we began winning, communicationally. It was we that started to inform people. . . . That they could oppose certain processes. (Personal interview, May 15, 2013)

The strategy of the project developer was initially to communicate as little as possible, on a national level at least, seemingly to avoid conflict, and to focus its communications in the communities local to the project area in Aysén. In the face of HydroAysén's low profile, *Patagonia sin Represas* initiated a communications campaign it called a national "citizen's alert." Urtubia explained how creating "visibility" in this way was crucial for *Patagonia sin Represas*:

> We believed that visibility was key. You have to be visible, and not establishing this kind of project, shall we say, behind the backs of the public. (Personal interview, May 15, 2013)

From 2006 to 2008 the social movement began to make HidroAysén visible on a national level, through a campaign including billboards and posters, public meetings, speaking engagements, "image events" (DeLuca 1999)—like a ten-day protest horse ride across Patagonia—and, for a limited time, a television advertisement. At this time, certain frames began to dominate media reporting on HidroAysén. Key commentators and conservative-oriented media tended to construct the project as a reason for national pride, a "nation building project" (e.g., *El Mercurio* 25/04/2007). The social movement's dominant discourse at this time directly rejected this frame. In its television ads and campaign posters, it focused on Chile's "national image" and its "destruction," arguing that the dams and power line would destroy landscapes and ecosystems and undermine this national image, at home and abroad. Campaign images at this stage in the debate showed iconic natural landscapes with power lines superimposed, often accompanied by the caption: "The worst national image" and "Destruction is not the solution."

"A Closeness to the Authorities": HidroAysén and Elites

While apparently endeavoring to remain out of nationwide public discourse, it seemed that HidroAysén was carrying out important work lobbying decision-makers in the capital to favor the megaproject. One ex-employee of HidroAysén commented in an interview:

> I think that HidroAysén from its beginning, was a project that cultivated a low profile and tried to generate a closeness to the authorities. (Personal interview, August 7, 2013)

An executive from the company's former PR firm confirmed:

> There was an active lobby by the shareholders. This wasn't outsourced, there wasn't any external agency. The company's owners did this directly to the source of power. (Personal interview, December 16, 2013)

This interviewee explained that HidroAysén's executives and owners could have influence over decision-makers in government "because they knew each other" and shared social circles. Latta (2010) writes that HidroAysén's owners "undoubtedly had little difficulty securing an audience with the president" (14).

Although closed-doors negotiations are difficult to pick apart, precisely because they are intended to be invisible, there were hints of such meetings in aspects of the discourse surrounding HidroAysén. For example, the general manager of Endesa explained to a shareholders' meeting in April 2007, "We are collaborating closely with the regulating entity so that norms are established that are in accordance with our interests" (Rodrigo and Orrego 2007). Interviewees for this study also spoke of "invisible negotiations" (personal interview, June 12, 2013) and "subterranean negotiations" (personal interview, November 26, 2013) between HidroAysén's directors and authorities, intended to smooth the path of the approvals process for the megaproject. Online newspaper *El Mostrador* also reported the existence of a "secret weapon" in the form of "energy fast trackers" who "travelled the length and breadth of the country," meeting with and "pressuring" local authorities (*El Mostrador* 17/11/2008). Over time, HidroAysén's perceived "closeness to the authorities" became an important aspect of the discursive construction of the project, so that its embeddedness within an intergenerational elite in Chile and the social unfairness of elite closure became things with which the project was associated, and for which it was criticized.

"Sovereign Energy" and "Foreign-ness"

HidroAysén's strategic communications material also tapped into ideas of "sovereign-ness"and "foreign-ness" in relation to the megaproject. A presentation that HidroAysén made to interest groups emphasized that HidroAysén was "national energy," "a sovereign resource," and offered "greater independence from other countries" (HidroAysén 2009c). The company's national television publicity also portrayed the megaproject in nationalistic terms, calling it "Chilean energy" (HidroAysén 2009a) and "sovereign energy" (HidroAysén 2009b), reducing dependence on fossil fuel imports from neighboring countries. A series of television ads and billboards from late 2009 described the energy that would come from HidroAysén as "clean, renewable and *Chilean*," and this frame became the common anchorage for much of the company's advertising output over the whole trajectory of the project.

Patagonia sin Represas and HidroAysén used the "nationality" of the project in their own strategic discourse—the social movement by reminding that this was a foreign majority-owned project, not just a Chilean one, and the project developer (backed by much media coverage) by noting the "foreign" nature of protest against the project. An *El Mercurio* editorial, in early 2009 (04/01/2009), wrote of "*energía propria*"—in Spanish, "our own energy"—in relation to hydroelectricity and HidroAysén. *El Mercurio* also described HidroAysén as "national . . . indigenous" (Bustamante and Orellana, *El Mercurio* 16/04/2009), "native hydraulic potential" (*El Mercurio*, 23/06/2009), and as making use of "sovereign resources" (*El Mercurio*, 15/04/2010).

On the other hand, the label of "foreign-ness" was also applied to the project developers. Center-right newspaper *La Tercera* published interviews with opponents of the project in which HidroAysén was criticized for being neocolonial and a product of "overseas capital," "Spanish capital," and "Italian capital" (for example, *La Tercera* 09/05/2010). Discourse around "foreign-ness" also appeared in influential left-leaning online news outlet, *El Mostrador*. For example, *El Mostrador* (4/10/2010) interviewed HidroAysén's CEO who spoke critically of "foreign interests," "overseas funding," and "foreign people" to describe the anti-dams movement. Douglas Tompkins, as *Patagonia sin Represas*'s key funder, was similarly referred to in negative terms as "this foreign person" (*El Mostrador* 21/10/2010). Notions of the sovereign and its opposites, the foreign and transnational, were becoming closely attached not only to the project, but also to its *opponents*. Later, in 2013 through 2015, a concerted social media campaign would also be waged against Douglas Tompkins, specifically, in which he was demonized as an "outsider," "extremist," "fundamentalist," and "usurper."

HidroAysén and Climate Change

HidroAysén's strategic communications across the trajectory of the debate had described the project as "clean" and "renewable." In early 2009, the World Bank (2009) released a study that underscored the importance of developing non–fossil fuel energy resources in Chile, amongst them hydroelectricity.

In the wake of this report, activist discourses and mediatized debate focused more closely on the project's link with climate change. *La Tercera* reported that, while "not a perfect solution, because in terms of energy, perfect solutions don't exist," HidroAysén was "the most reasonable, the cleanest, and the most compatible in the fight against climate change" (Astudillo, *La Tercera* 15/04/2009) and was part of the "worldwide crusade against global warming" (1/04/2010). HidroAysén's strategic communications output, including a pair of television ads shown in 2010/2011, also sought in this context to undermine the capabilities of alternative renewables, like wind and solar. The company's ex-PR executive, interviewed for this study (16/12/13), explained this was a "deliberate strategy." Government voices quoted in mainstream media seemed to concur. An article in *La Tercera* (Marticorena, *La Tercera* 21/11/2009) quoting the energy minister of the time, was headlined: "Tokman Warns That If the Dams in Aysén Are Not Built, Coal Will Replace Them." Mainstream media outlets in the study echoed this sentiment, presenting a binary of "HidroAysén or coal." *Patagonia sin Represas*'s advertising material attempted to highlight this as a "false dichotomy." In a full-page newspaper advertisement, for example, it stated:

> The Minister for Energy Marcelo Tokman and the National Energy Commission insist on deceiving the public with the false dichotomy between dams in Aysén and coal-based, or even nuclear, power plants. This dilemma does not exist.

The construction of false dichotomies between energy forms and their attributes is reminiscent of McGaurr and Lester's (2009) writing on "complementary problems and competing risks." McGaurr and Lester have shown that climate change can be used as a "justification" (184) for promoting particular types of energy, that particular energy sources can be constructed as a "solution to climate change" (184), and that "when environmental risks compete, media framing of one problem can clearly serve to disguise the risks of another" (185). In the case of HidroAysén, presenting the project as a solution to climate change was a persistent discursive characteristic for most of the project's trajectory. Only later in the project trajectory did alternative renewables escape the pervasive negatively framed discourse, as I discuss further in this chapter's final section.

"Rendering Technical," Rendering "Political"

Early strategic communications output by HidroAysén was dominated by a focus on the *technical*—that is, on the magnitude of the project, its vast environmental impact study (or EIS, which amounted to ten thousand pages), and the brilliance of its engineering. Key media also followed this cue, highlighting the technical in reporting the project. Such a focus in discourse around development projects has been conceptualized by Murray Li (2011) as "rendering technical." This term, which comes from a context of international development, describes the way in which neoliberal systems, often from the Global North, impose their requirements for classification on peoples and environments of the Global South. Rendering technical requires that a place be "investigated, mapped, classified, documented, interpreted" (60) in order to be better governed or economically improved. The results of such processes are often lengthy documents and strategies for change that propose "technical solutions" (66) to all perceived problems.

The HidroAysén EIS, submitted in 2008, was in the approvals process for nearly three years. The company was repeatedly required to submit new information, and elected to suspend the approvals process several times. During this time, leading up to the project's approval in May 2011, discourse around HidroAysén changed significantly. Between 2008 and 2011, HidroAysén was consistently being referred to as "political." By late 2009, in the lead up to the 2010 presidential elections, *El Mercurio* reported that HidroAysén had "transformed directly into a campaign issue" (20/08/2009). In an editorial, *El Mercurio* wrote that "once the technical requirements have been taken care of, these decisions are, after all, political" and that there was "influence here with political motives" (30/10/2009). In the same month, *El Mostrador* (20/10/2009) stated that "this has become a political project." Tompkins later noted "political manipulation and pressure" around the project in an article published in *La Tercera* (29/12/2010).

The change from discourse around the technical to discourse on the political is significant. On the eve of the approval of HidroAysén in 2011, the project was seen as one in which politically motivated decision-making was more significant than a decision on the project based on its technical aspects. This "political project" now became the subject of citizens' action, and this was the turning point in the social movement struggle against the project.

HidroAysén and Inequality

In May 2011, HidroAysén was granted its environmental approvals. Street protests began in Chile's capital and several other cities, including Aysén's regional capital. For two months, intermittent protest took place in up to twenty cities across the nation, particularly in the capital Santiago, where the largest protests numbered one hundred thousand people. Protests against HidroAysén intermingled with those of striking students, who had taken to the streets to oppose profiteering in higher education. The anti-dams protesters and the students shared many of the same grievances—chief among them, as discussed, the sense that Chileans were being taken advantage of by social and economic elites, and that the neoliberal model of Chile's post-dictatorship development must now change. This sentiment now seemed to have become strongly associated with HidroAysén, both in discourse on social media networks, and to some degree in the traditional media. A post from *Patagonia sin Represas*'s Facebook page demonstrates, for example, how the issue of inequality was highlighted by the anti-dams movement, alongside environmental concerns. The post emphasized the wealth of the Matte family, Colbún's owners, with a link to an article by broadcaster Radio Bío Bío, titled, "Researcher Analyses Inequality in Chile: 'The Country's Wealth Is Concentrated among 4 Families,'" and the direct quote, "among them, the Matte brothers, one of the HidroAysén partners, with $10.4 billion."

While some media coverage demonized the protestors as "eco-terrorists" (*La Tercera* 11/05/2011) and "quite extreme groups" that "defend their visions to the death" (*La Tercera* 04/05/2011), *El Mostrador* explicitly connected citizen discontent at inequality and wealth concentration with the unrest over the dams. One commentator in *El Mostrador* wrote, for example:

> The wound that has been opened is not to do with energy, but rather society. . . . HidroAysén has just been an excuse to talk about the wealth concentration which keeps Chile at the very bottom of the OECD and keeps Chileans expressing a profound social discontent. (Marinovic, *El Mostrador* 18/05/2011)

Sociologist Marcelo Mendoza wrote in *El Mostrador*, "HidroAysén . . . represents much more than the construction of dams in Patagonia. Somehow, it symbolizes the disgust of the majority precisely for this fantasy of unbridled 'growth' and the abstraction known as 'the market,'" adding, "This rebellion symbolizes the stellar distance which exists between power and the interests of the business-political class, and ordinary people" (Mendoza, *El Mostrador* 21/05/2011).

In an interview for this research, deputy editor of *La Tercera*, Felipe Contreras (16/07/2013) also described HidroAysén in terms of what it represented to the Chilean majority: "an emblem, an icon of the neoliberal market model." Discourse linking the dams and social unfairness became a persistent feature of the mediatized dams debate from 2011 through 2014. Unable to effectively refute this discursive construction, the dams developer elected to withdraw from national publicity at this stage, and now concentrated its communications on a small scale in the communities in the potentially affected region.

HidroAysén and Energy Transformation

Patagonia sin Represas confronted the task of keeping the dams "visible" nationally between 2011 and 2014 with various "image events" (DeLuca 1999), and also by using public interest litigation against the project. HidroAysén still required one further level of approval, and, given intervening political machinations, litigation, appeals, and investigations (some the initiative of the anti-dams movement), this approval was not finalized until 2014. In June of that year, a committee of ministers (of environment, energy, mining, agriculture, and health) unanimously, and momentously, found HidroAysén's EIS had not complied with environmental law and withdrew the permissions granted in 2011. The day after environmental permits were withdrawn, *La Tercera* published a piece by then Energy Minister Pacheco in which he wrote of the decision as "symbolic" and of the necessity to "dehidroaysén-ize" the debate over energy (Pacheco, *La Tercera* 11/06/2014). Three years of further litigation followed. Although the company vowed to keep working on the project, Chile was now on the brink of an energy transformation.

By early 2017, energy debates—and the practical likelihood of HidroAysén being built—seem to have changed entirely. Mediatized discourse on energy seemed to credit the debate around HidroAysén, at least in part, for initiating this transformation. In the seven years to 2017, the price of photovoltaic power had fallen 75 percent (Vargas, *El Mercurio* 20/04/2017), and Chile, with its near-cloudless skies in the arid north, could now produce some of the cheapest solar power in the world. In 2017, *El Mercurio* reported as front-page news that by 2024, alternative renewables, primarily solar energy, would be supplying 58 percent of the energy that HidroAysén would have generated, and quoted the president of Enel in Chile, Herman Chadwick, as saying that it would now be "a monumental error to invest in 'traditional' power generation projects in Chile" (*El Mercurio* 26/04/2017). *El Mercurio* quoted the Minister of Energy, Andrés Rebolledo, as saying: "We have proven that it is possible to implement energy in a different way" (*El Mercurio* 17/11/2017a) while Environment Minister Marcelo Mena said, "Since we closed the door on HidroAysén, investment in energy has doubled, alternative renewables have increased five-fold and prices have fallen 75 percent" (*El Mercurio* 05/11/2017a).

On December 22, 2017, HidroAysén called an extraordinary meeting of shareholders to vote on dissolution of the project and the return of water rights to the Chilean state. *El Mercurio*'s editorial on that day stated, "In practical terms, the suspension of HidroAysén has been beneficial for the country, because it opened the possibility to further develop alternative renewables, which have experienced a spectacular growth, especially wind and solar, with a reduction in tariffs" (*El Mercurio*, 22/12/17). For a conservative newspaper which had almost invariably supported HidroAysén, this admission was surprising, and is perhaps a measure of how much the case of HidroAysén changed Chile. Even more telling is another piece in *El Mercurio*, which described the end of HidroAysén as "a citizens' achievement" and quoted an activist as saying:

> *Patagonia sin Represas* has been the most important socio-environmental cause in Chile's recent history. It has strongly influenced public policy in relation to water, social participation, and of course, energy. . . . It's something to be happy about, proud of, and grateful for, that the work of so many has achieved what seemed impossible on so many fronts. Now is the time to celebrate. (*El Mercurio* 18/12/2017)

As mediatized debate on Chile'e energy transformation continues to date, HidroAysén is often cited as a turning point. For example, energy minister Susana Jiménez, interviewed in *La Tercera* in 2018, cited the HidroAysén debate and noted "how important this discussion has been for the country" (Orellana and Santa María, *La Tercera* 18/03/2018). Chile has now become what the *New York Times* has described as "a global leader in countering climate change" (10/03/2018). Without HidroAysén, the country is moving toward a more diverse, less centralized, and lower carbon model of energy. This change may not have been as complete had the "clean" energy from HidroAysén been available.

CONCLUSION: HIDROAYSÉN DISCOURSES AND THE MEDIATIZED STRUGGLE FOR WATER JUSTICE

As Alexander (2006, 33–34) tells us, in "power conflicts" such as the mediatized conflict between *Patagonia sin Represas* and HidroAysén, "representation is critical." The protest campaign had fought since 2006 to influence mediatized representation of the HidroAysén megaproject, to sway public opinion, and to influence government decision-making on the project. The social movement had had some success with making the project "visible" (Thompson 2005) in the early years, while the project developers were reluctant to communicate on a national level. Media discourse still tended to cover HidroAysén more, however, and to reproduce its frames, arguably becoming an *actor* in the mediatized conflict of the HidroAysén debate, as Cottle puts it: "implicated within conflicts while disseminating images or ideas about them" (Cottle 2006, 9).

As a set of discourses built up around HidroAysén and accumulated over time, the HidroAysén project itself took on symbolic meaning. If protesters were, as mediatized discourse constructed them, "extreme," "foreign," and "eco-terrorists," blocking development and rejecting "clean" energy (which several scholars including Bratman, 2015; Da Silva and Rothman 2011; and Ozen, 2014 have pointed to as common discourses on anti-dams protesters), then HidroAysén was environmentally disastrous, "foreign" owned, profiteering, and a project for and by elites, which was ultimately unnecessary. As one *Patagonia sin Represas* activist put it, HidroAysén "confused the necessity of energy development in Chile with the energy business of a few" (Rodrigo, *La Tercera* 07/04/2009). As the massive street protests of 2011 showed, in the mediatized struggle for issue definition between the social movement and the project developer, it was ultimately *this* discursive representation of the HidroAysén project which attracted the attention of Chileans. Aligning HidroAysén with social unfairness was also a way to connect those in the remote far south of the country with people in Chile's cities and its capital. Here, where water-based injustice in the case of HidroAysén would not be experienced directly, inequality and social unfairness *were* a daily experience for many—and this galvanized Chileans to protest HidroAysén. With one hundred thousand people in the streets, media coverage was abundant, because, as Wolfsfeld (1997) tells us, media and protest are interacting, mutually dependent systems. It seemed, then, it was finally the association between the project and *inequality* in Chile that allowed the social movement to operate the "switches" that Arsenault and Castells (2008), Castells (2009), and Hutchins and Lester (2015) have theorized as the sites of influence and power in mediatized enactment of conflicts related to the environment.

Liberal justice theorists tell us that beyond simply unequal *distribution*, lack of recognition, lack of opportunities for participation, and lack of the capabilities necessary to enable individual lives to flourish are at the core of injustice. In short, the circumstances that cause inequality are also the circumstances that allow injustices to prevail. In questions of water justice specifically—particularly in the context of globalization and neoliberal economic policy—inequality is underscored in the ways that powerful national and often transnational interests accumulate water resources and rights to the detriment of the economically less powerful (Boelens, Perreault and Vos 2018; Zwarteveen and Boelens 2014). Such transfers of water from supposedly less to supposedly more productive use—in the case of HidroAysén, from maintaining Chile's natural ecosystems and local cultural integrity to providing power for the country's "development"—are legitimized and supported by neoliberal modes of thinking, acting, and speaking about water. The language we use for talking about and the ways that we represent knowledge about issues are constitutive of those issues, and controlling issue-definition is a position of power. Power and politics can therefore pervade both the allocation of water and ways of thinking and talking about water. But, as the case of HidroAysén shows, power—even the power to affect an energy transformation—can also be acquired by less resourced actors, and such actors can steer public discourse on an issue to achieve "socio-ecological justice" (Zwarteveen and Boelens 2014).

The case of HidroAysén, debated for eleven years and producing vast amounts of media coverage as well as a plethora of strategic communications aimed at influencing public opinion, is a rich source for the formulation of empirical insights into discourses over water justice. It points to the necessity to consider a *discursive* dimension to the political ecology of water; a place where definitions of water justice are formulated but also *contested,* and a place where the less powerful sometimes prevail.

REFERENCES

Alexander, Jeffrey C. 2006. *The Civil Sphere*. Oxford: Oxford University Press.

Anderson, Alison. 1997. *Media, Culture and the Environment*. London: Routledge.

Ansar, Atif, Bent Flyvbjerg, Alexander Budzier, and Daniel Lunn. 2014. "Should We Build More Large Dams? The Actual Costs of Hydropower Megaproject Development." *Energy Policy* 69: 43–56.

Arsenault, Amelia H., and Manuel Castells. 2008. "The Structure and Dynamics of Global Multi-Media Business Networks." *International Journal of Communication* 2: 707–48.

Bakker, Karen. 2010. *Privatizing Water: Governance Failure and the World's Urban Water Crisis*. Ithaca, NY: Cornell University Press.

Berkun, Mehmet. 2010. "Hydroelectric Potential and Environmental Effects of Multidam Hydropower Projects in Turkey." *Energy for Sustainable Development* 14, no. 4: 320–29.

Blake, David J. H., and Keith Barney. 2018. "Structural Injustice, Slow Violence? The Political Ecology of a 'Best Practice' Hydropower Dam in Lao PDR." *Journal of Contemporary Asia* 48, no. 5: 808–34.

Boelens, Rutgerd, and Jeroen Vos. 2012. "The Danger of Naturalizing Water Policy Concepts: Water Productivity and Efficiency Discourses from Field Irrigation to Virtual Water Trade." *Agricultural Water Management* 108: 16–26.

Boelens, Rutgerd, Tom Perreault, and Jeroen Vos. 2018. *Water Justice*. Cambridge: Cambridge University Press.

Borgias, Sophia L., and Yvonne A. Braun. 2017. "From Dams to Democracy: Framing Processes and Political Opportunities in Chile's Patagonia without Dams Movement." *Interface: A Journal on Social Movements* 9, no. 2: 300–328.

Bratman, Eve. 2015. "Passive Revolution in the Green Economy: Activism and the Belo Monte Dam." *International Environmental Agreements: Politics, Law and Economics* 15, no. 1: 61–77.

Cammaerts, Bart. 2015. "Movement Media as Technologies of Self-Mediation." In *The Routledge Companion to Alternative and Community Media*, edited by Chris Atton, 445–56. London: Routledge.

Carmona, Ernesto. 2002. *Los Dueños de Chile*. Santiago, Chile: Ediciones La Huella.

Carruthers, David, and Patricia Rodriguez. 2009. "Mapuche Protest, Environmental Conflict and Social Movement Linkage in Chile." *Third World Quarterly* 30, no. 4: 743–60.

CASEN. 2015. "Region de Aysén 2015." Ministerio de Desarollo Social. Accessed October 21, 2018. http://observatorio.ministeriodesarrollosocial.gob.cl/indicadores/docs/region/Aysen.pdf.

CASEN. 2013. "Region de Aysén 2013." Ministerio de Desarollo Social. Accessed July 7, 2015. http://www.ministeriodesarrollosocial.gob.cl/resultados-encuesta-casen-2013/%3E.

Castells, Manuel. 2007. "Communication, Power and Counter-Power in the Network Society." *International Journal of Communication* 1, no. 1: 238–66.

Castells, Manuel. 2009. *Communication Power*. Oxford: Oxford University Press.

Cordano, Mark, Stephanie Welcomer, Robert F. Scherer, Lorena Pradenas, and Víctor Parada. 2011. "A Cross-Cultural Assessment of Three Theories of Pro-Environmental Behavior: A Comparison between Business Students of Chile and the United States." *Environment and Behavior* 43, no. 5: 634–57.

Cottle, Simon. 2006. *Mediatized Conflict: Developments in Media and Conflict Studies*. New York: Open University Press.

Couldry, Nick, and Andreas Hepp. 2018. *The Mediated Construction of Reality*. Cambridge: John Wiley & Sons.

Da Silva, Patrícia Pereira, and Franklin Daniel Rothman. 2011. "Press Representation of Social Movements: Brazilian Resistance to the Candonga Hydroelectric Dam." *Journal of Latin American Studies* 43, no. 4: 725–54.

DeLuca, Kevin Michael. 1999. *Image Politics: The New Rhetoric of Environmental Activism*. New York: Guilford Press.

Elizondo, Gabriel. 2009. "Kaipaó Protest in Brazil against Dams." *Al Jazeera*, 29 October 2009. aljazeera.com/blog/americas/kaipa%C3%B3-protest-brazil-against-dams.

Fearnside, Philip M. 1997. "Greenhouse-Gas Emissions from Amazonian Hydroelectric Reservoirs: The Example of Brazil's Tucuruí Dam as Compared to Fossil Fuel Alternatives." *Environmental Conservation* 24, no. 1: 64–75.

Fearnside, Philip M. 2004. "Greenhouse Gas Emissions from Hydroelectric Dams: Controversies Provide a Springboard for Rethinking a Supposedly 'Clean' Energy Source." *Climatic Change* 66, no. 1: 1–8.

Fearnside, Philip M. 2015. "Tropical Hydropower in the Clean Development Mechanism: Brazil's Santo Antônio Dam as an Example of the Need for Change" *Climatic Change* 131, no. 4: 575–89.

Foucault, Michel. 1977. *Discipline and Punish: The Birth of the Prison*. Translated by Alan Sheridan. New York: Pantheon.

Franzen, Axel, and Dominikus Vogl. 2013. "Two Decades of Measuring Environmental Attitudes: A Comparative Analysis of 33 Countries." *Global Environmental Change* 23, no. 5: 1001–8.

Gamson, William A., and Gadi Wolfsfeld. 1993. "Movements and Media as Interacting Systems." *The Annals of the American Academy of Political and Social Science* 528, no. 1: 114–25.

Hall, Stuart. 1997. *Representation: Cultural Representations and Signifying Practices*. Vol. 2. London: Sage.

Hansen, Anders. 2010. *Environment, Media and Communication*. London: Routledge.

Harris, Leila M., Jacqueline A. Goldin, and Christopher Sneddon. 2013. *Contemporary Water Governance in the Global South: Scarcity, Marketization and Participation*. Abingdon: Routledge.

HidroAysén. 2009a. "HidroAysén Viento." Accessed December 12, 2018. https://www.youtube.com/watch?v=HhSqKZeF74g.

HidroAysén. 2009b, "HidroAysén Nieve." Accessed December 12, 2018. https://www.youtube.com/watch?v=TXTV_gmV0Pk.

HidroAysén. 2009c. *El Proyecto HidroAysén*. Santiago, Chile.

Hoffman, Steven M. 2002. "Powering Injustice: Hydroelectric Development in Northern Manitoba." In *Environmental Justice: Discourses. International Political Economy, Energy and Environmental Policy,* edited by John Byrne, Leigh Glover, and Cecilia Martinez, 147–70. London: Routledge.

Hutchins, Brett, and Libby Lester. 2011. "Politics, Power and Online Protest in an Age of Environmental Conflict." In *Transnational Protests and the Media*, edited by Simon Cottle and Libby Lester, 159–71. New York: Peter Lang.

Hutchins, Brett, and Libby Lester. 2015. "Theorizing the Enactment of Mediatized Environmental Conflict." *International Communication Gazette* 77, no. 4: 337–58.

Inglehart, Ronald. 1997. *Modernization and Postmodernization: Cultural, Economic, and Political Change in 43 Societies*. Princeton, NJ: Princeton University Press.

Inglehart, Ronald, Miguel Basanez, Jaime Diez-Medrano, Loek Halman, and Ruud Luijkx. 2000. *World Values Surveys and European Values Surveys, 1981–1984, 1990–1993, and 1995–1997*. Ann Arbor, MI: Institute for Social Research.

Islar, Mine. 2013. *Private Rivers: Politics of Renewable Energy and the Rise of Water Struggles in Turkey*. PhD diss., Lund University.

Joignant, Alfredo, Lucas Perelló, and Javier Torres. 2015. "Political Capital and the Unequal Career Origins of the Political Elite in Chile." In *Political Inequality in an Age of Democracy: Cross-National Perspectives*, edited by Joshua Kjerulf Dubrow, 87–94. London: Routledge.

Khagram, Sanjeev. 2004. *Dams and Development: Transnational Struggles for Water and Power*. Ithaca, NY: Cornell University Press.

Latta, Alex. 2010. "Generating Consent: Mega-Hydro and the Post-Neoliberal Corporation, the Case of HidroAysén." Paper presented at WATERLAT International Conference, São Paulo, Brazil, 2010.

Lester, Libby. 2010. "Big Tree, Small News: Media Access, Symbolic Power and Strategic Intervention." *Journalism* 11, no. 5: 589–606.

Li, Tania Murray. 2011. "Rendering Society Technical." *Adventures in Aidland: The Anthropology of Professionals in International Development* 6: 57–80.

Lutsey, Nicholas, and Daniel Sperling. 2008. "America's Bottom-Up Climate Change Mitigation Policy." *Energy Policy* 36, no. 2: 673–85.

Maeck, Andreas, Tonya DelSontro, Daniel F. McGinnis, Helmut Fischer, Sabine Flury, Mark Schmidt, Peer Fietzek, and Andreas Lorke. 2013. "Sediment Trapping by Dams Creates Methane Emission Hot Spots." *Environmental Science & Technology* 47, no. 15: 8130–37.

Martínez, Viviana, and O. L. Castillo. 2016. "The Political Ecology of Hydropower: Social Justice and Conflict in Colombian Hydroelectricity Development." *Energy Research & Social Science* 22: 69–78.

McCully, Patrick. 1996. *Silenced Rivers: The Ecology and Politics of Large Dams*. London: Zed Books.

McGaurr, Lyn C., and Libby A. Lester. 2009. "Complementary Problems, Competing Risks: Climate Change, Nuclear Energy and the Australian." In *Climate Change and the Media*, edited by Tammy Boyce and Justin Lewis, 174–85. New York: Peter Lang.

Núñez, Javier, and Leslie Miranda. 2011. "Intergenerational Income and Educational Mobility in Urban Chile." *Estudios de Economía* 38, no. 1: 195–221.

OECD. 2014. "Society at a Glance 2014 Highlights: Chile." OECD Social Indicators, accessed April 13, 2016. http://www.oecd.org/social/societyataglance.htm%3E.

Ozen, Hayriye. 2014. "Overcoming Environmental Challenges by Antagonizing Environmental Protesters: The Turkish Government Discourse against Anti-Hydroelectric Power Plants Movements." *Environmental Communication* 8, no. 4: 433–51.

Rodrigo, Patricio, and Juan Pablo Orrego. 2007. *Patagonia Chilena ¡Sin Represas!* Santiago, Chile: Ocho Libros Editores.

Rucht, Dieter. 2013. "Protest Movements and Their Media Usages." In *Mediation and Protest Movements*, edited by Bart Cammaerts, Alice Mattoni, and Patrick McCurdy, 249–68. Bristol: Intellect.

Sapelli, Claudio. 2011. *Chile Más Equitativo: Una Mirada Distinta a la Distribución del Ingreso, la Movilidad Social y la Pobreza en Chile*. Santiago: Ediciones Universidad Católica de Chile.

Schneider, Flurina, T. Buser, and Olivier Graefe. 2014. "Scales of Justice in Water Governance: Hydropower Controversies in Switzerland." *Water Policy* 16, no. S2: 137–54.

Siciliano, Giuseppina, and Frauke Urban. 2017. "Equity-Based Natural Resource Allocation for Infrastructure Development: Evidence from Large Hydropower Dams in Africa and Asia." *Ecological Economics* 134: 130–39.

Solimano, Andres. 2011. *Prosperity without Equality: The Chilean Experience after the Pinochet Regime.* International Center for Globalization and Development: Santiago, Chile.

Solimano, Andrés. 2014. *Economic Elites, Crises, and Democracy: Alternatives beyond Neoliberal Capitalism*. New York: Oxford University Press.

Solimano, Andrés, and Molly Pollack. 2006. *La Mesa Coja: Prosperidad y Desigualdad en el Chile Democrático*. Santiago: Centro Internacional de Globalización y Desarrollo, CIGLOB.

Swyngedouw, Erik. 2005. "Dispossessing H2O: The Contested Terrain of Water Privatization." *Capitalism Nature Socialism* 16, no. 1: 81–98.

Thompson, John B. 1995. *The Media and Modernity: A Social Theory of the Media*. Stanford, CA: Stanford University Press.

Thompson, John B. 2005. "The New Visibility." *Theory, Culture & Society* 22, no. 6: 31–51.

Thompson, John B. 2011. "Shifting Boundaries of Public and Private Life." *Theory, Culture & Society* 28, no. 4: 49–70.

Thompson, John B. 2013. *Political Scandal: Power and Visibility in the Media Age*. Cambridge: Polity Press.

Torche, Florencia. 2005. "Unequal but Fluid: Social Mobility in Chile in Comparative Perspective." *American Sociological Review* 70, no. 3: 422–50.

Tur, Antonio Aledo, Hugo García-Andreu, Guadalupe Ortiz, and Jose Andres Domínguez-Gomez. 2018. "Discourse Analysis of the Debate on Hydroelectric Dam Building in Brazil." *Water Alternatives* 11, no. 1: 125–41.

Wehrli, Bernhard. 2011. "Climate Science: Renewable but Not Carbon-Free." *Nature Geoscience* 4, no. 9: 585–86.

Wolfsfeld, Gadi. 1997. *Media and Political Conflict: News from the Middle East*. Cambridge: Cambridge University Press.

Wooden, Amanda E., and Jeanne Féaux de la Croix. 2015. "The 'Great Future of the Country'? Dams and Hydroelectricity Discourses in Kyrgyzstan." In *Environmental Crises in Central Asia*, edited by Eric Freedman and Mark Neuzil, 70–82. London: Routledge.

World Bank. n.d. "Chile Country Data 2018." https://data.worldbank.org/country/chile.

Zanotti, Laura. 2015. "Water and Life: Hydroelectric Development and Indigenous Pathways to Justice in the Brazilian Amazon." *Politics, Groups, and Identities* 3, no. 4: 666–72.

Zarfl, Christiane, Alexander E. Lumsdon, Jürgen Berlekamp, Laura Tydecks, and Klement Tockner. 2015. "A Global Boom in Hydropower Dam Construction." *Aquatic Sciences* 77, no. 1: 161–70.

Zwarteveen, Margreet. 2015. *Regulating Water, Ordering Society: Practices and Politics of Water Governance*. PhD diss., University of Amsterdam.

Zwarteveen, Margreet Z., and Rutgerd Boelens. 2014. "Defining, Researching and Struggling for Water Justice: Some Conceptual Building Blocks for Research and Action." *Water International* 39, no. 2: 143–58.

Chapter Three

Culture-Jam or Log-Jam?

Rhetorics of Spectacle Protest in the Free the Snake Flotilla

Casey R. Schmitt

In September 2017, they came in rafts and in kayaks, in motor boats, paddle-boards, and dugout canoes, drifting down the Snake River just west of Clarkston, Washington, for six miles to the tune of hoots, laughter, drumming, and song. Some floated freely and others paddled together, tethered to twelve giant red letters floating along the surface with a message visible from the air: "FREE THE SNAKE." In 2017, they were four hundred strong. In 2018, roughly six hundred joined the protest, the celebration, the party, the cause.

For four years now the annual Free the Snake Flotilla has grown in size, enacting a ludic and joyful yet pointed protest calling for the removal of four "deadbeat dams" along the lower Snake. Participants gather on a weekend to eat and dance, learn and discuss, and ultimately float together as one pack down the namesake waterway. The event is one part environmental activism, one part leisure afternoon. It's one part political and one part fun. It is a visual spectacle, raising awareness and media attention in the tradition of Greenpeace "happenings" and Earth First! monkeywrenching (Short 1991; Killingsworth and Palmer 1992), but also different from those radical instances for its performative elements, which depend on recreation rather than outright disruption or revolution.

As the event grows from year to year, it's apparent that it resonates with participants, including scientists, nonprofit activists, outdoor enthusiasts, and members of the Nez Perce, Colville, Kalispel, Coeur d'Alene, and Spokane tribes. The shared drive to remove the dams, to enact community sovereignty over the river's flow, is motivated by multiple crises. These include plummeting chinook and steelhead salmon populations, eliminating the primary food source for the North Pacific's southern resident killer whales (SRKW), decreasing economic benefits from barge traffic, and a demand for reparations by the tribes whose historical gathering sites and burial grounds were submerged when the dams were installed. The varied commitments unite participants, and the event at first glance exemplifies the "image politics" of environmentalist rhetoric identified by DeLuca (1999) and the "spectacle" of participatory display discussed by Procter (1990) and Halloran (2001).

Yet, at the same time, the Free the Snake Flotilla represents a notable subcategory within these fields of highly visible and intentionally creative demonstrations. Certainly, the Flotilla is a spectacle and certainly it is protest, but it is also similar in many ways to other nonpolitical and recreational crowd-scale boating, swimming, biking, and hiking events. Participants in the spectacle are not "culture jamming" (Harold 2004) in the radical sense of social disruption activism but simply gathering in a critical mass of leisure engagement.

This form of spectacle protest sees analogs in crowd-scale bike rides for ostensibly activist causes, including the World Naked Bike Ride and the appropriately named Critical Mass bike tour, or "dance-ins" at spaces already demarcated for public festival and celebration.

These leisure-structured spectacle protests could, potentially, be misinterpreted or divorced from their activist message by the passerby audience, and they demand less risk on the part of the participant than, say, ecotage and ecoactivism (Standing 2012) or mass demonstration in heavily policed city streets.

This chapter examines *leisure spectacle demonstration* as an emergent subgenre of spectacle protest, outside of deliberative or other sanctioned modes of protest, distinct from radical ecoactivist culture jamming but also from everyday acts of resistance (Senda-Cook and McHendry 2018) or slacktivism (Arora 2015). It applies rhetorical criticism to the material and image rhetorics of the annual event to explore whether the leisure-structured spectacle gatherings function as persuasive/disruptive culture-jam or, rather, are merely spaces of constitutive identity performance (Charland 1987) reaffirming an in-group sentiment for those who take part in the display/recreational activity.

The analysis suggests that the Flotilla invites some degree of confusion from outsiders, by depending on the material markers of simple leisure watersport (i.e., outdoor exercise gear, high-end kayaks, Patagonia labels) but also encourages insider identification with the collective cause, through playful inclusion of protest signs, orca artwork, and songs about salmon. I also trace the event as a notable moment in Native American resistance rhetoric and media spectacle, building on themes from Roosvall and Tegelberg

(2018) to suggest that such demonstrations are particularly open avenues for indigenous protest.

I argue that such events differ from radical culture jams but also from the "everyday" or "mundane acts of resistance" that Senda-Cook and McHendry (2018) argue can "rhetorically construct practices of *ecological habitual resistance*" (225) because the simple, leisure spectacle is limited to a once- or twice-a-year event. The chapter seeks to contribute to critical critique of the recreation- or leisure-structured protest, engaging recent work by Arora (2015) and Newlands (2018), as well. It engages questions of justice through its relationship to Bullard's (2005) central questions for environmental justice and equity (i.e., "What groups are most affected by a specific environmental problem? Why are they affected? Who caused the problem? What can be done to remedy it?") (29). Along the way, it grapples with questions of vulnerability, need, and responsibility (Walker 2012, 46). Within the broader aims of this collection, the chapter illuminates a novel and creative rhetorical form in current water sovereignty campaigns.

IMAGE POLITICS, HAPPENINGS, AND CULTURE JAMMING

The Free the Snake Flotilla and other such highly visible social/political demonstrations, disrupting status quo flows of space, discourse, and movement with visual displays of unusual size or incongruity are what DeLuca (1999) identifies as *image events*. The unusual event serves as spectacle, demanding attention and forcing conversation in spheres and in publics wherein attention to and conversations about specific topics of advocacy might not otherwise happen. DeLuca highlights how environmental advocacy groups have frequently used such image events to campaign for environmental justice and other environmental causes when deliberative, legal, and mass media avenues have failed to prioritize or even attend to environmental concerns. The image event disrupts and demands attention through visual spectacle, to "contest social norms and deconstruct the established naming of the world" (59). The disruption draws media attention and coverage, and the coverage forces the conversation. Among groups like Greenpeace and Earth First!, he notes, the events make the conversation happen and thus are literally termed "happenings."

Image events draw attention and function rhetorically as *spectacle* in the sense discussed by Farrell (1989), Procter (1990), and Halloran (2001). Spectacles are as much material constructions of an event as they are "transformations" of that event and thus rhetorical, "because the construction of the event by interest group rhetorics encourages the body politic to: (1) recognize some issue or event as salient and (2) perceive the issue or event from the rhetor's perspective" (Procter 1990, 119). Farrell (1989) attends to the performative and ideologically constructed aspects of spectacle events, referring to spectacle events as "choreographed displays" (161) infused with "social knowledge assumptions" (p. 180), which, Procter (1990) notes, "include explicit or implicit characterizations of heroes and villains, accounts of acceptable and unacceptable behaviors, and statements of appropriate and inappropriate values" (119). Procter adds, "Spectacles, then, are dramatic accounts of material experience which occurs beyond our personal purview and which we receive only through symbols developed by some interest group" (119). And Halloran (2001) extends the concept by emphasizing the role of audience in the event. Spectacle, as Halloran defines it, is "a public gathering of people who have come to witness some event and are self-consciously present to each other as well as to whatever it is that has brought them together" (5). The sensory engagement of spectacle—that which is both *spectated* and *spectacular*—invites interpretation through direct experience that viscerally guides viewers to particular subject positions. Through circulation among the community of spectators, spectacular cases encourage communities to rally around collaborative interpretations, with judgments formed through shared, lived experience of the memorable event.

So, on one level, the spectacle image event engages the communities immediately present to witness, or "spectate" the happenings. On another, though, DeLuca notes, such events are designed explicitly to catch the attention of *mass media*.

Newlands (2018) has recently explored the relationship between ecoactivism and mass, commercial media, applying critical discourse analysis to media and political representation of ecoactivism. She notes that in any public controversy or event, the core conflicts of story coverage come from media outlets as "*primary definers*" (33), and traces how, over the past fifty years, the role of primary definer in environmental reporting has shifted "from protestors to police, politicians, and social commentators." Mainstream reporting, she argues, has removed "the activists from the narrative and replaced [them] with comments from official spokespersons who speak on or about, but not within movements" (63). In recent decades, ecoactivists have increasingly been defined as a threat, through language framing ecoactivism as a form of terrorism. Newlands notes how many ecoactivist and advocacy groups initially met this media shift with blackout silence, but that they have since become "media movements," strategically mobilized to engage media messaging through savvy baiting and guiding of mass and social media attention (126).

Image events, in this respect, become a tool for seizing media and public attention, and setting the agenda for ensuing public conversations about the environment. Such tactics have recently proved especially effective for indigenous ecoactivism, in communities historically ignored by or actively rejected from mainstream political and social discourses. Roosvall and Tegelberg (2018) have, for instance, explored how *indigenous communities*, unevenly impacted by the effects of climate change, have campaigned for transnational climate justice by using spectacle to correct misframing and enact self-representation. Moore (2019), too, has used framing analysis to explore how profit-driven news media coverage has historically ignored environmental justice campaigns and how members of the Standing Rock Sioux Tribe achieved media and political recognition through media spectacle surrounding protests against the Dakota Access Pipeline.

In these cases, as in others, part of the image event/spectacle rhetoric is the location of the display or demonstration itself. Indigenous activists from around the world marching in the streets of Paris (Roosvall and Tegelberg 2018) and Sioux protestors facing attack dogs and water cannons on historically potent tribal land (Moore 2019) enact what Endres and Senda-Cook (2011) have identified as both *place-based argument* and *place-as-rhetoric*. Image events disrupt the everyday flow of movement through spaces and places and, in doing so, make their appeals through place and placement. "Place-based arguments," they write, "discursively invoke images or memories of a place to support an argument, such as summoning the melting of the arctic as a reason to stop global warming, and make salient that dominant place meanings are sometimes linked to systems of power that discourage protest" (258). "Place-as-rhetoric," conversely, "refers to the material (physical and embodied) aspects of a place having meaning and consequence, be it through bodies, signage, buildings, fences, flags, and so on" (265) and "assumes that the very place in which a protest occurs is a rhetorical performance that is part of the message of the movement" (258–59). "(Re)constructing the meaning of place, even in temporary ways," they write, "can be a tactical act of resistance along with the tactics we traditionally associate with protest" (258), and in ecoactivist demonstrations, the environment is always part of the spectacle itself, dynamic and reconstructed through and with the bodies moving within it, "making and unmaking the possibilities of protest" (258).

These different elements in combination—image event, spectacle, mass media attention and framing, local, marginalized and/or indigenous communities, and reconstructions of place/space—are central to contemporary ecoactivist rhetorics and campaigns like the Free the Snake Flotilla. They all also, in combination, begin to fit the definition of a radical activist "culture jam."

"Culture jamming" is subversive cultural expression, mixing and remixing cultural symbols and practices to disrupt the status quo, question mainstream assumptions, and challenge institutional power structures (Harold 2004). It is often anticonsumerist and satirical, and culture jamming "pranksters" use cultural disruption and incongruent action to critique the very structures in which they perform.

Harold (2004) notes the term is "based on the CB slang word 'jamming' in which one disrupts existing transmissions" and "usually implies an interruption, a sabotage, hoax, prank, banditry, or blockage of what are seen as the monolithic power structures governing cultural life" (192). He presents, as examples, the Barbie Liberation Organization, which critiqued mass market gender norms by swapping the electronic vocal chips of in-store Barbie and G. I. Joe figurines, and the Biotic Baking Brigade, which challenges global monetary power structures by hitting influential figures with literal pies in the face. These playful sabotages and pranks are designed not only to disrupt norms but also explicitly to garner media attention, coverage, and dissemination.

Culture jamming is a mainstay of the environmental advocacy image politics explored by DeLuca, including Greenpeace and Earth First! Inspired largely by Edward Abbey's 1975 novel *The Monkey Wrench Gang*—in which a ragtag band engages in a series of playfully disruptive and increasingly destructive acts of sabotage to protect wild spaces—these groups have frequently relied on culture jamming disruption. Earth First!'s first major "happening" was, for instance, the installation of a three-hundred-foot black ribbon on the Glen Canyon Dam (the same dam targeted for demolition by Abbey's fictional gang) to create the illusion of an enormous crack in the structure (Short 1991).

Ecological culture jamming can vary in form, ranging from dressing as grizzly bears and wandering about Yellowstone National Park and camping in the canopy of trees scheduled for harvest (Short 1991) to trespassing and turning energy pipeline shutoff valves or sabotaging forestry equipment (Nicholson 2019), but while it is always disruptive, it is never violent against living beings. Rather, Short writes, it is "guerilla theater" meant to enact an agitative "rhetoric of confrontation" (172). It is designed to (1) draw attention and (2) push mainstream audiences—including more mainstream environmentalists—to respond to controversial issues (Short 1991).

Culture jams are inherently nonrational appeals. They are ludic, hyperbolic, ridiculous, and bizarre, and that's the very point. They are meant to push boundaries and conversation, and, McEdwards (1968) argues, such agitative rhetoric is used to gain public attention, without which reasoned appeals and debates would not begin and could not exist.

Arora (2015) has identified a possible extension of culture jamming in recent years, as *spaces of public leisure*—both physical and digital—have become especially notable spheres for protest, dissent, mobilization, and agitative action. Urban parks and leisure spaces like Tahrir Square, Zuccotti Park, and Gezi Park have become high-profile spaces of protest, but, moreover, she writes, "we see protest taking on more creative, play-like forms of expression, creating new rituals of communication among citizens and the state" (2), wherein public demonstrations and protests frequently take on a celebratory and/or festival feel. Arora posits that "the 'relaxed framing' of such sites create an inclusive public space for political mobilization and expression" (4), but the end result is a kind of culture jam, mixing and transgressing public and private, leisure action and political action.

This *ludic recombination of space* in demonstration is exactly what Endres and Senda-Cook (2011) address in their discussion of place-as-rhetoric, noting that social movements use place through "(1) building on a pre-existing meaning of a place, (2) temporarily reconstructing the meaning of a place, and (3) repeated reconstructions that result in new place meanings." (266). Holding a mass demonstration like the Free the Snake Flotilla (1) directs public attention to the Snake River itself, (2) temporarily reconstructs the meaning of river recreation as environmentalist and political statement, and (3) aims to reconstruct the place through "short-term fissures in the dominant meanings" of what the river is meant to be (259). In their work on (PARK)ing Day, Endres, Senda-Cook, and Cozen (2014) find a similar phenomenon, in a public image event that likewise perverts and merges public and private, urban and natural. The (PARK)ing Day movement temporarily transforms city parking spaces into miniature parks, thus functioning as "tactical reconstruction of place" to argue for alternative possibilities (123). "PARK(ing) installations," they write, "are temporary places in protests, or temporary tactical disruptions of normalized spatial practices that seek to spur change in thinking" (123). They argue that "many protests enact *ephemeral fissures* in the meaning of place" to transgress expectations "by positing an alternate vision" (Endres and Senda-Cook 2011, 268), and that "ephemeral spatial arguments can temporarily challenge collective, normalized understandings of places and spaces" (Endres, Senda-Cook, and Cozen 2014, 124).

For its playful and recreation-based nature combined with serious advocacy and cultural critique, its mixing and matching of symbolic elements, its festival aura, its location in a leisure sphere, and its temporary fissure in the Snake River as place, the Free the Snake Flotilla might fit the broad definition of culture jam activism. Yet in the next section, I argue that such leisure spectacle demonstrations differ from culture jamming and other forms of ecoactivism in notable ways.

A GROWING SUBGENRE: LEISURE SPECTACLE DEMONSTRATION

While the Free the Snake Flotilla uses a mass of boats and other recreational rivercraft, there are other notable forms of public mass demonstration that make their case through other forms of public recreation. Perhaps on the simplest and most common level, the "run/walk for charity" form has seen widest proliferation in the United States, Canada, and Europe. Including, for instance, the American Cancer Society's Relay for Life, the Susan G. Komen Race for the Cure, the American Heart Association's Heart Walk, the Leukemia and Lymphoma Society's Light the Night Walk, and others, these participatory events invite attendees to run, walk, or otherwise exercise in support of a cause, personally enacting their commitment but also having a good time in a celebratory atmosphere. In recent years, to attract new crowds and ensure originality, other organizations have sponsored bike rides, color runs, dance-a-thons, sporting or gaming tournaments, and more, all in the name of a cause. Frequently, public leisure or recreation events that before had no explicit social or political commitments are linked to a cause of one kind or another.

Such events are, of course, fundraisers, and participants either pay a fee to join (which is used to raise funds for the cause) or find sponsors who themselves agree to donate a certain amount for every lap, mile, and so forth, completed by the participant recreator/activist. The recreation/leisure format is a way to motivate participants to join and participation fees ensure that the event is also a productive one, creating capital that can then promote more traditional forms of activism and advocacy.

But not all public leisure events are fundraisers. Those like the Free the Snake Flotilla are demonstrations of solidarity or protest, enacting free expression in creative ways but not explicitly tethered to a fundraising structure.

Take, for instance, Critical Mass, a once-per-month public bike ride through city streets in the name of raising awareness for bicyclists and reducing automobile pollution. The informal movement has no official organizing structure but exists monthly in more than three hundred cities worldwide as a spectacle of social protest (Newlands 2018). In a "temporary reconstruction of the meaning of the city streets" hundreds of cyclists clog the roadways to disrupt status quo flows of movement, allowing bikes, not cars, to dictate pace and direction (Endres and Senda-Cook 2011, 270).

The public statement is enacted through temporary fissure of space, and that fissure is achieved through activity that would, in other spaces or contexts, appear as purely leisure or recreational activity.

Similar biking events include the World Naked Bike Ride (a nudist bike excursion held worldwide in cities each June and often campaigning loosely or explicitly for social causes, like opposition to war, body pride, LGBTQ+ rights, and legalization of recreational drugs) and the Full Cycle Supper, documented by Senda-Cook and McHendry (2018) as an experiential, ephemeral enactment of locavore food ethics.

In the leisure space protests documented by Arora (2015), public demonstrations also sometimes take the form and appearance of mass leisure and recreation in the name of purpose or cause. *Dance-ins and dance protests* have coupled with groups chanting and singing to create celebratory atmospheres in these leisure spaces of protest. Groups like the Raging Grannies (Narushima 2004) and Werk for Peace (Zorthian 2017) couple song and dance with public mass demonstrations to add a playful, recreational element to heated social/political dissent. Kutz-Flamenbaum (2007) argues that such "performance activism" is "striving simultaneously to attract and hold attention and challenging the understandings and expectations of fellow protesters and the general public while, paradoxically, staying within the boundaries of these commonly held understandings and expectations" (91).

Yet despite their apparent growing popularity, such leisure spectacle demonstrations have some significant rhetorical shortcomings in a broader public space and lend themselves uniquely to misinterpretation or failure compared to other forms of protest and advocacy.

That is to say, simply: if a demonstration looks and sounds like simple play, leisure, and/or recreation, it's possible that broader audiences may mistake it as simple play, leisure, and/or recreation and entirely miss the underlying social/political message.

This is similar to the critique of image events more broadly made in a book review of DeLuca's work on the concept. Delicath (2001) writes, "While I applaud the optimism that infuses his exploration of the transformative potential of dramatic acts of protest, I am left wanting more in the way of reasons why audiences are likely to make sense of image events in ways favorable to those groups who employ them" (108), adding, "DeLuca settles for locating the power of image events in their heteroglossic potentialities rather than any specific possibilities for shifting the terrain of environmental argument in fundamentally radical ways" (109).

Lawrence Prelli (2009), too, critiques the shortcomings of what he calls "*demonstrative protest rhetoric*," a "kind of enacted performance that anticipates the presence of both participating and witnessing audiences" (93). When effective, he argues, such rhetoric "is inventive; it shows, manifests, and exhibits something previously unseen or concealed" (99) by forcing a conversation in the public attention; but he also argues that all too often such demonstrations fail to "*demonstrate* urgency through enacting proof of power that constrains both demonstrators' subsequent words and deed and political leaders' subsequent decisions to remain consistent with the cause's leading moral and political presumptions" (93). Critiquing the Step It Up climate change awareness events in Boston, he writes, the "urgency of the climate situation" was "defused when speakers and participants talked and acted in ways that minimized the situation's gravity," or when they argued that "saving the planet could be great fun," such that "symbolic actions, both discursive and nondiscursive, were enacted throughout the day that conveyed '*having fun*' as a theme of the rally." As a result, he argues, "one could experience the rally more as a fun-filled Saturday afternoon at the park than as a protest in response to an urgent crisis" (96). Ultimately, he concludes, "These efforts were good-hearted and cheerful overall; they often seemed more entertaining than political, more funny than poignant" (98). Prelli thus argues that public demonstrations need to "manifest" their central crisis "before a wide audience in some strikingly tangible way" (86), as "invention of new or previously unknown or concealed perspectives or facts, meanings, relationships, or actions hinges on the capacity to 'prove' or 'demonstrate' their existence through situated discourse and action" (86). He adds that future demonstrations should explicitly provide attendees and audiences with explicit next steps for action (90–91) and be more confrontational and disruptive (102), as "supporters, onlookers, potential adversaries, and other witnesses of the performance of protest are left anything but mildly disposed toward the situation; rather, the drama leaves them no middle grounds for decision but instead confronts them with a stark choice that puts their own commitments and values at issue" (104).

This is a fair critique, but it's even more pointed if applied to the leisure spectacle demonstrations discussed above and deserves attention as we interrogate events like the Free the Snake Flotilla.

Imagine, for instance, you are walking along a public street and see a mass of hundreds of bikers moving past in the other direction, singing, laughing, and appearing to have a good time. There's a fair chance that, absent explicit labels or signs or other information, you would interpret the group as part of a recreation event, and not a social/political statement. The same would be true for dancing, singing, boating, or recreating in other ways in public spaces.

And, if the mass of apparent recreators blocked your path of movement, for work or for play, you might become frustrated as you waited for the jovial crowd to pass. Critical Mass's "corking" or blockage of traffic has frequently been met with angry shouts from motorists (Endres and Senda-Cook 2011; Newlands 2018). The Flotilla down the Snake River might appear to some audiences as less of a culture-jam and more of a simple, human-made logjam, blocking everyday passage down the river.

Endres and Senda-Cook (2011) make a similar critique in their analysis of a Step It Up climate change demonstration in Salt Lake City. The playful, celebratory atmosphere is appealing and attractive to engage those already committed to the cause, making the work of social protest invigorating and enjoyable, but also, they write, disorienting and not entirely coherent rhetorically. The event included, for instance, a yoga sun salutation in Liberty Park, "a typical city park with playgrounds, walking and running paths, [and] picnic areas," frequently a site for leisure but rarely for protest (273). The yoga acted on some level, they claim, as temporary fissure, but did not entirely disrupt the park as leisure space, making the yoga act at once as both protest and as simple, quotidian park recreation. As the demonstration continued at another city park, Endres and Senda-Cook note a "disconnect between the stated purpose of a protest and the festival-like feel of the event" (273). They explain: "In an effort to reach a wider audience than those who typically attend environmental protest events in Salt Lake City, one of the organizers told [them] they specifically attempted to evoke a Woodstock-esque feel to the protest by scheduling music interspersed with political speeches, an outdoor setting, and a gathering of like-minded organizations" (273). The event was advertised not as a protest but as a "Free Concert" on flyers and in press release materials. The physical layout of the place, including fences, security entrances, and concessions, "also emphasized the festival qualities as opposed to protest qualities" (273). They call the experience "jarring" (274), as the presence of children in a bounce house, hula hoops, stilt walkers, and festival foods failed to enact a protest atmosphere over a leisure one. They concluded that "the dominant meanings of some places may limit the potential of a protest to create a fissure in meaning" (277), and I argue that this is as true for dominant meanings of *actions* within certain places.

If leisure spectacle demonstrations perform more radical fissures, like leading bike rides through indoor or institutional spaces, or dance parties in government halls, there is greater potential for their culture jamming disruption to register. Yet even then, such demonstrations may not be fully accessible to the broader public who is not already participating in the spectacle—the "moveable middle" on any social controversy that Newlands (2018) identifies as the primary target of mass mediated activism.

Moreover, the leisure spectacle demonstration plays into a mass media narrative of activists as informal and colloquial, and thus more casual and disorganized than the State and more formal institutions (Newlands 2018, 63). This interpretation of activists as disorganized or less serious makes it easier for media and broader publics to dismiss activist causes as childish, reckless, or silly. It contributes to a broader media trend of framing activism as conflict or chaos instead of as caring or concerted action (Newlands 2018, 64). And, under what Arora (2015) calls the "relaxed framing" of demonstration structures themselves (4), such protests and events run the risk of being dismissed as *slacktivism*, requiring minimal personal risk and investment in a cause, for the way participants need only gather to recreate—often in leisure modes that they may enact for personal fulfillment completely separate from any broader social/political cause (4). Arora writes on this tension between the appeal and potential limits of leisure in protest, "Both material and virtual leisure platforms have evoked common reactions: either enthusiasm toward them, seen as a significant expansion of democratic public space, to the more dismal view of being prime spaces to disarm and manipulate the masses through their seemingly unregulated leisurely character" (2).

For these reasons—the potential for misinterpretation, the limits of fissuring space through recreation, and the current critiques of casual activism—leisure spectacle demonstrations like the Free the Snake Flotilla and the others mentioned here do not function fully or successfully as radical activist culture jamming. Yet they do not entirely fail rhetorically, either. Their messages and enactments do achieve a rhetorical end, and for that reason I suggest we treat *leisure spectacle demonstration* as its own unique subgenre of culture jamming image event rhetoric.

While it may at first glance appear to fit the definitions of other activist/demonstration genre forms, it differs from each of them in notable ways.

Leisure spectacle demonstration is:

- *NOT* DELIBERATION NOR TRADITIONAL PUBLIC PROTEST in that it does not directly engage with legislators or policy makers through argumentation or confrontation. It takes place in places separate from halls of deliberation and/or legal authority and consists merely of social enactment which may or may not ever be seen by those who can make final decisions. It appears to be more for the ephemeral, embodied experience of the demonstrators than for engagement of public argument.
- *NOT* RADICAL CULTURE JAMMING in the Earth First! sense, for while it certainly creates temporary fissures in space and action by playfully combining cultural symbols and practices to critique culture and policies, it is also subtle enough to go unnoticed or interpreted as mere leisure, even by those who live and work in the space and culture under critique. It lacks the disruptive satirical bite of ecoactivist culture jamming in other forms.
- *NOT* MUNDANE ACTS OF RESISTANCE, even if it shares similarities to the concept proposed by Senda-Cook and McHendry (2018) to describe the Full Cycle Supper bike ride. Senda-Cook and McHendry argue that the rides, consisting of bikers with shared ecological commitments, "rhetorically construct practices of *ecological habitual resistance"* to status quo practices by pushing "against marginalizing structures" (225) while "also advancing alternative practices that allow people to act in accordance with a variety of political concerns" (226). These "mundane acts of resistance" can articulate a "web of environmental micro-practices" (225) of resistance that become "concomitant with everyday life" (226). But leisure spectacle demonstration, as described in the examples above, is by definition not everyday or mundane; it is spectacular and unusual, and part of its appeal comes from the fact that it is a singular or, at most, annual or monthly event. That novelty and exclusivity lend to its rhetoric as image event/spectacle. And yet, it is also . . .
- *NOT* SLACKTIVISM. The leisure spectacle demonstration may indeed lend itself to critique when put beside everyday mundane acts of resis-

> tance, for the comparably minimal or only occasional investment of time and body that the participant must make. However, it does involve investment and, at times, legitimate risk. The spectacle participant does personally engage with and meditate on the cause, so while it may be casually organized and even fun, it is not "slacktivism" proper.

If leisure spectacle demonstration is distinct from any of these forms of protest, then it should not be critiqued for failing to function as they do.

I argue that while leisure spectacle demonstration (1) may not always engage policy makers, (2) may sometimes fail to disrupt the thinking and experience of the broader public of passerby audiences, (3) may not enact a change in the everyday lived experiences of participants and/or places, and (4) may not require sacrifice or great investment of the participant, it does certainly perform one significant rhetorical function seen in other forms of spectacle and image event: it performs and reifies *commitment and community* for those who participate. In divided and partisan times, this can be a crucially important connective function, whereby individuals cement their personal identity and position with respect to a broader social controversy. One-time participation in a leisure spectacle event creates an embodied, experiential performance of this commitment and community that creates coalition without the requirements of long-term shared interactions and is likely to guide actions and discourse with respect to the core cause when it appears in other spheres.

The leisure spectacle demonstration serves as a *constitutive event* (Charland 1987), performing a shared community identity into being. Those sharing in the event as participants and witnesses become consubstantial and co-identify in the moment of performance and that shared community identity and commitment is established as personal, experiential memory thereafter. It allows such events to act as connective rhetorics, establishing community among widely varied participants.

It serves as *spectacle*, then, in Halloran's (2001) sense, blurring the role of audience and rhetor as "a lived experience" wherein "members of the audience become rhetors through their visible and audible reactions, transforming the event as it transpires into an enactment of their social order" (6). The participants, he writes, as both performers and witnesses to the spectacle event, become "a gathering of people who may know nothing more of each other than that they have a common desire to see and hear something" (5) and "in that shared experience is the germ of a public" (6). And like Procter's (1990) *dynamic* spectacle, the event becomes a "nucleus of community identity" and "touchstone for community-building" (119). It "works to build community by: (1) casting the material event into a symbol of a communal past, (2) converting the event into a rhetoric of the community's ideology, and (3) transforming the event into a motive for community action" (120).

It serves, also, like Short's (1991) *agitative rhetoric and litmus test to group identity*. Willingness to participate in the ludic, embodied, and highly visible event "serves as a touchstone for measuring their individual level of commitment to the movement and how far they will go to purify the system" (175).

And it serves, thirdly, as *vernacular rhetoric*, with performance that "is *polyvocal, evades detection, interrogates authority*, and *performs power*" (Hauser and McClellan 2009, 26) in "*mundane, often unnoticed ways*" (3). Hauser and McClellan write, "The ability of vernacular rhetoric to affect existing power structures with consequences for a diverse population of marginalized and oppressed peoples is undeniable," and "This rhetorical form, at once liberating and constraining, inclusive and exclusive, accepting and rejecting, defines those who participate in it as a community whose vernacular discourse expresses bonds of affiliation and aspiration. It also excludes those who do not understand from participating in the discourse community that defines their dissident social imaginary" (33).

Newlands (2018) suggests that such protests and demonstration can act as Foucauldian "*heterotopia*," enacting a space that "acts as a mirror to the real world, made from ideas in the real" yet "transposes neither a true nor utopian reality, but an alternative" (178), or as a "*temporary autonomous zone*" that "does not engage directly with the State" but rather like a guerilla operation "which liberates an area (of land, or time, or imagination) and then dissolves itself to re-form elsewhere/elsewhen before the State can crush it" (179). As a *liminoid*, she argues, the "protest becomes a unifying space that, regardless of external factors, is united through a common objective" (182).

Even Prelli (2009) acknowledges a potential merit to such spectacle demonstrations of "having fun," noting that participants "felt positive and optimistic about the situation, with few exceptions. The fact that they attended the rally left them feeling 'encouraged,' 'hopeful,' 'better informed,' 'like you're part of something,' and 'happy to participate'" (98).

In this way, image events can be community and coalition-building moments. Participation allows an avenue for identification with the cause for those alienated by more confrontational and direct action, or "who do not see themselves as extreme" (Hall et al. 2009, 56). Participation and reflection on the encounter "offers the potential for expanding the original community of concerned citizens and galvanizing previously unconcerned participants" (Hall et al. 2009, 56–57). This opportunity to establish and reinforce identification with the cause allows varied and dispersed groups and participants "to identify simultaneously with other movement participants and against mainstream culture outside the movement" (Feldpausch-Parker et al. 2009, 25). Such image events, in spite of their shortcomings, can, through experience, "help people see themselves in movement actions (Sprain, Norton, and Milstein, 2009, 301).

Moore (2019) and Roosvall and Tegelberg (2018), in their recent work, have both documented how such rhetorical events and happenings serve to establish shared identity across cultural lines among various global and transnational indigenous communities. The liminal spectacle demonstration exemplifies constitutive community and identity-building rhetorics in various ways, likely to extend beyond local or cultural groups. Before explicitly listing the form's merits and constraints, then, it may be useful to consider a case example at length.

DAM REMOVAL AND THE FREE THE SNAKE FLOTILLA

The relationship between image events, happenings, spectacle, and dam removal is well established. The very concept of ecoactivist culture jamming begins with the "monkeywrenching" inspired by Abbey's 1975 novel, in which the protagonists target the Arizona's Glen Canyon Dam for sabotage and destruction (Short 1991). After Earth First!'s installation of the three-hundred-foot crack at the dam, other dam removal activists have used image events and creative spectacle to make their case—including, for instance, the well-publicized 2011 graffiti marking on California's obsolete Matilija Dam, whereby a dotted line and enormous scissors encouraged policy-makers to "cut here" to destroy the dam, opening the way for steelhead trout migration (Chawkins 2011).

While campaigning against the installation of dams dates back to the beginning of the modern environmentalist movement—what with John Muir and the Sierra Club's ultimately unsuccessful work to prevent installation of the O'Shaughnessy Dam in the Hetch Hetchy Valley (Oravec 1984)—the topic of dam *removal* is a relatively recent and rising trend in environmentalist discourse.

In the first half of the twentieth century, dam installation was frequently associated with industrial progress, technological advancement, and national pride, touted as a means of facilitating flood control, irrigation, and hydropower. In the United States, from the end of World War II to the end of the century, more than forty-four thousand dams were installed at an estimated price of $2 trillion (Stanley and Doyle 2003, 15). But with time, the structures deteriorated and required costly upkeep. While environmental consciousness rose in the general population and Congress passed legislation like the 1968 Wild and Scenic Rivers Act, ensuring the preservation of free-flowing rivers for future generations, dams gradually became less appealing to the general American public. Dam construction in the United States declined through the 1970s, and construction was surpassed by decommissioning of dams by 1998. In 2000, the World Commission on Dams "concluded that, although dams have significantly contributed to human development and the benefits derived from dams have been considerable, the economic, social, and environmental price has been unacceptably high" (Stanley and Doyle 2003, 15).

Gradually the term "deadbeat dams" crept into environmental activist vocabulary, referring to the increasingly useless and even harmful aspects of the aging structures. In 1992, the Elwha River Dam in Washington state was one of the first major large dams in the United States to be federally marked for removal and, with the 1999 Federal Energy Regulatory Commission's removal of the Edwards Dam on the Kennebec River in Maine primarily for ecological reasons, a precedent for deadbeat dam removal was set.

Conservation and other ecoactivist groups increased appeals and campaigns to remove deadbeat dams, spurred on by the 2014 documentary film *DamNation*. *DamNation* was funded by the outdoor clothing company Patagonia (which, of course, benefits from the outdoor recreation market), and its reach was both wide and swift (Moscato 2016). Deep-pocketed allies like Microsoft cofounder Paul Allen credited the film with their interest in dam removal and began funding dam removal campaigns.

The dam removal argument rests largely on promoting environmental and social justice, supporting native species, protecting and restocking migratory fish populations, and restoring river ecosystems to free-flowing states (Hychka and Druschke 2017), though they are also largely promoted for economic purposes, as the costs of restoration and upkeep of aging dams is beginning to outweigh the costs of removal.

It's worth noting that the debate over which course of action—restoration or removal—is more cost effective and environmentally beneficial is still very much in progress among engineers and natural scientists. Dam removal could hypothetically restore ecosystems, but it is also potentially dangerous, and removing a large dam could release uncertain levels of sediment, which could just as likely alter or destroy ecosystems currently in relative balance, making rich ground for invasive species (Stanley and Doyle 2003). While advocates for removal often argue that removal will restore fish migration numbers, there is no guarantee that this would be the case. Stanley and Doyle urge policy makers to consider the removal debate a "trade-off" with substantial variables, as "some of the results may be considered beneficial, while others are costly. Some management goals will be achieved quickly and easily, but others will be more elusive," reminding readers, "It is unrealistic to assume that removal will simply and rapidly reverse the suite of conditions created by a dam's construction, and it may be foolhardy to minimize or ignore the fact that some outcomes are likely to represent environmental setbacks" (20).

As the debate plays out, advocates for removal build their appeals on the claims that removal will restore ecosystems and revitalize economies. Removal is framed as a benefit to *life* (attributing recent declines in plant and animal populations to changes in river flow), *energy* (claiming new clean energy alternatives can outproduce aging hydropower), *economy* (asserting that tax payers will avoid upkeep and repair costs), and *justice* (reminding publics that colonized and indigenous populations have been unfairly subjected to the negative effects of damming and that, in many cases, reparations are due for dams constructed on or flooding ceremonial lands). Those opposing removal point frequently to the uncertainty of economic predictions, and voice discomfort with change to a generationally familiar landscape.

Recent cases have shown that there is occasionally room for deliberation and traditional legal channels in the debates. Notably, in 2010, more than twenty stakeholder groups from various sides of the debate met repeatedly to "set aside conflict-based litigation, single-interest political advocacy, and media wars, and adopt a collaborative approach to natural resources governance" (Gosnell and Kelly 2010, 365), ultimately reaching an agreement on the removal of four large dams in the Klamath Basin running between Oregon and California. But in other places, more radical activism and advocacy are still warranted to push the conversation and campaign for change.

Attention to the Snake River, running through Idaho to join the Columbia River in Washington state, and its four lower dams—Ice Harbor, Lower Monumental, Little Goose, and Lower Granite—has grown especially as the Puget Sound orca population has struggled and waned (Barker 2018a). The dams stabilize river flow to allow barge transport of Idaho and Washington grains to Portland, supplement electrical power grids, and facilitate sporting recreation on the Snake (Francovich 2018). However, barge traffic has declined in recent years, and studies suggest the energy produced by the dams could be replaced by other clean energy sources (Francovich 2018). The plight of the orcas, however, changed the conversation. In 2018, concerns over the North Pacific's southern resident killer whales (SRKW) gained international attention when the population hit a thirty-year low and one orca mother was found pushing the corpse of her deceased and malnourished newborn for seventeen days (Francovich 2018). While threats to orcas include inbreeding, disease, pollution, vessel noise, and competition from other marine mammals, many reports linked the declining orca health and population to diet. Chinook and steelhead salmon make up roughly 50 percent of the SRKW diet and since dam installation, populations of the migratory fish have been in steady decline. Roughly 40 percent of historically documented Pacific Northwest chinook runs are locally extinct, and the rest are threatened or endangered (Mapes 2019).

On the Snake, the campaign to remove the dams has become a campaign to save the salmon and, thus, save the orcas, ensuring the return of a regular diet of adequate quality.

Those against removal have argued that breaching the dams would eliminate the economic benefits of barge transportation of wheat, make it harder for local farmers to irrigate, and forfeit the benefits of hydroelectricity, suggesting that alternatives for chinook repopulation exist, including habitat recovery, hatcheries, predator controls, and reduction of commercial fishing (Barker 2019). A scientific debate argues as to whether the Columbia's chinook population is as crucial to Puget Sound and Salish Sea orcas as other stocks, and whether the crucial stock for orcas comes from fall or spring feeding times (Barker 2018a). Those who believe the fall chinook, arriving before winter, are more crucial to orca well-being argue against dam removal, claiming current levels are sufficient and the problem lies in a lack of diverse sources of food, while those who believe the spring chinook, arriving after the winter, are more crucial argue that dam removal will ensure higher populations at that crucial time (Barker 2018a). The National Oceanic and Atmospheric Administration (NOAA) notably claims data on the importance of the spring chinook is inconclusive (Barker 2018a).

In response, in late 2018, Washington Governor Jay Inslee included $750,000 in his two-year, $54.3 billion state budget to fund a cost/benefit study on whether removing the dams would actually restore salmon populations (Camden 2018). Political opponents like Washington state U.S. Representatives Cathy McMorris Rodgers and Dan Newhouse issued a news release statement of protest within two hours of the budget's release, stating, "We commit to do everything in our power to save our dams" (Camden 2018, News 8). Regardless, and despite noted contradictions in Inslee's vow to commit to fossil fuel alternatives while also removing sources of hydropower, the state legislature passed the budget in 2019, allowing a "neutral third party" to run the study on the potential impact of dam removal for salmon and orcas but also for farmers, shippers, and utility providers (Barker 2019).

At roughly the same time, a U.S. District Court agreement required the dams, under the jurisdiction of the Bonneville Power Administration, to allow increased spillover during the spring period of downstream salmon migration, allowing more water to pass over the dam during daily periods of lower power demand for the next three years (Bernton and Mapes 2018). This short-term agreement between Bonneville Power, Oregon, Washington, and the Nez Perce Tribe acknowledges that current data on the dam's impact on salmon is insufficient but also ensures the chance to collect better information (Barker 2018b). While the requirement will only be in effect from April to mid-June, when spring chinook head to oceans, McMorris Rodgers and Newhouse derided the agreement as "worse than useless," with a possible $38 million loss in potential energy profits and no proof that its findings will resolve the dispute (Probert 2019).

Meanwhile, amid and alongside the broader state debate, the Nez Perce, Colville, Kalispel, Coeur d'Alene, and Spokane tribes have consistently voiced concerns for the river and its animal life and campaigned on the side of removal. For members of the Nez Perce Tribe, for instance, "the dams represent an historical injustice: When built, they flooded many traditional Nez Perce gathering sites and burial grounds" (Francovich 2018). Julien Matthews, a Nez Perce tribal member and a board member of Nimiipuu Protecting the Environment, an environmental organization that focuses on upholding tribal treaty rights, was quoted in local media, saying, "We talk about the injustice, it's ongoing. . . . Our relatives are buried there under that water" (Francovich 2018).

Amid the U.S. twentieth-century period of dam building and expansion, "local people, especially indigenous, [were] often excluded from dam planning or construction and many have protested dam operations and/or called for their decommissioning and removal" (Gosnell and Kelly 2010, 365). Gosnell and Kelly note, "In the Columbia river basin in the US Pacific Northwest, for example, a number of tribes historically dependent on migrating salmon have lost access to traditional fishing spots due to the construction of more than 75 federal, non-federal, and Canadian dams in the basin since the 1930s" (365). Yet, they add, "The ability of indigenous peoples to effectively influence decisions around dam construction and management varies tremendously from region to region depending on the laws in place defining indigenous rights" (366).

It is no surprise that the Free the Snake Flotilla, growing in size since the release of *DamNation*, has attracted a coalition of environmental advocates and activists, biological scientists, progressive political allies, and indigenous communities.

The Flotilla began in 2015, heavily sponsored by Patagonia, local river advocates (Save Our Wild Salmon, Nimiipuu Protecting the Environment, Friends of the Clearwater, Sierra Club, Wild Steelhead Coalition, Eco Depot, Earth Ministry, Idaho Rivers United) and local outdoor recreation companies (Mountain Gear, ROW Adventures), as a means of campaigning for dam removal (Idaho Rivers United 2018a). Attendees arrive with their own leisure watercraft, including canoes, kayaks, paddleboards, inner tubes, rafts, and more, and put in at Chief Timothy Park, an island on the lower Snake, outside of Clarkston, Washington. Together, the Flotilla participants paddle roughly six miles, arriving near the Lower Granite Dam. Along the way, they create, in the words of event organizers, "a display of solidarity and a resounding call to FREE THE SNAKE!" (Free the Snake n.d.).

In 2018, the Flotilla had expanded to a two-day event on September 7 and 8, advertised with "outdoor activities, educational opportunities, and music" (Free the Snake n.d.). Attendees and participants were urged to "Stick around the whole weekend to camp out with fellow river soldiers and salmon advocates" (Free the Snake n.d.), with assurance that "there will be plenty of room: more than 50 RV sites and plenty of tent camping" (Idaho Rivers United 2018a). While leading with the recreational aspects of the weekend, in 2018, advertisements also stressed "small guided paddles around Chief Timothy, field trips, and educational talks" (Idaho Rivers United 2018a). Environmental and indigenous rights activist Winona LaDuke appeared as keynote speaker at the group camp following the Flotilla, and Democratic candidate for Idaho governor Paulette Jordan spoke, verbally committing "if elected, to fight to decommission the four lower Snake River Dams" (Francovich 2018).

Also in 2018, organizers touted an increased indigenous contingent and participation, and a "focus on tribal land and tradition." LaDuke's talk centered on dam removal "as a tribal treaty rights issue of great significance to indigenous communities the world over" (Free the Snake n.d.). The 2018 Flotilla began with a "contingent of traditional tribal canoes" beginning "their paddle in the morning a few miles upstream on the Clearwater River, joining the flotilla of kayaks, canoes, rafts and motor boats on the lower Snake River for a ceremony and rally in the early afternoon" (Idaho Rivers United 2018a).

The event was clearly a leisure spectacle demonstration, but much of its organizers' promotional materials focused on the spectacle as community-building event, putting tribal members and concerns in direct communion and conversation with "hundreds of non-tribal salmon advocates" (Free the Snake n.d.), outdoor recreation enthusiasts, and people from all walks of life, united if only in their shared interest in a free-flowing lower Snake.

Notably, though, the community sentiment and identity *is not universal.* Local media also covered how, on the same day, "just miles upriver from the Free the Snake Flotilla, a pro-dam event was held." The counterprotest, River Fest, was hosted by the Port of Clarkston and also enacted a festival/cookout atmosphere, instead celebrating the dams and asserting that removing them would be "inconceivable." Here, too, the event was wrapped in identity assertion and performance. One attendee interviewed asserted the dams' importance to community, claiming, "This whole town is built on these dams" (Francovich 2018).

In examining the verbal, visual, material, and experiential rhetorics of the Free the Snake Flotilla, it's clear that it is open to the same critiques as Endres and Senda-Cook's Step It Up festival or to Delicath's and Prelli's skepticism of image events in general. The emphasis on leisure spectacle and participation makes the deliberative aspects and policy arguments of the event obscure and even inaccessible to those who do not participate.

Sure, if you attend to the event promotions and/or take part in the event, the dam removal message is clear and potent. But for outsider audiences—like those attending River Fest or simply those driving along a riverside road on September 8, 2018, the Flotilla and its events look remarkably like any other group of outdoor enthusiasts out for a good time—a large group, to be certain, but not an explicitly political and activist one. The emphasis on camping—in less-than-eco-friendly RVs, no less—presents the event as festival and recreation. And the visuals of the event are not explicitly activist either. The vast majority of participants arrived in generic outdoor recreation gear—Patagonia or otherwise, but often relatively high-end—the same as they would don for any other weekend water excursion. While a few crafts within the Flotilla did contain protest signs, the majority did not and those that were visible were somewhat vague about their argument (reading, for example, "IT'S ABOUT DAM TIME," and, simply, "FREE THE SNAKE!"). Of course, holding an oar or paddle makes it difficult to also carry a traditional protest sign, but the visual rhetorics of the event fail to disrupt perceptions of spectacle as simply a leisure event, like a fun run or kayak-enthusiast gathering.

In recent years, inflatable orcas and hand-made salmon have occasionally joined the Flotilla, but the display is not constructed for immediate audiences or passersby. It is intended for (1) mass media and social media coverage, as an image event, constructed to garner media attention (this is evidenced in part by the floating "FREE THE SNAKE" letters in the Flotilla, visible only from the air and thus only accessible to an elevated photographer) and (2) the participants themselves, who have the only ready access to the visceral, embodied, ephemeral, and experiential rhetorics of thrill and exhilaration as one joins hundreds of others in a ludic and unusual mass disrupting an open waterway.

Outside audiences may perceive the Flotilla as casual, disorganized, juvenile, slacktivist, or silly, but the reading of the event as leisure spectacle demonstration instead of as deliberative protest, radical ecoactivist culture jam, or everyday practice still allows it to be a constructive form. What's important in such leisure spectacle rhetorics is the ensuing sense of community and commitment to the cause experienced by those who take part in the event.

In this respect, all that matters for a general audience is that the Flotilla looks fun—that it seems like the kind of thing that you or others might want to try for yourselves next year. If it looks fun and appealing, the event can grow in size and, thus, so can the community and cause of dam removal. And the promotional materials for the Flotilla seem to acknowledge and embrace this idea, boasting the Flotilla not as a meaningful protest of environmental activism, but rather as "a popular fall event for people from around the Pacific Northwest (Idaho Rivers United 2018a).

This chapter has sought to contribute to critical discussions of leisure- and recreation-based culture jams and social/political demonstration, particularly as they grow in popularity in environmental justice and advocacy spheres.

It argues that such demonstrations, including the Free the Snake Flotilla, should be analyzed separately from other image events, spectacles, and culture jams for the ways they engage broad nonparticipant audiences and participant audiences differently.

Certainly, they disrupt status quo assumptions in highly spectacular, media-attracting ways, creating temporary fissures in public experience and, thus, discourse. But the end result (and perhaps the intentional end goal) of such demonstrations is not to persuade policy makers or broader audiences who may see these leisure spectacle demonstrations as merely ridiculous if jubilant festivals that clog city streets and jam local waterways. Rather, it is to give the participants *hope* and *hype* for a cause that might otherwise seem demoralizing and hopeless, given the way mainstream media and political channels might dismiss and ignore it.

Leisure spectacle demonstration gives participants a reason to get and stay excited about a marginalized or largely ignored cause. Even if the cards are stacked against the core argument and the logistics of something like, say, dam removal, are slow and complex, the leisure spectacle participant can be excited about the fun they will have personally through their participation. And, when gathering to have fun, they will be assured and reminded that they are not alone in their sentiments—that, to contrary, there is a mass of energized, high-spirited others who share that sentiment, fostering coidentification and potential optimism for the cause.

The shared experience performs communal identity, and opens the door to networking and future collaboration. Participants might be added to a mailing list or social media feed, creating an opportunity to mobilize collective action through petitions or other, more traditionally deliberative forms of engagement. Indeed, the Free the Snake Flotilla does exactly this by urging participants to sign petitions in the months following the event (Idaho Rivers United 2018b).

The danger of such demonstrations is, of course, rejection by the broader population. Some observers of agitative and disruptive rhetorics can reject the tactics as irrational, coercive, and even antidemocratic (Haiman 1967, 102). As Cathcart (1980) notes, such rhetoric "challenges the values of a political system, producing a counterrhetoric that in turn recognizes movement protesters and their potential threat to the established order" (Short 1991, 174). This can be productive in challenging the status quo, but it can also be grounds for rejection of a movement or spectacle. As with all vernacular rhetorics, the effects are "directly tied to a vernacular rhetoric's interpretation by a particular discourse community" (Hauser and McClellan 2010, 33). As Hauser and McClellan write, "Vernacular rhetoric, therefore, must presuppose some form of literacy; what is meaningful must be spoken and/or understood by a particular group in a particular way" (33). If the mode of leisure or recreational engagement is not accessible, appealing, or sensible to an audience, it indeed can become "exclusive" and "rejecting" as Hauser and McClellan warn. When they write that such rhetorics can exclude "those who do not understand from participating in the discourse community that defines their dissident social imaginary" (33), it's hard not to think of the Flotilla's River Fest counterprotest group. In especially partisan times and climates like those of the current day, playful modes of civic protest can potentially be divisive and contribute to tribalism and lack of meaningful dialogue.

At the same time, the benefits of such demonstrations are also clear, providing a space for hope and hype but also for networking, connecting a critical mass of allies for future mobilizations, and fostering identification across diverse and otherwise separate groups. This is especially notable for widely dispersed ecoactivist and indigenous activist communities, as Roosvall and Tegelberg (2018) highlight how "recently, a shift from *collective action,* including the reinforcement of group identity and the development of share frames, towards *connective action,* which entails a more personalized engagement within loose networks, has been noted as a key feature of the increasing intersection of social movements" (14).

At the time this collection was in print, the Free the Snake Flotilla was undergoing telling and intriguing changes. Suddenly, in late August 2019, social media posts announced that the Flotilla had been rebranded, as the "Nimiipuu River Rendezvous." This new name speaks to growing involvement in the event from local indigenous peoples—especially the Nez Perce, or Niimíipuu, people—but it is at present unclear how else the enactment of the event will or will not change. Either way, I hope attention to such gatherings continues, and that other writers continue to examine other LSDs (an acronym that I recognize has some delightful connotations and parallels to the recreational, ludic, and culture jamming mischief at hand). Up close and personal will be the best way to more thoroughly understand the embodied, ephemeral, and experiential rhetorics at play.

REFERENCES

Arora, Payal. 2015. "Usurping Public Leisure Space for Protest: Social Activism in the Digital and Material Commons." *Space and Culture* 18, no. 1: 55–68.

Barker, Eric. 2018a. "Snake River Salmon Key to Orcas' Health." *The Spokesman-Review,* November 26, 2018.

Barker, Eric. 2018b. "Salmon Deal to Add Spillage at Region's Dams." *The Spokesman-Review,* December 20, 2018.

Barker, Eric. 2019. "Reaching Out on Breaching." *The Spokesman-Review,* May 2, 2019.

Bernton, Hal, and Linda V. Mapes. 2018. "Changes to Dams on Columbia, Snake Rivers Benefit Salmon, Hydropower and Orcas." *The Spokesman-Review,* December 19, 2018.

Bullard, Robert D. 2005. *The Quest for Environmental Justice: Human Rights and the Politics of Pollution.* San Francisco: Sierra Club Books.

Camden, Jim. 2018. "Dams Also Generate Controversy." *The Spokesman-Review,* December 14, 2018.

Cathcart, Robert S. 1980. "Defining Social Movements by Their Rhetorical Form." *Central States Speech Journal* 31, no. 4: 267–73.
Charland, Maurice. 1987. "Constitutive Rhetoric: The Case of the *Peuple Québécois*." *Quarterly Journal of Speech* 73, no. 2: 133–50.
Chawkins, Steve. 2011. "On a Divisive Barrier, a Snippy Bit of Graffiti." *Los Angeles Times*, September 19, 2011. https://www.latimes.com/archives/la-xpm-2011-sep-19-la-me-dam-scissors-20110919-story.html.
DeLuca, Kevin Michael. 1999. *Image Politics: The New Rhetoric of Environmental Activism*. New York: Guilford Press.
Delicath, John. 2001. Review of *Image Politics: The New Rhetoric of Environmental Activism*, by Kevin Michael DeLuca. *Quarterly Journal of Speech* 87, no. 1: 107–9.
Endres, Danielle, and Samantha Senda-Cook. 2011. "Location Matters: The Rhetoric of Place in Protest." *Quarterly Journal of Speech* 97, no. 3: 257–82.
Endres, Danielle, Samantha Senda-Cook, and Brian Cozen, B. 2014. "Not Just a Place to Park Your Car: PARK(ing) as Spatial Argument." *Argumentation and Advocacy* 50, no. 3: 121–40.
Farrell, Thomas B. 1989. "Media Rhetoric as Social Drama: The Winter Olympics of 1984." *Critical Studies in Mass Communication* 6, no. 2: 158–82.
Feldpausch-Parker, Andrea M., Damon Hall, Christi Horton, Jodi Minion, Anna Munoz, Israel Parker, and Tarla Rai Peterson. 2009. "Step It Up in the Lone Star State: How Identity and Myth May Impact a Movement." In *Social Movement to Address Climate Change: Local Steps for Global Action*, edited by Danielle Endres, Leah Sprain, and Tarla Rai Peterson, 23–51. Amherst, NY: Cambria Press.
Francovich, Eli. 2018. "More Than 600 Turn Out for Snake River Protest Saturday." *The Spokesman-Review*. September 9, 2018. https://www.spokesman.com/stories/2018/sep/09/more-than-600-turn-out-for-snake-river-protest-sat/.
Free the Snake. (n.d.). Retrieved from https://www.freethesnake.com/.
Gosnell, Hannah, and Erin Clover Kelly. 2010. "Peace on the River? Social-Ecological Restoration and Large Dam Removal in the Klamath Basin, USA." *Water Alternatives* 3, no. 2: 361–83.
Haiman, Franklyn. 1967. "The Rhetoric of the Streets: Some Legal and Ethical Considerations." *Quarterly Journal of Speech* 53, no. 2: 99–114.
Hall, Damon M., Leigh A. Bernacchi, Tema O. Milstein, and Tarla Rai Peterson. 2009. "Calling All Artists: Moving Climate Change from My Space to My Place." In *Social Movement to Address Climate Change: Local Steps for Global Action*, edited by Danielle Endres, Leah Sprain, and Tarla Rai Peterson, 53–79. Amherst, NY: Cambria Press.
Halloran, S. Michael. 2001. "Text and Experience in a Historical Pageant: Toward a Rhetoric of Spectacle." *Rhetoric Society Quarterly* 31, no. 4: 5–17.
Harold, Christine. 2004. "Pranking Rhetoric: 'Culture Jamming' as Media Activism." *Critical Studies in Media Communication* 21, no. 3: 189–211.
Hauser, Gerald A., and Erin Daina McClellan. 2010. "Vernacular Rhetoric and Social Movements: Performances of Resistance in the Rhetoric of the Everyday." In *Active Voices: Composing a Rhetoric for Social Movements,* edited by Sharon McKenzie Stevens and Patricia Malesh, 23–46. Albany: State University of New York Press.
Hychka, Kristen, and Caroline Gottschalk Druschke. 2017. "Adaptive Management of Urban Ecosystem Restoration: Learning from Restoration Managers in Rhode Island, USA." *Society & Natural Resources* 30, no. 11: 1358–73.
Idaho Rivers United. 2018a. "Free the Snake Flotilla to Draw Hundreds to Lower Snake River Sept. 7 and 8." *Idaho Rivers United*, July 30, 2018. https://www.idahorivers.org/newsroom/2018/7/30/flotilla.
Idaho Rivers United. 2018b. "Call to Action: Tell Idaho's New Governor We Value Our Wild Salmon and Steelhead." *Idaho Rivers United*, November 7, 2018. https://www.idahorivers.org/newsroom/2018/11/6/call-to-action-tell-idahos-leaders-why-we-value-our-wild-salmon.
Killingsworth, M. Jimmie, and Jackie Palmer. 1992. *Ecospeak: Rhetoric and Environmental Politics in America*. Carbondale: Southern Illinois University Press.

Kutz-Flamenbaum, Rachel V. 2007. "Code Pink, Raging Grannies, and the Missile Dick Chicks: Feminist Performance Activism in the Contemporary Anti-War Movement." *NWSA Journal* 19, no. 1: 89–105.
Mapes, Linda V. 2019. "The Decline of Salmon Adds to Orcas' Struggle." *The Spokesman-Review*, April 1, 2019.
McEdwards, Mary G. 1968. "Agitative Rhetoric: Its Nature and Effect." *Western Speech* 32, no. 1: 36–43.
Moore, Ellen. 2019. *Journalism, Politics, and the Dakota Access Pipeline: Standing Rock and the Framing of Injustice*. New York: Routledge.
Moscato, Derek. 2016. "The Brand behind the Activism: Patagonia's DamNation Campaign and the Evolution of Corporate Social Responsibility." *Case Studies in Strategic Communication* 5: 99–116.
Narushima, Miya. 2004. "A Gaggle of Raging Grannies: The Empowerment of Older Canadian Women through Social Activism." *International Journal of Lifelong Education* 23, no. 1: 23–42.
Newlands, Maxine. 2018. *Environmental Activism and the Media*. New York: Peter Lang.
Nicholson, Blake. 2019. "Companies Decry Oil Pipeline 'Valve Turners.'" *The Spokesman-Review,* March 10, 2019.
Oravec, Christine. 1984. "Conservationism vs. Preservationism: The 'Public Interest' in the Hetch Hetchy Controversy." *Quarterly Journal of Speech* 70, no. 4: 444–58.
Prelli, Lawrence J. 2009. "Demonstrative Protest Rhetoric and the Boston Step It Up Campaign." In *Social Movement to Address Climate Change: Local Steps for Global Action*, edited by Danielle Endres, Leah Sprain, and Tarla Rai Peterson, 81–116. Amherst, NY: Cambria Press.
Probert, Cameron. (2019). "Opponents Decry New Dam Agreement to Help Salmon." *The Spokesman-Review*, December 24, 2019.
Procter, David E. 1990. "The Dynamic Spectacle: Transforming Experience into Social Forms of Community." *Quarterly Journal of Speech* 76, no. 2: 117–33.
Roosvall, Anna, and Matthew Tegelberg. 2018. *Media and Transnational Climate Justice: Indigenous Activism and Climate Politics*. New York: Peter Lang.
Senda-Cook, Samantha, and George F. McHendry Jr. 2018. "Embodying Resistance: A Rhetorical Ecology of the Full Cycle Supper." In *Tracing Rhetoric and Material Life: Ecological Approaches*, edited by Bridie McGreavy, Justine Wells, George F. McHendry Jr., and Samantha Senda-Cook, 223–50. London: Palgrave Macmillan.
Short, Brant. 1991. "Earth First! and the Rhetoric of Moral Confrontation." *Communication Studies* 42, no. 2: 172–88.
Sprain, Leah, Todd Norton, and Tema Milstein. 2009. "Step It Up and Image Politics in the Pacific Northwest." In *Social Movement to Address Climate Change: Local Steps for Global Action*, edited by Danielle Endres, Leah Sprain, and Tarla Rai Peterson, 281–307. Amherst, NY: Cambria Press.
Standing, Sarah Ann. 2012. "Earth First!'s 'Crack the Dam' and the Aesthetics of Ecoactivist Performance." In *Readings in Performance and Ecology,* edited by Wendy Arons and Theresa J. May, 147–55. New York: Palgrave Macmillan.
Stanley, Emily H., and Martin W. Doyle. 2003. "Trading Off: The Ecological Effects of Dam Removal." *Frontiers in Ecology and the Environment* 1, no. 1: 15–22.
Walker, Gordon. 2012. *Environmental Justice: Concepts, Evidence and Politics.* London: Routledge.
World Commission on Dams. 2000. *Dams and Development: A New Framework for Decision-Making: The Report of the World Commission on Dams*. London: Earthscan.
Zorthian, Julia. 2017. "LGBT Activists Had a Dance Party Protest in Mike Pence's Neighborhood." *Time*, January 19, 2017. https://time.com/4639152/mike-pence-queer-dance-party/.

Chapter Four

Reimagining Dam Removal to Resist Settler Colonial Logics

Emma Lundberg, Caroline Gottschalk Druschke, and Alicia Lehrer

There are compelling reasons to see dam construction—at its peak in the United States in the mid-twentieth century (Ho et al. 2017) but currently increasing throughout the world (Biello 2009)—as an example of the physical manifestation of settler colonial logics on the landscape (Schneider 2013; Griffith 2017). As articulated by indigenous scholars, settler colonialism is a dominant social or political structure where a colonizer comes to a place, lays claim to it, and disenfranchises its people by stripping their autonomy as a means to exploit the land for monetary gain (Wolfe 1999; Tuck and Yang 2012; Arvin, Tuck, and Morrill 2013). Settler colonialism is not an event in time; it is an ongoing process (Wolfe 1999) and "a matter of life-world reordering . . . associated with death and the end of the world" (Boswell 2015, 1).

A settler colonial lens offers theoretical language for understanding the connections between the arrival of Anglo-European settlers in North America and the damming of waterways. For instance, in New England, Anglo-European populations began to intensify in the eighteenth century due to the advances of the Industrial Revolution, initially made possible by the use of small mill dams that harnessed the power of water, many of which still punctuate the landscape. The construction of in-stream structures designed to extract the power of flowing water physically restructured and symbolically reimagined North American rivers and many human relationships to those systems. As Eve Tuck and K. Wayne Yang (2012) argue, "Within settler colonialism, the most important concern is land/water/air/subterranean earth [. . .] what is most valuable, contested, required" (5). In the case of New

England's dammed landscape, water—and access to it for the sake of generating power—became a primary concern; water power was valuable, contested, required. While most New England mill dams no longer generate power, their long-term persistence on the region's landscape serves as an ongoing and consequential scar from settler colonial actions.

If dam building in New England is a settler colonial manifestation, then dam removal—currently on the rise in the United States (Ambers 2007; Gosnell and Kelly 2010)—would seem to offer a remedy. River restoration practices, including dam removal—a process that attempts to return degraded aquatic ecosystems to healthier conditions through the removal of human-made barriers—have become an increasingly common strategy for repairing river connectivity and natural riverine processes (Graf 1999; Wohl et al. 2007; Bernhardt et al. 2007; Palmer et al. 2007). In addition to improving ecological function and supporting native species (Bellmore et al. 2019), aquatic restoration projects offer potential to strengthen social justice efforts by providing accessibility to green spaces, improving public health, and reconnecting human and nonhuman communities in riverine corridors, particularly in urban areas (Palamar 2008; Wolch, Byrne, and Newell 2014; Hychka and Druschke 2017). Thus, dam removals offer a potential solution to the settler colonial logics of dam construction: seeming to undo the damage of New England's fourteen thousand small and large dams that harnessed and privatized river flow for the financial benefit of Anglo settlers.

And yet, we argue here that dam removals in the United States may be just as likely to reinscribe and reinforce dominant settler colonial logics of domination and control, acting as a palliative but not a remedy to settler colonial river management practices. As we will explain, the management discourse and logics that drive dam removals and fish passage projects in the U.S. Northeast often seem to repeat the same logics of efficiency and disconnection employed in dam construction. We also briefly work to describe how such restoration projects are often easiest to accomplish in economically privileged and politically savvy communities.

To offer an alternative, we build from one author's work directing the Woonasquatucket River Watershed Council in Providence, Rhode Island. The Woonasquatucket River Watershed Council is a nonprofit organization with a vision for community-based river restoration that includes dam removal and fish passage projects, and does so in ways that amplify the voices of marginalized community residents and encourages the reconnection of human and nonhuman relationships. We suggest that this creative, collaborative model of dam removal and river restoration might usefully undermine—rather than reinforce—settler colonial logics, while achieving pragmatic goals for rivers and for the human and nonhuman communities they support.

DAM CONSTRUCTION IN THE SETTLER COLONIAL CONTEXT

Dams make present a literal, physical instantiation of settler colonial logics. In the U.S. Northeast—where we have researched, taught, and practiced—dams are an ubiquitous and often unquestioned feature punctuating the landscape. The National Inventory of Dams (NID) estimates there are more than eighty thousand dams in the United States, which amounts to roughly one dam built per day since the signing of the Declaration of Independence (Lieb 2015). New England boasts the highest concentration of dams in the United States. The region features the single county with the highest number of dams in the United States (Worcester, Massachusetts) and hosts the oldest surviving U.S. dam: Connecticut's Mill Pond Dam, built in 1677 (Graf 1999). With the staggering number of dams built from European contact to present, it is not an exaggeration to say that Anglo-European colonization of the United States goes hand in hand with the damming of its free-flowing waters.

The U.S. Northeast hosts a staggeringly dense concentration of dams, estimated at fourteen thousand (Fox, Magilligan, and Sneddon 2016). While it is true that humans have used the force of free-flowing water to advance societal development for thousands of years (Kluune 2012), the demand for water power in the United States increased exponentially during and after the Industrial Revolution to meet settler communities' increasing demands (Gleick 2003; Magilligan et al. 2016; Magilligan et al. 2017). By the 1940s, hydropower dams provided approximately 75 percent of all electricity generated and consumed in the growing U.S. Pacific West, and constituted approximately one-third of total U.S. electrical energy production (Kluune 2012). The damming history of waterways is a pointed example of colonial logics validating the understanding of the natural world as something to be controlled and harnessed. From this perspective, resource extraction often occurs at the expense of public well-being and ecological health, and without regard for indigenous people's connection to place and relation to water (Desbiens 2013).

The damming of U.S. rivers is not an isolated enactment of settler disruption. To the north of the U.S. border, Canada boasts more than fifteen thousand dams, almost one thousand of which are considered "large" dams (either 15m and greater, or 5–15m and impounding more than three million meters cubed of water) (Canadian Dam Association, n.d.). Those dams have wide scale ecological and social impacts. As Caroline Desbiens (2013) describes in *Power from the North: Territory, Identity, and the Culture of Hydroelectricity in Quebec*, hydroelectricity from dams in northern Quebec has ideologically and materially supported French Canadian nationalism, while marginalizing Aboriginal territories by way of dispossession and capitalization of place-based natural resources. In Newfoundland and Labrador, meanwhile, the Muskrat Falls Project, a proposed hydroelectric dam on the Churchill River, has been embroiled in controversy between the hydroelectric company, Nalcor Energy, and the Government of Nunatsiavut, representing the Inuit people of Labrador, who have concerns about possible methylmercury contamination (Fitzpatrick 2012; 2015; 2016). As the president of Nunatsiavut, Sarah Leo, claimed in 2015, "The Muskrat Falls development will put mercury into Lake Melville and negatively affect our people, our land, and the fish and wildlife we depend on" (Fitzpatrick 2015). Members of the Nunatsiavut have expressed frustration over the years about their concerns seeming to take a backseat to economic arguments in favor of the project (Fitzpatrick 2012; 2015; 2016), which would largely support the demand for energy in the U.S. Northeast. These examples support our claim that U.S. and Canadian past and present offer evidence of Anglo-European populations dislocating and marginalizing indigenous peoples through a variety of means, including large and small projects to control and harness the power of free-flowing rivers.

The influences of settler colonialism on river development and the monopolization of water power are not isolated within the North American context; the development of large-scale dams and infrastructure projects are current and controversial, particularly in the Global South (McCully 2001; Braun 2011). Dam development projects globally, especially in the Global South, can be accompanied by social and ecological hardships and unfulfilled promises of community and economic development (McCully 2011) that reveal ongoing settler colonial structures and logics. Many large-scale dam projects come to life through significant domestic and foreign investments of capital and resources and are justified through neoliberal ideologies that promise economic returns (McCully 2001). Although projects promote economic returns to local residents, fulfillment of these promises are limited, as researchers have detailed in South Africa (Braun 2011), India (Chattopadhyay 2014), Myanmar (Simpson 2013), and elsewhere (McCully 2001). Indigenous peoples and ethnic minorities tend to be more frequently impacted by the forty-five thousand dams built globally since World War II (Johnston 1997). Though reported to be regularly excluded from development decision-making related to dams (Grumbine, Dore, and Xu 2012; Manorom, Baird, and Shoemaker 2017), indigenous groups are often on the front line of resistance movements fighting for environmental protection and protesting large-scale development projects (Fisher 1994; Fulmer et al. 2008). Despite some of the very real benefits dams can offer for human development, including power, flood control, nutrient retention, and economic capital, dams present a variety of human and ecological drawbacks (Bednarek 2001; Ambers 2007), and their ubiquity on the global landscape demonstrates the power and extent of the colonial reach.

RIVER RESTORATION AS ONGOING SETTLER COLONIALISM

As we have detailed, ongoing settler colonialism renders itself visible on the New England landscape through the region's fourteen thousand small and large dams, put in place to harness the energy of free-flowing rivers that, in part, supported an industrial and more broadly capitalist agenda. It would seem, then, that the current U.S. shift from dam construction to removal would help to correct the inequities posed through dam development. Construction of dams in the United States has been on the decline since the 1970s, and the U.S. rate of decommissioning overtook the rate of construction in 1998 (WCD 2000; Gosnell and Kelly 2010). Dam removal now plays an important role as a tool for ecosystem restoration in the United States (Bednarek 2001; Ambers 2007; Bellmore et al. 2019). But we suggest that the recent trend of dam removal seems to echo a similar sentience as historic dam building throughout Anglo-European development of New England waterways. Rather than acting as a tool for dismantling the settler impact on and control of rivers, we detail the ways that dam removal often serves, like dam construction, as a means for perpetuating the logics and physical spaces of management, control, and efficiency.

The support and extension of settler colonial logics is perhaps most visible in the official, managerial language of dam removal and river restoration. In Rhode Island, specifically, language that aims to protect the state's waterways and aquatic organisms employs verbiage that tends to distance restoration managers from the systems they manage, fitting river systems and aquatic organisms into frameworks of dominance and control. Efforts to support Rhode Island migratory fish passage with respect to the legacy of mill dams that impede fish's seasonal upstream migrations, for instance, are supported through a complex network of interlocking agencies at the watershed, state, and federal scales, but largely reside under the guidance of the Rhode Island Department of Environmental Management's "Strategic Plan for the Restoration of Anadromous Fishes to Rhode Island Coastal Streams" (Erkan 2002). The introduction to the Strategic Plan actually opens by pointing to the logics that drove the expansion of industrial era mill dams throughout the region, and describes their impacts on migratory fish populations in the state:

> The dams of the Industrial Revolution were placed in rivers for an inexpensive and reliable source of power. Little thought was given to the effects on native fish populations. Fish passage efforts at the time met with little success. The end result was the extirpation of local populations of American shad, alewives, blueback herring, Atlantic salmon, and other species. As the Industrial Revolution waned, so has the need for the dams, but they remained. (Erkan 2002, 6)

We argue that settler colonial impulses offer the subtext and foundation for these industrial, managerial logics, where a concern for powering Anglo-European industry with "an inexpensive and reliable form of power" elided the needs of the river's resident and migratory fish, much less the needs of the Narragansett, Pequot, Niantic, and Wampanoag peoples native to the region now known as "Rhode Island."

Still, where the document does point to the ongoing impacts of these coupled colonial and industrial logics on fish species (but not humans), it seems to simultaneously articulate and perpetuate these same logics. Practically speaking, the solutions offered, while useful and important for restoring migratory fish populations in Rhode Island rivers and streams, largely consist of technical solutions that include constructing engineered fish passages, stocking of migratory fish species, and ongoing population monitoring efforts without questioning or undermining the very logics that drove dam construction in the first place.

These state-based logics echo federal perspectives on fish passage restoration efforts, evidenced in part in 2016 federal interagency guidelines for the construction of fish passages (Turek, Haro, and Towler 2016). As that technical memorandum details, "Both NMFS [the National Marine Fisheries Service] and USFWS [U.S. Fish and Wildlife Service] along with USGS [U.S. Geological Service] seek to advance engineering design and technology in providing safe (from both physical injury and predator avoidance), timely, and effective upstream and downstream passage for all diadromous species targeted for restoration" (2). Again, though useful and timely at the level of practice, this language of management, safety, timeliness, and efficiency of fish movement past dams bears the trace of settler colonial logics. These discourses, as they play out in everyday discussions, management documents, and state policies, further support, extend, and normalize settler colonial logics and the structures they support.

Crucially, the Strategic Plan and federal policy discourse sets an agenda for funneling available resources toward priority watersheds that, as the Rhode Island document describes, offer "the potential to restore, establish, or enhance anadromous fish populations through upstream passage for migrating adults and downstream passage for juveniles" (Erkan 2002, 7). As the Plan details, "In both cases, the primary goals are to minimize passage-induced mortality allowing expansion into unutilized and underutilized habitats with the most cost-effective method available. Reintroduction of spawning broodstock is another critical component of the restoration efforts" (Erkan 2002, 7). Technical/managerial logics are evident in the language of mortality rates, cost-effectiveness, and fish stocking, and these logics guide restoration efforts throughout the state. While migratory fish species like Atlantic salmon, alewives, blueback herring, and shad were all but extirpated in New England due to the rise of early industrial capitalist construction of some fourteen thousand dams region-wide, the solution to restoring these species is likewise envisioned through human control of natural processes and landscapes: through breeding programs, stocking, and ongoing monitoring and management of populations driven by perspectives on cost-efficiency. These impulses fundamentally illustrate the settler colonial logics present in natural resource management (Acheson 2006; Palmer 2006; Schneider 2013; Boswell 2015).

On the federal level, policy discourse supports the prioritization of watersheds and water bodies; rivers are categorized by management agencies from low- to high-priority systems, often with limited involvement from local communities in making these categorizations (Geist and Galatowitsch 1999; Petts 2007). Site selection for restoration projects is often associated with colonial value hierarchies, supported by inflexible institutional processes that limit meaningful trans-species relationships, while reinforcing problematic structures of settler power and privilege.

Most of the 1,400 dams in the United States deliberately removed since the 1970s have been less than 5m tall and are aging, obsolete dams that pose potential threats to human health and safety, while impeding river connectivity (Ambers 2007; Magilligan et al. 2016; Bellmore et al. 2019). Crucially, the highest frequency of dam removals in the United States occurs with small dams (WCD 2000), many of which, thanks to the ongoing legacy of the Industrial Revolution, occur in densely populated regions, and, as we discuss further, occur where communities have access to monetary resources and where the restoration sites are connected to swaths of "pristine" land (Pow 2009).

While not always the case, some evidence seems to suggest that communities with economic means and powerful political voices are those that have the knowledge and clout to initiate, direct, or support restoration activities (Hychka and Druschke 2017). Crucially, that trend can exclude communities that lack economic means and know-how from dam removal and river restoration efforts. Kristen C. Hychka and Caroline Gottschalk Druschke (2017) describe a variety of barriers to urban restoration initiatives that include dam removal, especially in under-resourced communities, including a lack of availability of funds in so-called "degraded," nonpristine river systems and a lack of capacity for spending among many small, under-resourced organizations. Among the interviews with restoration managers and community advocates highlighted in their work, Hychka and Druschke (2017) feature a restoration manager who summed up the funding challenge of working in urban areas: "You can get maybe a quarter of an acre of restoration in, where if you go down to [a more rural part of the state] the same money will buy you 20 acres. . . . If you have $2 million, you can save a huge pristine [area]. Or you can restore a five-acre contaminated site in an inner city. And that is the challenge" (6). Urban restoration projects, including dam removals, in under-resourced communities are harder to justify, harder to fund, and often present more technical obstacles due to legacy effects of industrial pollution (Hychka and Druschke 2017). Some communities either do not have the educational empowerment to know about restoration activities, or do not know how to move forward in advocating for socially and environmentally just river management (Hychka and Druschke 2017).

Settler colonial logics have laid the foundation for and continue to sustain a particular management discourse within dam removal projects, including fish passage around dams, that, in many ways, distances humans from fish, while prioritizing efficiency and managerialism over human with more-than-human relationships. The status quo terminology of fisheries research and management offers fish as "units" or "products" that are stocked, planted, closely monitored, and even reared through industrial modes and logics, while the status quo terminology of river restoration prioritization privileges literal spaces of power and exclusion.

Despite the dominance of settler logics in dam construction and removal, we consider how to work toward the deconstruction of these structures within river restoration discourses and practices by using an unsettling framework, in solidarity with decolonization. As white, female settlers whose work does not bring "about the repatriation of indigenous land and life" (Tuck and Yang 2012, 1), we aim to support decolonization without metaphorizing by using unsettling means. Unsettling work disrupts colonial structures and is attuned to the ways that privilege-based ideologies are perpetuated, instilled, and reinforced through social and scientific discourses. Below, we offer a practical case example, drawn from one author's work as director of the Woonasquatucket River Watershed Council in Providence, Rhode Island, of local community members and technical scientists collaborating with fish and other river dwellers.

COUNTERING SETTLER COLONIAL LOGICS: THE WOONASQUATUCKET RIVER WATERSHED COUNCIL

To highlight this alternate perspective on river restoration, we feature an ongoing urban restoration initiative in Providence, Rhode Island, that employs a variety of restoration techniques. The initiative, which includes dam removal projects, aims to restore aquatic connectivity and improve water quality, but also to increase the visibility of shared community resources, rebuild community connections to place, direct resources toward under-resourced communities of color, and amplify the voices of community members. This case example, Rhode Island's Woonasquatucket River Watershed Council (WRWC), was one of the earliest urban river restoration projects in the United States. The WRWC model has the potential to inspire other communities working to restore degraded, urban aquatic ecosystems by reconnecting and redeveloping human-community relationships to rivers. The work of WRWC demonstrates the possibilities for a group of creative advocates to contest 140 years of industrial domination of a river system that many city dwellers had forgotten even existed, thereby engaging in dam removal projects that offer a model for productively interrupting settler colonial logics. Important to note is that although WRWC is not actively returning land to the Narragansett Tribe, the indigenous peoples of this region, they do offer a model for undertaking restoration projects that support the initial unsettling steps toward decolonial work more broadly.

WRWC's restoration work continues to be accomplished in part through physical restoration of the river corridor—by way of removing dams and constructing fish ladders, green spaces, and a bike path—but also through the increasing reorientation of the lower Woonasquatucket's underresourced human community toward the river itself and the reclaiming of habitat by the river's once nearly extirpated migratory fish species. As WRWC inspires human community members to reconnect to the Woonasquatucket River, this work—and the narratives they co-create about that work and about each other—redraws relationships with the river and bolsters the importance of trans-species connections that are mutually beneficial for diverse human communities, river organisms, and rivers themselves.

We point to the work of the Woonasquatucket River Watershed Council here because of the ways that their work seems to, at least partially, deconstruct dominant restoration narratives related to river management, restoration hierarchies, and place-based relationships to more-than-human worlds and species. We suggest that the WRWC model may prompt future dam removal projects to consider employing ecologically and socially equitable discourses and actions. Projects aiming to support decolonization efforts would benefit from extending the WRWC model to include repatriation of land and fostering collaboration among indigenous leaders, settlers, and scientific communities in the design and implementation of restoration projects.

Working within an Existing Settler Colonial Structure

Crucially, we argue that WRWC is such a useful model because it conducts its creative work within the very settler colonial structures we have described. For instance, WRWC has to work against the fact that the Woonasquatucket watershed is deprioritized in Rhode Island's "Strategic Plan for the Restoration of Anadromous Fishes to Rhode Island Coastal Streams" because, while it is cited as having "local interest in establishing fish passage," the "poor water quality and high number of dams make other watersheds more attractive alternatives" (Erkan 2002, 25). In other words, the state management plan documents local desire for river restoration projects like dam removal in the Woonasquatucket watershed that might reconnect human and piscine residents, but the system's "poor water quality," its current degradation—the consequence of colonial industrial transformation of the waterway from the 1700s through the 1970s (Greenwood, n.d.)—excludes it from the mainstream restoration process. In the technical/managerial logics of effective and efficient fish passage around dams, the Woonasquatucket fails to offer a compelling argument for sustained state or federal attention, a stalemate that conspires to worsen the ecological health of the river. Further complicating the matter are the ongoing and extensive impacts of Superfund-level legacy pollution in the lower reaches of the watershed. Due to this complication, it is not likely that dam removal can offer anything like the sorts of recoveries that are touted as the outcomes of river restoration and dam removal, what a key environmental nonprofit in Rhode Island, Save the Bay, describes as bringing a river "back to life," by ushering in a "return to their former channel and habitat function" (Save The Bay, n.d.).

A Watershed-Based, Systemic Approach for Supporting Human-Fish Interaction

And yet, dam removal and engineered fish passage solutions have become an increasing point of emphasis in the Woonasquatucket watershed. As WRWC describes on its web site:

> Fish have once again become a common sight in the tidal portions of the Woonasquatucket. The Woonasquatucket River Watershed Council is working with NRCS and other partners to provide fish passage at the first five dams on the Woonasquatucket to allow fish such as shad and herring, which migrate from saltwater to fresh to spawn, and eels, which migrate from fresh to salt to spawn, to move up and down the lower river. (WRWC Fish Passage)

That work involves a systemic approach to fish passage at five sequential dam sites with various restoration approaches: Rising Sun Fish Ladder, Paragon Dam, Atlantic Mills Fish Ladder, Dyerville Dam, and Manton Dam. In contemporary river restoration, basin-wide repair efforts are becoming more common but are still often the exception to the rule (Opperman et al. 2011; Magilligan et al. 2016). Recognizing that each dam location is site-specific yet related to the others within the watershed is an important component of restoration that can positively influence strategic restorations aimed at broad-scale recovery. Place-based solutions aimed at balancing riverine resources and reconnecting human-river actors cannot be achieved on an individual project scale (Opperman et al. 2011). Therefore, this turn toward a system-wide approach, as demonstrated on the Woonasquatucket, provides evidence for a shift in restoration discourse and implementation that aims to improve connectivity at the watershed scale.

Beyond this watershed-scale approach to improving fish passage at existing dams, WRWC restoration efforts have worked to open opportunities for public interaction with the river and its annually returning migrating fish. The WRWC's restoration efforts that have aimed to expand beyond the physical river have fostered the visibility of trans-species relationships that we argue are essential for working to dismantle settler colonial structures. For instance, the Rising Sun Fish Ladder, completed in 2007 at the Rising Sun Mills complex, incorporates a viewing platform where community members can observe migrating river herring and American eel and count returning fish throughout the spawning season. Atlantic Mills Fish Ladder in Riverside Park, completed in 2009, also includes public engagement features such as a walk-on platform to look down into the ladder at migrating fish. The uppermost dam on the river—just over a half mile upstream of Dyerville and two miles downstream of the Centredale Manor Superfund Site—is Manton Dam, where the WRWC has worked for several years to secure funding for a nature-like fishway. The Manton Project, completed in late 2016, allows migratory fish access to Manton Pond's nine acres of spawning habitat upstream and brings the total available spawning habitat in the system from 4.6 to 5.3 river miles. In many ways, the Manton Project closes the loop on a vision for migratory fish passage past dams on the Woonasquatucket that seemed impossible in the framing of the 2002 Rhode Island anadromous fish report and its prioritization scheme.

Listening to Members of Under-Resourced Communities

The bulk of the work done by WRWC on the Woonasquatucket—including these dam removal and fish passage projects—takes place in the lower reaches of the watershed, largely in the Olneyville neighborhood of Providence. In the early 1990s, stakeholders from the City of Providence, led by community advocate Jane Sherman, began working with neighborhood residents on reintroducing greenspace and recreational opportunities in abandoned industrial lands and public spaces along the Woonasquatucket River in Olneyville. The success of the environmental and community restoration movement was achieved through a blend of community networking and multistakeholder communication from planning to execution to ongoing monitoring and involvement, in coordination with community members, Rhode Island School of Design, the Army Corps of Engineers, the Narragansett Bay Commission, the Natural Resources Conservation Service, and many others.

Olneyville's demographics are shifting—potentially due, in part, to gentrification, a consequence of increased access to greenspace, visibility of ecological assets, increased housing costs and property values, and improvements to local community and environmental health, all of which have been reported elsewhere (Wolch, Byrne, and Newell 2014)—but the neighborhood has long been home to a large population of undocumented and transitional residents. Community activists have pointed to frequent resident turnover and political disconnection as challenges for organizing, as well as the frequent overlooking of the neighborhood as a political priority because it straddles two districts and does not offer any pristine habitat for restoration efforts (Hychka and Druschke 2017). Nevertheless, WRWC continued to build trust locally and worked toward community consensus on mutually identified goals. Though it was difficult because of a lack of political and financial power in the neighborhood and the general invisibility of the river due to its channelization behind and sometimes underneath abandoned industrial sites, one of the early central goals of WRWC was to listen to the concerns and interests of local residents and reposition the Woonasquatucket River as a neighborhood asset.

Public engagement was essential to these early discussions about neighborhood improvement and river restoration, and WRWC emerged out of that momentum and support for shared community objectives. Founding Executive Director Jane Sherman started her work in the neighborhood by attending existing neighborhood meetings, listening to the concerns and interests of neighborhood residents. She held block parties in parks that were then better known as sites for drugs and prostitution. Early public meetings with designers and architects from Rhode Island School of Design allowed residents to discuss opportunities for largely abandoned areas adjacent to the river. These meetings were bilingual, educational, and open-ended, designed to meet the needs of the Olneyville population, which the 2000 Census reported was 57.4 percent Hispanic, 13.6 percent African American, 7.4 percent Asian, and 30 percent foreign born (Providence Plan, n.d.). By design, according to Sherman, there was no master plan to be edited or policy to be followed. Instead, community members were given tours of newly renovated urban parks in Providence, existing greenspaces in the upper watershed, and derelict open spaces and hidden natural resources in the Olneyville neighborhood. The community was asked how the available land in Olneyville should be restored. A plan was developed out of these community meetings, which prioritized safe places for children to play and a corridor for safe transportation.

The project slowly started to generate funding by the same method of providing tours and soliciting suggestions from local actors. Sherman brought anyone who would go—members of the Southern New England Conservation Group, City Parks Department, Planning Department, Department of Transportation, Department of Environmental Management, and others—on tours around the Olneyville area showing the current conditions of public spaces as well as the hidden natural resource of the river and models for potential change. She started with small projects—sharing the burden of a seemingly insurmountable task across many shoulders and garnering confidence from other, potential funders—and eventually received several large pools of funding to establish a riverside greenway. The development of the greenway, now known as The Fred Lippitt Woonasquatucket River Greenway, was aimed at fulfilling the goals of increasing access to recreational opportunities and visibility of environmental assets, restoring the river and surrounding land, and working toward neighborhood stabilization.

By extending opportunities for engagement and education to many stakeholders and actors (both inside and outside the neighborhood) and enlisting community residents in all stages of planning and design, the project generated support from the ground up that resulted in local, state, and federal support. The watershed council in its current form emerged out of this work and has become a nationally recognized leader in watershed stewardship, contributing to the designation in 1998 of the Woonasquatucket as a federally recognized American Heritage River.

The suite of projects along the Woonasquatucket River in Olneyville resulted in the fruition of plans envisioned by community members during public meetings. The Fred Lippitt Woonasquatucket River Greenway project—a bike route spanning from Johnston, Rhode Island, to Waterplace Park in Providence—travels through Olneyville thanks to funding from a public bond issue and the Brownfield Showcase Community designation. Also, thanks to the latter, Riverside Park has been renovated from a neglected, high-crime space into a usable public good for the Olneyville community featuring greenspace, river access, a functional fish ladder, and a community bike hub. Restoration projects in the Woonasquatucket watershed offer lessons about the importance of educational access, networking, and the development of strong community support and subsequent political strength—all moves that we see as potentially countering the settler colonial logics typically present in dam removal discourse. As Sherman described of her early work in the watershed:

> Everybody said, "Nothing good will happen here. We don't have environment in this community. Anything you do in this community they'll destroy." But what happened was, as you kept talking to more and more people about it you kept getting sort of indications, "Well, I'll support you. We won't take on the whole thing but I can do a little bit." And, so it was a matter bringing all the different elements together in a way that no one felt that they were bearing the total burden. . . . And people came by and said, "Where did you get the river?" They didn't know it was there.

Further, it seems likely that those greenspace and restoration projects helped to galvanize a community-wide interest in related environmental projects, including the dam removal and fish passage projects.

Fish Passage and Human-Fish Relations on the Woonasquatucket

As the work of the WRWC progressed, one of the crucial ways of connecting neighborhood residents with the formerly hidden river was through restoring passage for migratory fish. While the state considered fish passage on the Woonasquatucket River a low priority, these projects have been integrated as an important component of community engagement and river restoration on the Woonasquatucket. As a WRWC staff member described in 2013, these fish passage efforts are integral to WRWC's work:

> We also have fish restoration projects happening at the river, where we're trying to bring back anadromous fish or migratory fish. The fish that live both in the river and in the open ocean for part of their life cycle. Because this was such an urban industrialized, dammed up river, we haven't had those kind of migratory fish coming through the river for, since 1860. And in 2008 we opened our first fish ladder at the first dam on the river. And that has allowed the herring to return to that dam and since then we've gotten through the first four dams on the river and we're trying to get through the fifth one. And once we get through that fifth one, we will have enough spawning habitat or reproduction space for about 40,000 adult herring, which are one of the target species we're trying to bring back to the river, and they have lived here historically.

Importantly, that work has enhanced the natural resources of the watershed for human community members, even if ongoing water quality issues in the watershed hamper long-term, significant changes. As WRWC staff described back in 2013, the goal of the organization has to do with the fact that:

> [The river]'s pretty degraded, and, in fact, it's the only real natural resource for a lot of communities or a lot of neighborhoods. And my goal or, the organization's goal is the make the resources usable, available, enjoyable, safe, for the people and the wildlife that live here.

As a result, as that staff member explained, she's always asking: "How can we improve the water quality? How can we improve the sediment quality? How can we improve the land around it so that it protects the water better and to make it more accessible, usable, recreational, fun?"

This is an agenda of incremental change against daunting impossibility. The WRWC knows the disastrous realities it is up against. As WRWC staff described:

> We constantly see people catching the fish down here and we're constantly like please don't eat those. This is not safe, because this is below the Superfund Site. They shouldn't really even be in the water. There's so many sewer overflows at this point going into the river. There's 19 that any time there's a half-inch of rain there's raw sewage pouring into the river. So that's fortunate-

> ly being fixed by the Narragansett Bay Commission right now. . . . So water quality will be improving very much, but until we get that Superfund site cleared up somewhat, it's still always going to be dangerous.

Despite those real challenges, returning fish passage to the river, among other efforts, as this staff member explained, "Was really to like, revitalize, restore a community by revitalizing the natural resources." The current and past work of WRWC has been about taking a river that was "shrouded," according to this staff member, making it more visible, and highlighting river connectivity. This focus on connectivity has allowed residents to see they were actually connected to healthy areas upstream and to Narragansett Bay downstream via the river, and to the creatures within the river. Dam removal and fish ladder construction became integral to that visibility—inviting community members to witness the benefits of connectivity, like observing migrating fish, padding the river, inviting them to view maps of connectivity, and taking them on field trips upstream.

BUILDING FROM WRWC RESTORATION EFFORTS

We see WRWC utilizing an unsettling, countercolonial approach to dams and restoration. This unsettling approach attempts to understand and account for the positions and needs of impacted persons and more-than-humans within the hegemonic dominance and control paradigm. Importantly, unsettling work aims to make visible and disrupt settler logics through processes that redistribute power and privilege, while recentering community-based needs in place of institutional priorities and desires. Activities like watershed-scale dam removal and fish passage, reconnecting human residents with migratory fish, paddling the river together, offering technical assistance, listening to and amplifying resident voices in state political discussions, and working in legal arenas to address the ongoing impacts of the Superfund site on the river, have encouraged river restorationists and community members alike to understand their relationships with each other and with the river differently. This has been done through recognizing power inequalities associated with class, race, gender, and river restoration prioritization processes, in addition to addressing both the exclusion and needs of under-resourced groups and the environmental conditions in which they live.

WRWC can prime us to learn from indigenous scholars and communities in ways that challenge us to think of and examine New England dams differently, placing issues of equity, race, and privilege at the foreground of community engagement and natural resource management. WRWC restoration efforts aimed to be inclusive of the wide variety of potential community-river interactions, a move toward dismantling long-instilled settler logics. Providing mechanisms for marginalized community members to reclaim voice and space is crucial to creating inclusive, sustainable, and just restoration projects.

We suggest there is an acute need for place-based work that adopts feminist and trans-species approaches critical of settler colonialism to dam development and removal, like the work we offer here. We attempt to enter what we see as a relative silence in the literature related to human dimensions of dam construction and removal, where gendered and critical feminist analyses of hydropower in the Global North are largely absent, in places where dam decisions have transitioned into questions of dam removal rather than continued development of free-flowing rivers for power generation, navigation, and economic benefits. But these discourses of dam removal also need to be explored with a critical lens.

River restoration strategies, including dam removal, can perpetuate ongoing settler colonialism by privileging managerial logics and reinforcing a narrow vision of success, without considering the impacts of river restoration projects on rivers and both their in-stream and streamside inhabitants, both human and nonhuman. The Woonasquatucket River Watershed Council offers an example of river restoration that does not conform to the structure of completely reversing anthropogenic action, but, instead, reimagines what the community's relationship to the river is and can be. The Woonasquatucket River restoration projects were aimed at both reimagining the river and redeveloping the natural resources of the community. Through this reimagining and redeveloping, communities along the banks of the Woonasquatucket River were able to begin to realize the potential of their mutually beneficial relationships with the river and see the river as a vital asset previously forgotten. By refusing to uphold unrealistic expectations for restoration—like returning the river to pre–European contact conditions—community members and restorationists were able to co-create new relationships with each other and with the river, while also making obvious gains for the ecological and residential health within the watershed.

REIMAGINING DAM REMOVAL IN RESISTANCE TO SETTLER COLONIAL LOGICS

In this chapter, we argued that a settler colonial lens can spark the deconstruction of problematic management and restoration narratives and structures. We hope this piece will contribute to the global retelling of dams and river restorations that amplify the voices and concerns of impacted peoples, impacted aquatic populations, and the impaired river itself. And we offer this WRWC case study as a way to begin the unsettling of New England waterways in solidarity with and in support of decolonization efforts spearheaded by indigenous communities worldwide. As Tuck and Yang (2012) have powerfully articulated, "Decolonization in the settler colonial context must involve the repatriation of land simultaneous to the recognition of how land and relations to land have always been differently understood and enacted; that is, *all* of the land, and not just symbolically. This is precisely why decolonization is necessarily unsettling, especially across lines of solidarity" (7). Dam removal, in and of itself, is an unsettling disturbance for humans and more-than-humans, disrupting what has stood in place in recent memory (Stanley and Doyle 2003). But we take seriously Tuck and Yang's (2012) insistence that "decolonization is not a metaphor" (1), and we offer this work as a call for the literal dismantling of the colonial structures and logics within river management and restoration, and do not suggest that this work is a model for decolonization. We encourage our readers to embrace unsettling work that supports indigenous-led projects and decolonial actions, and, even when uncomfortable, to continue incorporating and amplifying marginalized voices in restoration projects. Otherwise, we risk perpetuating the same dominance and control paradigm that has led to widespread ecological and social degradation of the world's rivers.

We also encourage water researchers, restoration managers, and water workers to consider the settler colonial mechanisms at play in the ways we communicate about these topics. An early step to dismantling colonial structures within dam, hydropower, and dam removal discourses is to recognize that the ways we communicate information, prioritize spaces, and design management plans for human and more-than-human organisms in impacted systems can support colonial and distancing processes that limit trans-species relationships among terrestrial and aquatic realms. For instance, scientific discourses used to produce technical reports and develop management regimes frequently perpetuate narrow, Western, colonial perspectives that contribute to ongoing colonialism. When language is used that allows for distancing, it becomes easy to "other" organisms, systems, and peoples that do not fit within the dominant narrative. This material form of dominance has the power to further strip autonomy, disregard inherent value, and situate human and nonhuman organisms and systems into the dominance and control paradigm. In this positioning, institutional power degrades marginalized groups' ability to organize, make decisions, and have their voices and needs heard and taken seriously.

Here we have utilized a settler colonial lens to critique both dam construction and dam removal discourse and practice, and to complicate what we see as a prominent restoration practitioner perspective that extols the benefits of river restoration in ways that can actually perpetuate an unjust status quo (Dhillon 2015). We have attempted to destabilize hegemonic dam removal discourses and highlight a long-term, real-life, watershed-based, trans-species approach to restoration that attempts to resist colonial influences in favor of grassroots community-river-species reconnection and trans-species repair, a model that might offer important pathways and strategies for future decolonizing projects. We hope readers will join us in this ongoing work.

REFERENCES

Acheson, James M. 2006. "Institutional Failure in Resource Management." *Annual Review Anthropology* 35: 117–34.

Ambers, Rebecca K. R. 2007. "Effects of a Small, Century-Old Dam on a Second-Order Stream in the Virginia Piedmont." *Southeastern Geographer* 47, no. 2: 181–201.

Arvin, Maile, Eve Tuck, and Angie Morrill. 2013. "Decolonizing Feminism: Challenging Connections between Settler Colonialism and Heteropatriarchy." *Feminist Formations* 25, no. 1: 8–34.

Bednarek, Angela T. 2001. "Undamming Rivers: A Review of the Ecological Impacts of Dam Removal." *Environmental Management* 27, no. 6: 803–14.

Bellmore, J. Ryan, George R. Press, Jeffery J. Duda, Jim E. O'Connor, Amy E. East, Melissa M. Foley, Andrew C. Wilcox, Jon J. Major, Patrick B. Shafroth, Sarah A. Morley, Christopher S. Magirl, Chauncey W. Anderson, James E. Evans, Christian E Torgersen, and Laura S. Craig. 2019. "Conceptualizing Ecological Responses to Dam Removal: If You Remove It, What's to Come?" *BioScience* 69, no. 1: 26–39.

Bernhardt, Emily S., Elizabeth B. Sudduth, Margaret A. Palmer, J. David Allan, Judy L. Meyer, Gretchen Alexander, Jennifer Follastad-Shah, Brooke Hassett, Robin Jenkinson, Rebecca Lave, Jeanne Rumps, and Laura Pagano. 2007. "Restoring Rivers One Reach at a Time: Results from a Survey of US River Restoration Practitioners." *Restoration Ecology* 15, no. 3: 482–93.

Biello, David. 2009. "The Dam Building Boom: Right Path to Clean Energy?" *Yale Environment 360*. February 3, 2009. http://e360.yale.edu/features/the_dam_building_boom_right_path_to_clean_energy.

Boswell, Anna. 2015. "The Sensible Order of the Eel." *Settler Colonial Studies* 5, no. 4: 363–74.

Braun, Yvonne A. 2011. "Left High and Dry: An Intersectional Analysis of Gender, Dams and Development in Lesotho." *International Feminist Journal of Politics* 13, no. 2: 141–62.

Canadian Dam Association. n.d. "Dams in Canada." Canadian Dams Association. Accessed May 1, 2017. http://www.imis100ca1.ca/cda/MainEN/Dams_in_Canada/Dams_in_Canada/MainEN/CDA/Dams_In_Canada.aspx?hkey=5abec4f0-5033-4ab6-9802-313c1dece13e.

Chattopadhyay, Sutapa. 2014. "Postcolonial Development State, Appropriation of Nature, and Social Transformation of the Ousted Adivasis in the Narmada Valley, India." *Capitalism Nature Socialism* 25, no. 4: 65–84.

Desbiens, Caroline. 2013. *Power from the North: Territory, Identity, and the Culture of Hydroelectricity in Quebec*. Vancouver: University of British Columbia Press.

Dhillon, Jaskiran K. 2015. "Indigenous Girls and the Violence of Settler Colonial Policing." *Decolonization: Indigeneity, Education & Society* 4, no. 2: 1–31.

Erkan, Dennis E. 2002. "Strategic Plan for the Restoration of Anadromous Fishes to Rhode Island Coastal Streams." Rhode Island Department of Environmental Management.

Fisher, William H. 1994. "Megadevelopment, Environmentalism, and Resistance: The Institutional Context of Kayapó Indigenous Politics in Central Brazil. *Human Organization* 53, no. 3: 220–32.

Fitzpatrick, Ashley. 2012. "Nunatsiavut Government Claims Right to Consultation." *The Telegram*, November 29, 2012. http://www.pressreader.com/canada/the-labradorian/20121203/281651072408047.

Fitzpatrick, Ashley. 2015. "Nunatsiavut Appealing on Mercury Concerns." *The Telegram*, February 10, 2015. http://www.southerngazette.ca/news/regional/2015/2/9/nunatsiavut-appealing-on-mercury-concern-4037344.html.

Fitzpatrick, Ashley. 2016. "Clear Muskrat Falls Reservoir, Protect People: Nunatsiavut Leaders." *The Telegram*. April 19, 2016. http://www.pressreader.com/canada/the-compass/20160426/281724088736665.

Fox, Coleen A., Francis J. Magilligan, and Christopher S. Sneddon. 2016. "'You Kill the Dam, You Are Killing a Part of Me': Dam Removal and the Environmental Politics of River Restoration." *Geoforum* 70: 93–104.

Fulmer, Amanda M., Angelina Snodgrass Godoy, and Philip Neff. 2008. "Indigenous Rights, Resistance, and the Law: Lessons from a Guatemalan Mine." *Latin American Politics and Society* 50, no. 4: 91–121.

Geist, Cathy, and Susan M. Galatowitsch. 1999. "Reciprocal Model for Meeting Ecological and Human Needs in Restoration Projects." *Conservation Biology* 13, no. 5: 970–79.

Gleick, Peter H. 2003. "Water Use." *Annual Review of Environment and Resources* 28, no. 1: 275–314.

Gosnell, Hannah, and Erin Clover Kelly. 2010. "Peace on the River? Social-Ecological Restoration and Large Dam Removal in the Klamath Basin, USA." *Water Alternatives* 3, no. 2: 361–83.

Graf, William L. 1999. "Dam Nation: A Geographic Census of American Dams and Their Large-Scale Hydrologic Impacts." *Water Resources Research* 35, no. 4: 1305–11.

Greenwood, Richard E. n.d. "A Brief Assessment of the Historical Significance of the Woonasquatucket River Valley." Woonasquatucket River Watershed Council. http://www.wrwc.org/history.php.

Griffith, Jane. 2017. "Hoover Damn: Land, Labor, and Settler Colonial Cultural Production." *Cultural Studies ↔ Critical Methodologies* 17, no. 1: 30–40.

Grumbine, R. Edward, John Dore, and Jianchu Xu. 2012. "Mekong Hydropower: Drivers of Change and Governance Challenges." *Frontiers in Ecology and the Environment* 10, no. 2: 91–98.

Ho, Michelle, Upmanu Lall, Maura Allaire, Naresh Devineni, Hyun Han Kwon, Indrani Pail, David Raff, and David Wegner. 2017. "The Future Role of Dams in the United States of America." *Water Resources Research* 53, no. 2: 982–98.

Hychka, Kristen, and Caroline Gottschalk Druschke. 2017. "Adaptive Management of Urban Ecosystem Restoration: Learning from Restoration Managers in Rhode Island, USA." *Society & Natural Resources* 30, no. 11: 1358–73.

Johnston, Barbara Rose. 1997. *Life and Death Matters: Human Rights and the Environment at the End of the Millennium*. Walnut Creek, CA: AltaMira Press.

Klunne, Wim Jonker. 2012. "Hydropower." In *Encyclopedia of Energy*, edited by Morris A. Pierce, 651–55. Hackensack, NJ: Salem Press.

Lieb, Anna. 2015. "The Undamming of America." NOVANEXT. Public Broadcasting Service. http://www.pbs.org/wgbh/nova/next/earth/dam-removals/.

Magilligan, Frank J., Brian E. Graber, Keith H. Nislow, Jonathan W. Chipman, Christopher S. Sneddon, and Coleen A. Fox. 2016. "River Restoration by Dam Removal: Enhancing Connectivity at Watershed Scales." *Elementa: Science of the Anthropocene* 4, no. 1: 000108.

Magilligan, F. J., C. S. Sneddon, and C. A. Fox. 2017. "The Social, Historical, and Institutional Contingencies of Dam Removal." *Environmental Management.* 59, no. 6: 982–94.

Manorom, Kanokwan, Ian G. Baird, and Bruce Shoemaker. 2017. "The World Bank, Hydropower-Based Poverty Alleviation and Indigenous Peoples: On-the-Ground Realities in the Xe Bang Fai River Basin of Laos." *Forum for Development Studies* 44, no. 2: 275–300.

McCully, Patrick. 2001. *Silenced Rivers: The Ecology and Politics of Large Dams*. New York: Zed Books.

Opperman, Jeffrey, Joshua Royte, John Banks, Laura Rose Day, and Colin Apse. 2011. "The Penobscot River, Maine, USA: A Basin-Scale Approach to Balancing Power Generation and Ecosystem Restoration." *Ecology and Society* 16, no. 3: 7.

Palamar, Collette R. 2008. "The Justice of Ecological Restoration: Environmental History, Health, Ecology, and Justice in the United States." *Human Ecology Review* 15, no. 1: 82–94.

Palmer, Lisa. 2006. "'Nature,' Place and the Recognition of Indigenous Polities." *Australian Geographer* 37, no. 1: 33–43.

Palmer, Margaret, J. David Allan, Judy Meyer, and Emily S. Bernhardt. 2007. "River Restoration in the Twenty-First Century: Data and Experiential Knowledge to Inform Future Efforts." *Restoration Ecology* 15, no. 3: 472–81.

Petts, Judith. 2007. "Learning about Learning: Lessons from Public Engagement and Deliberation on Urban River Restoration." *The Geographical Journal* 173, no. 4: 300–11.

Pow, Choon-Piew. 2009. "Neoliberalism and the Aestheticization of New Middle-Class Landscapes." *Antipode* 41, no. 2: 371–90.

Providence Plan. n.d. "Olneyville in Depth: Background." The Providence Plan. http://local.provplan.org/profiles/oln_bk.html.

Save the Bay. n.d. "River Restoration and Dam Removal." Save the Bay. http://www.savebay.org/rivers.

Schneider, Lindsey. 2013. "'There's Something in the Water': Salmon Runs and Settler Colonialism on the Columbia River." *American Indian Culture and Research Journal* 37, no. 2: 149–64.

Simpson, Adam. 2013. "Challenging Hydropower Development in Myanmar (Burma): Cross-Border Activism under a Regime in Transition." *The Pacific Review* 26, no. 2: 129–52.

Stanley, Emily H., Martin W. Doyle. 2003. "Trading Off: The Ecological Effects of Dam Removal." *Frontiers in Ecology and the Environment* 1, no. 1: 15–22.

Tuck, Eve, and K. Wayne Yang. 2012. "Decolonization Is Not a Metaphor." *Decolonization: Indigeneity, Education & Society* 1, no. 1: 10–40.

Turek, James, Alex Haro, and Brett Towler. 2016. "Federal Interagency Nature-Like Fishway Passage Design Guidelines for Atlantic Coast Diadromous Fishes." Interagency Technical Memorandum.

Wolch, Jennifer R., Jason Byrne, and Joshua P. Newell. 2014. "Urban Green Space, Public Health, and Environmental Justice: The Challenge of Making Cities 'Just Green Enough.'" *Landscape and Urban Planning* 125: 234–44.

Wohl, Ellen, Paul L. Angermeier, Brian Bledsoe, G. Mathias Kondolf, Larry MacDonnell, David M. Merritt, Margaret A. Palmer, N. LeRoy Poff, and David Tarboton. 2005. "River Restoration." *Water Resources Research* 41, no. 10.

Wolfe, Patrick. 1999. *Settler Colonialism and the Transformation of Anthropology: The Politics and Poetics of an Ethnographic Event.* New York: Cassell.

World Commission on Dams. 2000. *Dams and Development: A New Framework for Decision-Making: The Report of the World Commission on Dams.* Earthscan.

Section II

Access and Water Wars

In the United States, the average home uses about one hundred gallons of water each day. In the poorest nations, a household may need to make do with only five gallons a day, or less (National Geographic Society 2010). Each year, inadequate access to water and sanitation kills an estimated 780,000 people (Subramanian 2019). In arid and landlocked regions, families and communities vie against nation-states, corporations, and other powerful interests for the same fresh water. As each party seeks sufficient fresh water for its own needs, tensions rise into "water conflict" and "water wars," antagonistic struggles over the resource. As populations rise and climate changes, the struggle for equity in water access enacts Hardin's (1968) tragedy of the commons before our eyes.

At the time this collection was being drafted and in print, the world grappled with some of the largest water shortage crises and water access conflicts ever recorded. In Chennai, capital of the southern Indian state of Tamil Nadu and home to more than seven million people, the four major reservoirs of clean drinking water ran dry (Sengupta 2019). It was at least partially a human-made crisis; while Chennai had seen 55 percent less rainfall than usual during its rainy season and then seen two hundred days without rain (Subramanian 2019), the shortage was also largely the result of rampant industrial growth and building projects. Fields and lakes that once acted as natural reservoirs, catching and preserving fresh water, had, over the past several decades, been filled in and built over to meet the demands of a rapidly growing population. As one commentator wrote, "Land is too expensive to be left fallow" (Sengupta, 2019). Moreover, within that burgeoning population grew burgeoning class divisions and desires; as annual water reserves dwindled, those who could afford to do so drilled private bore wells to ensure their own water access, then took beyond what was needed (Subramanian, 2019; Hardin 1968). The residents of Chennai must now ration water use and seek fresh water from novel sources, turning to daily shipments of fresh water from out of the city (dependent on fossil fuel consumption for tankers traveling from afar), investing in industrial desalination projects (which themselves consume material resources in mass), and simply drinking water that is not safe for consumption. Around India, twenty-one other major cities are also facing extreme shortages, when "municipal water sources are unable to meet demand" (Subramanian 2019). In Cape Town, South Africa, this point has been given a name: "Day Zero," the day when city taps would be turned off and replaced with daily water rations. In 2018, Cape Town residents were warned of the imminent water apocalypse, and the term entered the global vocabulary, as cities from London to Tokyo to São Paulo have experienced their own water crises (Laudicina 2018).

In this new millennium, economists have already begun to see potable water as an investment opportunity (Gray 2019), fueling a rush toward water privatization. The water wars of the current century have begun in cities like Cape Town, where residents cut back while corporate wells sat untapped. In countries like the United States, where farm and industrial use accounts for roughly 93 percent of water consumption (National Geographic Society 2010), individual and household water access risk becoming only minor concerns for major markets.

But here, as with the injustices surrounding water sovereignty, activists, policy makers, and communities have mounted a rhetorical response. In Chennai, this has taken the form of public demonstration—hundreds have been arrested for sometimes violent protests—but also the form of public advocacy and campaigning for local, practical solutions that prevent reliance on further industrial development megaconstruction and resource extraction from afar. The water wars are being waged with words and frames and symbols as much as they are with power, influence, and pocketbooks.

This section examines such rhetorical invention in four cases of water access activism and deliberation that attempt to reframe and reorient public conversations over who deserves water and why, how to conserve water, and how to ensure everyone gets equitable access to the water they need. Each chapter highlights a rhetorical concept useful to conceptualizing public campaigns for water access.

In chapter 5, Theresa R. Castor examines the *meanings of "public" and "community"* across two water diversion requests in southeast Wisconsin, attending to frames in the 2008 Great Lakes Compact. Given that "public" and "community" are a part of the Compact's key criteria, analysis of stakeholder use of these terms provides insight on the parameters of the Compact in regulating water diversions. Castor applies Latour's (2005) actor-network theory and Cooren's (2010) ventriloquial approach to guide a discourse analytic examination of water diversion public hearing transcripts. The findings show general agreement on the *importance* of the Compact and the "public" and "community" good, but key disagreement in the *application* of the Compact and how "public" and "community" are constituted. The chapter contributes to the understanding of how fresh water access problems are defined and the applications and limitations of water policies.

In chapter 6, Jordan Christiansen applies thematic analysis to Kansas Governor Sam Brownback's appeals to economic growth and freedom as rhetorical response to the depletion of the Ogallala Aquifer, the "lifeblood of the Great Plains agriculture economy." Christiansen argues that Brownback's *greenwashing justifications* for conserving the Ogallala Aquifer worked to maintain the conflicting practices of water conservation and upholding the viability of the Kansas economy, and concludes that these strategies relied less on washing away the negative effects of current water usage, and instead greenwashed Brownback and Kansas farmers as simultaneously environmentally responsible and economically viable. In conducting its analysis, the chapter extends understandings of greenwashing to include individuals and political parties.

In chapter 7, Kelsey Mesmer, Mostafa Aniss, and Rahul Mitra examine municipal water shutoffs in Detroit, Michigan, as a case study in communicative meaning-making of water access and scarcity among both urban and suburban residents, exploring the constitution of privilege and difference that works to "other" low-income and, often, black residents of the city. They focus on city-sanctioned shutoffs in 2014 that led to a spike in waterborne illnesses, an increase in tax foreclosures, and a declaration by the United Nations describing the shutoffs as an unprecedented human rights travesty. Despite media coverage of these events and grassroots furor, suburban residents in the region and leaders remained ambivalent or even outright hostile to the plight of fellow citizens barely a few miles away. The chapter puts into conversation theoretical frameworks from environmental communication, social justice law, and anthropology to unpack how relatively privileged residents of metro Detroit constitute their "relations-in-place" (Milstein et al. 2011) with those residents denied water access. Building from interviews with various community and institutional stakeholders, it proposes new discourses and meaning-making devices to enable grassroots activists and progressive policy makers to "*preach beyond the choir*," and lay a foundation for long-term community-building and policy-making around vital natural resource infrastructures.

In chapter 8, Christopher Oldenburg takes a more global scope, with an examination of Pope Francis's 2018 papal message on the "World Day of Prayer for the Care of Creation," which focused specifically on the *topos* of water. In the address, Francis reaffirms water as a social justice issue and that "access to safe drinkable water is a basic and universal human right" frequently denied to the poorest and most vulnerable global communities. Oldenberg examines how Francis invites his audience to ecological conversion by forging a confluence between social justice and water, and through a rhetorical strategy of *hydro-anthropomorphism*. Drawing from classical rhetorical theory, the chapter defines hydro-anthropomorphism as an eco-trope that functions to embody the presence, rights, and excluded expressions of both the biophysical Other (water) and the poor, serving as a countercorrective "voice" to the epistemological constitution of water.

REFERENCES

Cooren, François. 2010. *Action and Agency in Dialogue: Passion, Incarnation and Ventriloquism*. Amsterdam, Netherlands: John Benjamins.

Gray, Tim. 2019. "As Fresh Water Grows Scarcer, It Could Become a Good Investment." *New York Times*, July 13, 2019.

Hardin, Garrett. 1968. "The Tragedy of the Commons." *Science* 162, no. 3859: 1243–48.

Latour, Bruno. 2005. *Reassembling the Social: An Introduction to Actor-Network-Theory*. Oxford: Oxford University Press.

Laudicina, Paul. 2018. "Water Day Zero Coming to a City Near You." *Forbes,* June 7, 2018. https://www.forbes.com/sites/paullaudicina/2018/06/07/water-day-zero-coming-to-a-city-near-you/#70bf421c1738.

Milstein, Tema, Claudia Anguiano, Jennifer Sandoval, Yea-Wen Chen, and Elizabeth Dickinson. 2011. "Communicating a 'New' Environmental Vernacular: A Sense of Relations-in-Place." *Communication Monographs* 78, no. 4: 486–510.

National Geographic Society. 2010. "Special Issue: Water: Out Thirsty World." *National Geographic*, April 2010.

Sengupta, Somini. 2019. "Life in a City without Water: Anxious, Exhausting and Sweaty." *New York Times*, July 11, 2019.

Subramanian, Meera. 2019. "India's Terrifying Water Crisis." *New York Times*, July 15, 2019.

Chapter Five

Water for the "Community" Good

Contested Meanings of Stakeholder Interests in Great Lakes Water Diversion Controversies

Theresa R. Castor

The Great Lakes of North America are fortunate in having a bountiful, freshwater supply, containing 21 percent of the world's surface freshwater (EPA, n.d.). The Great Lakes region is home to thirty million residents of the United States and Canada and is an important economic driver for the region (NOAA 2019). However, only 1 percent of the waters of the Great Lakes are replenished each year. Thus, for a variety of reasons, there is a great deal of care and attention devoted to protecting Great Lakes' waters for environmental and economic reasons.

In his discussion of the Aral Sea disaster, Peter Annin in his book, *The Great Lakes Water Wars* (2006), provides a cautionary tale of what could happen to the Great Lakes if the waters are not taken care of in a sustainable way. Because of diversions of water from the rivers that fed the Aral Sea to farmland starting in the 1960s and over a roughly fifty-year period, the waters of the Aral Sea were depleted by 90 percent. Former sea bottom became areas of dry land; fishing villages were no longer viable. It was an environmental and public health disaster. Great Lakes water policies and water diversion proposals have a long history (see Annin 2006 for background). For this project, I focus on the Great Lakes–St. Lawrence River Basin Water Resources Compact, often referred to as the Great Lakes Compact or Compact.

In 2008, the U.S. government enacted the Great Lakes Compact to safeguard the Great Lakes against water diversions. While the Compact limits diversions, there are exceptions allowed in the case of "straddling communities." This historic legislation was tested recently in 2013 when the city of Waukesha, located in the southeast corner of Wisconsin, successfully requested a water diversion. This controversial case was watched closely due to concern regarding the precedent that it could set for the application of the Compact. Waukesha's diversion request was eventually approved, but only after modifications requested by the Great Lakes–St. Lawrence River Water Resources Regional Body (Regional Body) were adopted into Waukesha's proposed diversion plan.

Approximately five years later, in 2018 and roughly forty miles away, the city of Racine requested a water diversion of its own to supply water for the village of Mount Pleasant and in anticipation of the planned development of a Foxconn electronics manufacturing facility in Mount Pleasant. While there are differences across the Waukesha and Racine water diversion cases, both have been very controversial for some overlapping as well as different reasons. One feature they have in common, in part as required by the Great Lakes Compact, is a focus on "community."

This chapter contributes to the understanding of the discursive construction of problems and specifically to how freshwater issues surrounding the Great Lakes are defined. Given the value of the Great Lakes and the role of the Great Lakes Compact for protecting the Great Lakes, it is important to understand the key terms and their meanings under which water has been diverted from the Great Lakes. This work can contribute to understanding the applications and limitations of water policies.

The purpose of this project is to analyze the meanings of "community" across two water diversion requests in southeast Wisconsin. "Community" is a part of key criteria for approval of a Great Lakes water diversion, and "community" has been a controversial idea in public discourse related to the diversion requests. This project analyzes the use of the term "community" in public discourse in order to understand challenges in applying the Compact. The project draws from Cooren's (2010) ventriloquial approach which is a communication perspective that highlights how, when one speaks, it is often on behalf of another. Acts of ventriloquism provide a basis for authority (Vasilyeva et al. 2020) and the formation of associations in community controversies (Cooren 2001).

The primary data for this project consists of public hearings on the diversion requests. Public hearings provide an ideal site for study given the presence of diverse stakeholders. A discourse analytic perspective is applied to analyzing public hearing transcripts. This data was supplemented through participant-observation and the collection of organizational and official documents as well as media accounts.

This chapter is organized as follows. First, background information will be provided on the Great Lakes Compact. Second, additional information and explanation will be provided on the theoretical framing. Then, the two water diversion requests, from the cities of Waukesha and Racine, will be analyzed. In applying a ventriloquial perspective across the two water diversion requests, I identify the challenges of applying federal water policy as it intersects with state policies and localized community concerns. The analysis explicates the rhetorical uses of the term "community" in justifying divergent positions on a water access issue; the ventriloquial perspective helps to explain the different, authoritative bases for such divergent positions that each support "community."

THE GREAT LAKES AND THE COMPACT

The Great Lakes were formed approximately fourteen thousand years ago at the end of the last ice age as glacial sheets moved across North America, carving out large basins that would eventually fill with melted glacial water to form the five Great Lakes: Lake Michigan, Lake Superior, Lake Huron, Lake Erie, and Lake Ontario. Combined, the Great Lakes are the largest source of freshwater on the planet. However, only 1 percent of Great Lakes water is replenished each year through rainfall with the remaining 99 percent a leftover from the original glaciers that formed the Great Lakes basin. The Great Lakes has had a long history of controversy related to water rights and conservation given that it is such a large source of freshwater (Annin 2006; Egan 2017). Because of its limitation on replenishment, there is a great deal of concern related to conserving Great Lakes water and in preventing diversions where water would be transferred out of the Great Lakes area.

The Great Lakes Compact was approved and signed into law in 2008 after several years of negotiations. It was prompted in 1998 by the prospect of a Canadian company bottling Great Lakes water to ship to Asia (Annin 2006). The Compact is a multistate federal agreement among the eight Great Lakes states (Illinois, Indiana, Michigan, Minnesota, New York, Ohio, Pennsylvania, and Wisconsin) pertaining to the restriction of water diversions from the Great Lakes. It has international connections in that the Compact is the United States's fulfillment of the Great Lakes–St. Lawrence River Basin Sustainable Water Resources agreement with the Canadian provinces of Ontario and Quebec. There are several terms and details of the Compact.

The Compact prohibits diversions of water from the Great Lakes but does allow for some exceptions. A diversion is defined as when water is removed from one water boundary area and transferred to another. Section 4.9 (subpart 1[a]) of the Compact, however, mandates that diverted water must be returned to the Great Lakes, minus "an allowance for Consumptive Use." The Compact, therefore, was designed to recognize that while preservation of Great Lakes water is important, the Great Lakes are also a resource that has been and should continue to be used by humans. This approach relates to

Tarlock's (2006) discussion of five meanings of Great Lakes water. Whereas some views (i.e., Great Lakes as perpetual gift, and as world heritage ecosystem) focus on preservation and the minimization of human impact, others emphasize how Great Lakes water can be used for the benefit of human endeavors (i.e., Great Lakes as natural resource, as commodity, and as constrained natural resource). In these latter views, human-initiated changes for economic and development gains are acceptable. Tarlock argues that the Compact is an example of the constrained natural resource perspective in that "consumptive and non-consumptive uses are legitimate but that the former should be carefully assessed and conditioned to minimize interference with the ecosystem" (35).

For the Great Lakes, the basin line demarcates where, when water falls, that water can eventually flow back into the Great Lakes. As explained through an Environmental Protection Agency (EPA) resource: "Most of the basin is defined by hydrology; watersheds that drain into the Great Lakes and their connecting channels are in the Great Lakes basin." However, there are areas included as a result of "engineering and politics" (n.d.). Additionally, the Compact relies on the surface basin line, whereas how waters actually flow and connect underneath the basin differs.

Of note in how the line of the basin is drawn is that there is not a consistency between the basin line and distance from a Great Lake. For example, while nearly the entirety of Michigan is within the Great Lakes basin, only small "corners" or portions of Illinois and Pennsylvania are within the basin. Adjacent to two Great Lakes (Lake Michigan to the east and Lake Superior to the north), Wisconsin is unusual in a number of ways with respect to its relationship to the Great Lakes basin. The entirety of Wisconsin's eastern border consists of Lake Michigan shoreline with 20 percent of Wisconsin residing within the basin. However, the basin line crosses the state unevenly, extending as much as one hundred miles into the state in some areas or as little as seven miles in other areas, notably, in the southeast corner of the state.

Although the Compact bans water diversions, there are some exceptions to this ban which involve addressing some form of "community": (1) a "straddling community" or (2) a "community in a straddling county" may request an exception to the ban. A "straddling community" is one that lies partly within the Great Lakes basin and partly outside of it. Because it is partly contained within the basin, and thus within a municipality or water service area that already interacts with Great Lakes water, an exception is allowed under a less rigorous review process (and typically, with much less controversy). An example of such a diversion request is from the southeast Wisconsin community of New Berlin, a city in the eastern portion of Waukesha County. The dividing line of the Great Lakes basin meanders through the city itself. Thus, it would be possible for one neighbor to be within the basin while the neighbor next door is outside of the basin.

A "community in a straddling county" consists of a community in which the entirety of that community is outside of the Great Lakes basin line, but it is located in a county that is partly within the Great Lakes basin. Within the Compact, a community within a straddling county is defined as: "any incorporated city, town or the equivalent thereof, that is located outside the Basin but wholly within a County that lies partly within the Basin and that is not a Straddling Community."

Given that technically the requesting community is outside of the basin, the exception request is a much more complicated process that involves an international review by the Great Lakes–St. Lawrence River Water Resources Regional Body and public hearings, with a final decision being made by the Great Lakes–St. Lawrence River Basin Water Resources Council known more informally as the Compact Council. An example of a diversion request under this exception is from the city of Waukesha. Although Waukesha County itself is partly within the basin, the city of Waukesha is entirely outside of the basin.

A term related to community is "public." Habermas defined the public sphere as "a virtual or imaginary community which does not necessarily exist in any identifiable space" (see Soules 2007), making a connection between that which is in the public realm and communal interests. The Great Lakes Compact specifies in essence that water obtained through a diversion must be for a "public good" in that it must serve a *public water supply purpose*:

> Public Water Supply Purposes means water distributed to the public through a physically connected system of treatment, storage and distribution facilities serving a group of largely residential customers that may also serve industrial, commercial, and other institutional operators. Water Withdrawn directly from the Basin and not through such a system shall not be considered to be used for Public Water Supply Purposes. (https://www.congress.gov/bill/110th-congress/senate-joint-resolution/45/text)

While the Compact does recognize that water may be used by and for non-public purposes, the emphasis is on providing a public water supply to residents of an area.

On the surface, the Compact guidelines seem to provide an appropriate solution for helping to ensure that residents can have access to safe and clean water. However, as Asen (2017) notes, "a public good is dynamic and mobile, operating at different levels of society, and open to contestation and reformulation" (332). In applying the Great Lakes Compact, definitions of community and public are intertwined with economic interests such that it is at times unclear if fresh, clean water is primarily for the benefit of business and corporate interests (i.e., expansion and development) and/or for broader community interests (i.e., fresh, safe water). This dynamic can be seen in how "public" and "community" are contested notions in the public discourse and deliberation on the water diversion proposals, as will be expanded on later.

Since the approval of the Compact, to date, two diversions have been requested through the "straddling community" exceptions (from the cities of New Berlin and Racine), and one diversion has been requested through the "community in a straddling county" exception (from the city of Waukesha). All of these requests were eventually approved. All of these requests, coincidentally, hail from the southeast corner of Wisconsin. For the following analysis, I focus on the requests from Waukesha and Racine given the specific and heated controversies associated with each of them.

THEORETICAL BACKGROUND

This work draws upon Cooren's (2010) ventriloquial approach which identifies how, when people speak, it is often on behalf of other beings, including humans, texts, objects, nonhumans. Cooren's approach draws from several language analysis and discourse analytic traditions in addition to the science, technology, and society studies work of Bruno Latour (e.g., 2005). The ventriloquial approach has been used to explain how organizations are communicatively constituted or constructed through communication (Schoeneborn et al. 2014), as well as more specific concepts such as authority (Vasilyeva et al. 2020), leadership (Clifton 2017; Fairhurst and Cooren 2009), and crisis management (Bergeron and Cooren 2012). A ventriloquial perspective is well-suited for this project given the insights it can provide on issues related to the use of texts for competing actions (Castor 2018) and framing crises (Bergeron and Cooren 2012).

Cooren (2010) identified several features of ventriloquism including the upstream and downstream implications of ventriloquial acts. The concept of upstream highlights the 'drivers' of ventriloquizing or the precursor situations and activities that lead to ventriloquizing. For example, when an employee says to a customer, "It is our policy to stop service at 9:00 p.m.," that employee is speaking on behalf of the policy (presumably, a written document) as well as on behalf of the company. It would be an entirely different matter if the employee had said "I don't want to serve you," in which case a customer could retort by saying, "Your sign says you are open for business." The latter statement highlights a ventriloquial act in which the customer is speaking on behalf of a physical sign that is posted to speak on behalf of or represent the policy of the business. "Speaking on behalf of" another is intertwined with the construction of authority (Bencherki, Matte, and Cooren 2020) in that when one speaks and expresses a sentiment, makes a request, or justifies an action, it is not just because it is what the other wants, but it is because there is another (e.g., policy, institution, organization, value) that is driving or motivating the request *as well as* authorizing and legitimizing the request. Upstream actions implicate downstream outcomes. In the ventriloquial perspective, the concept of downstream refers to the outcomes or products of ventriloquizing such as signs, policies, additional texts.

Cooren's ventriloquial approach, although initially, dominantly was applied to organizational communication contexts, more generally it has been used to examine institutional contexts, issues of authority and power, and environmental issues. While not applying a ventriloquial perspective specifically, Cooren (2001) applied Greimas's semio-narrative theory to analyze the formation of coalitions through associations among groups on different sides of a Canadian river controversy. That analysis demonstrated the connection between narratives related to environmental disputes and the different alignments of groups with different stances. Focusing on the text of the Great Lakes Compact in a water diversion controversy, Castor (2018) examined the multiplicity of ways that a specific text as well as multiple texts can be applied to support divergent stances on the issues of water and sustainability. Also focusing on a water diversion controversy, Castor (2016) focused on the constitution of "relational ontologies" or conceptions of how people relate to "material" entities such as water and the Great Lakes. What is common across the aforementioned projects is the identification of associations across the various entities involved with a water dispute. These entities include social collectives and nonhuman entities such as texts (e.g., policies) and water.

This project adds to the prior work in examining a key motivation with respect to water, that of the community good. This is not meant to imply that water issues should only be about benefits for people—this is actually one perspective among many regarding water and how people relate to water (see Tarlock 2006)—rather, it acknowledges a key component of the Great Lakes Compact as identified in the preceding section and identifies how community has been problematized in the discourse associated with water diversion controversies.

ANALYSIS: WAUKESHA WATER DIVERSION REQUEST

The Waukesha request for a water diversion was the first of its kind in being issued from a community in a straddling county under the terms of the Great Lakes Compact. Given its precedent-setting status, it was also very controversial and involved the review and decision-making of an international body, the Great Lakes–St. Lawrence River Water Resources Regional Body. The story of Waukesha's relationship to water is ironic. In the late nineteenth century, Waukesha was famous for its spring waters with ongoing attempts to buy and export Waukesha spring water and rumors of attempted theft of said water (Annin 2006). The springs have since dried up, and Waukesha's underground water supply is contaminated with dangerously high levels of radium that is twice the federal limit. In 2003, Waukesha officials signed a state agreement to commit to addressing and solving their water problem by 2006. Waukesha's choice of solution in this matter was to request to divert twenty million gallons of water per day from Lake Michigan.

In 2010, the city of Waukesha began its official, Compact-guided request to divert water from Lake Michigan. Their proposal involved purchasing water from the city of Oak Creek, piping that water approximately seven miles inland to Waukesha. And, after usage and cleaning, returning most of the water back to Lake Michigan by way of the Root River. Their proposal was met with a great deal of controversy that included general concern with diverting Great Lakes water, criticisms of Waukesha's water conservation efforts, and suspicions that the requested water was driven by the desire to support Waukesha business growth.

For this project, I conducted field research related to the public input and decision-making surrounding Waukesha's water request. This included attending public hearings held in summer 2015 on the diversion proposal, attending the meetings of the Regional Body and Compact Council (all of which were open to the public), collecting various news media articles, and attending the public tour offered by the city of Waukesha for viewing their water facility processing plant. For this analysis, I focus on the three public hearings that were held in August 2015 in the city of Waukesha, and the nearby cities of Milwaukee and Racine. Milwaukee is the largest city in Wisconsin and situated in between Waukesha County and Lake Michigan. Racine, to the southeast of Waukesha is where the return flow from Waukesha would go through as part of returning the water to Lake Michigan. Each public hearing lasted for approximately two to three hours with nearly two hundred people speaking at each hearing, with many more also in the audience listening.

Within Waukesha, who or what constitutes a community was under dispute. At issue was the fact that when Waukesha proposed its diversion, it was not just for the city of Waukesha, but also included a provision to expand its existing water supply service area to include most of the *town* of Waukesha, and portions of the neighboring cities of Delafield and Pewaukee as well as a portion of the town of Genesee. Moreover, there were areas in this "expanded service area" that included lands with the potential to be developed. Meaning, water needs were calculated based on projections rather than currently existing needs. The ventriloquial approach highlights how definitions of community are intertwined with prior, "upstream" interactions.

Community as Service Area

During the Waukesha public hearing, one of the first presenters was the mayor of Waukesha who highlights the various upstream interactions relevant to how community is constituted:

> In addition, it is frustrating that opponents claim the state's service area law is inconsistent with the Compact. Governor Doyle's administration, who helped write the Compact, also wrote the service area law. When the Compact was adopted, it was expected that Waukesha's application would include the proposed service area.

In his characterization, the mayor is ventriloquizing the prior Governor Doyle administration, highlighting the role of that administration in both contributing to the Compact and developing the service area law. This multifaceted authorship role is used to explain how the community of concern is constituted by the "proposed service area." Thus, community is defined through a legal, state-sanctioned association ("service area law"). This definition is sanctioned through the authority of a prior state government. Community, therefore, is a connection that is mediated through legal ties and sanctions rather than (strictly) geographical proximity. The mayor's characterization of how the Compact is compatible with the notion of service area, however, is disputed by another speaker who was a state negotiator for the Compact.

The speaker quoted below self-identified as a former water division administrator for the Wisconsin Department of Natural Resources and one of the state's negotiators of the Great Lakes Compact:

> The statutory construct of the phrase "city, town or equivalent thereof," was discussed at length over many meetings and phone calls as the Great Lakes Compact was developed. The whole purpose of the discussion was to capture those entities that exist somewhere in the Great Lakes Basin, they're not cities or towns, but have equivalent meaning in those states or provinces. . . . The whole discussion leading up to the inclusion of the term "equivalent thereof," was to find a phrase for those communities that were not already cities or towns.

A contrast between the mayor's ventriloquial move and that of the aforementioned speaker is that the latter is speaking on behalf of his own experience and involvement in negotiating the Compact. This speaker shifts the notion of community away from a public service area to an issue of terminology: because communities may be labeled differently depending on state and national guidelines (e.g., city, town, village), the broadness of label was intended to capture a terminology difference. In this speaker's characterization, the idea of community is more a matter of semantics rather than policy.

Another way that community was problematized relates to the intertwining of material conditions of water and water systems. In advocating for the inclusion of the town of Genesee within Waukesha's application, the town chair for Genesee explained during the Waukesha public hearing that:

> Groundwater does not stop at municipal boundaries. And what Waukesha needs to do to protect their water supply affects municipalities all over the, you know, the area.

Genesee's town chair's comments highlight the blurring of community boundaries. In this case, the boundary of municipalities is secondary to boundaries (or lack thereof) created by water flows of groundwater as Genesee's water is inextricably interconnected with Waukesha's water.

One of the challenges in defining "community" relates to the disparate ways it has been defined across various governmental documents, as explained by the speaker quoted below:

> So, the Compact and Wisconsin's implementing legislation both require that to apply for a diversion, the community within the straddling county needs to show that they are without adequate supplies of potable water and also that there is no reasonable water supply alternative within the basin in which the community is located. And this would include conservation of existing water supplies. So, evaluated against these core tenants of the Compact, the City of Waukesha's application we feel is deficient in several critical areas in relation to the community requirements. . . . The City of Waukesha's definition of "community need" is inconsistent throughout its application. When calculating how much the water—how much City water—how much water the City will need in the future, Waukesha includes portions of neighboring communities in the definition of their "community."

In the preceding comments, the speaker is in essence claiming that Waukesha's water diversion application is not actually speaking on behalf of (ventriloquizing) a community and that the water needs calculated and described in the application are inaccurate in speaking on behalf of the needs of Waukesha.

What "counts" as a community became blurred as boundaries became intermixed with definitions of city and village borders as well as water service area lines. In addition, in Waukesha's application, it is not a *currently* existing service area, but rather, a *future proposed* or currently nonexisting service area. Therefore, the calculations related to water needs are based on the assumptions of a "community" that does not exist, but rather, one to be intentionally created. An "upstream" impact of the Waukesha proposal is to enable the development of the new water service area to satisfy its projected need. In a way, Waukesha's need for water is being created through the diversion request itself.

Downstream Effects and Community

In Waukesha's proposed water diversion plan, water would be taken to Waukesha through a pipeline from the city of Oak Creek, utilized within the Waukesha water service area, treated at a Waukesha water utility plant, and then returned to Lake Michigan via the Root River that flows through the city of Racine. "Community" as described in the Compact focuses on the community in need of clean water. However, other communities are impacted overall through the process of water diversion. Thus, when applying a ventriloquial perspective to Waukesha's water diversion request, literally and analytically, Racine is a part of the downstream effects of the water diversion. Many presenters across all three public hearings that took place in Waukesha, Milwaukee, and Racine expressed concern for Racine and the downstream impact of the return flow.

There were several areas of concern related to the impact of the water diversion for Racine and for the Root River that included the extent to which pollutants and contaminants such as phosphorus and chloride would be removed and how river water flow and temperature (and therefore, its surrounding ecology) would be impacted by the additional treated water. As summarized by the speaker below:

> The Great Lakes Compact does not concern itself with possible—possible consequential damages to other ecosystems, and I'm afraid that's what's happening with us.

As a note, as part of a diversion proposal, there is required monitoring of the return flow. What is of relevance for this analysis is the way that community is communicatively constituted: there are multiple communities, and the community of Racine is constituted as a passive recipient.

Community Concerns as Contextually Situated

Not surprisingly, the public hearings which occurred in each of the three cities of Waukesha, Milwaukee, and Racine, each emphasized different themes. All of course addressed the importance of the Great Lakes and the Compact as well as specific issues related to Waukesha. However, there was a tendency among public hearing presenters to emphasize different themes that ventriloquized key interests related to the geographic area of their city.

Using Waukesha as a baseline, presenters addressed the water concerns of that area as well as matters pertaining to economic and business development. Some of these issues related to the radium water contamination problem and the need for a safe water supply for Waukesha residents. Some presenters also addressed the positive impact in terms of property values and supporting business growth.

In Milwaukee, the city that lies in between Waukesha and Lake Michigan, some presenters emphasized race, economics, and social justice issues. As a Milwaukee resident outlined:

> In the larger context of a city being more endangered, more and more managed from the capital, from education to highways to freedom to bear guns and underpinned by a translation into action by white flight, this Root River solution is another nail in the plantation fence desired by a few powerful people in Waukesha. I submit to the DNR that there is a racial component to the determination to get Lake Michigan water without allowing access to jobs and development, which I believe is the spark that pushes Waukesha to add a new fence.

In the preceding quotation, the speaker can be said to be ventriloquizing what Waukesha and its water diversion proposal are speaking on behalf of, which is a separation in terms of access to jobs and the continued segregation of communities: Milwaukee is a large urban center with a substantial black population with a high unemployment rate; Waukesha is a generally white, suburban community that is more economically stable. This sentiment is emphasized through phrases such as "white flight" and "plantation fence."

As another public hearing presenter explained (who also spoke as a representative from the Sierra Club):

> And the reason for this is that Waukesha is asking for water from Lake Michigan for areas well beyond the City limits specifically for future economic and industrial development for a projected population growth, and this water is coming from or via the City and County of Milwaukee, which has a lot of access for water, but has a lot of empty factories, and also has a lot of people who are unemployed, especially people of color, and I think this is an important consideration. So, to promote this water diversion will do more than just exacerbate urban sprawl, but will further divide the two cities. It used to be that the expressways were the promoters of urban sprawl and would divide peoples, but here we have water. We've got a new source of division that will divide the two cities further, both economically and racially.

As with the preceding quotation, this public hearing presenter creates associations between water, economic development, and business/industrial development as well as contrasts such as economic growth versus unemployment. A theme in these associations and divisions is that of race and economic social justice. Waukesha's use of Lake Michigan water would enable business growth in Waukesha, away from (racially diverse) Milwaukee. There is an issue of access and jobs where the water diversion would enable businesses to develop in an area that would be difficult for economically disadvantaged, mostly minority people of Milwaukee to get to.

As noted in the preceding section, presenters in the city of Racine emphasized the downstream impacts of the water diversion process vis-à-vis the return flow through the Root River. As conveyed rather dramatically by one speaker, a mother, as she held up what appeared to be a home-computer-printed image of her daughter playing by the Root River:

> My name is XXXX, and I would like to speak to you as a child who grew up playing on the shores of Lake Michigan, as did the five previous generations of my Wisconsinite family. I am speaking to you as a mother, a teacher, a home-owner in the City of Racine, a taxpayer, and a representative of my four-year-old daughter, who is one of the children that all these speakers are talking about. She's four years old, and I do not want her having an increased risk of cancer, fertility problems as an adult, or any other health problems that we don't know about because we are taking people at their word that we can't prove. She is the one who will pay the real price of your mistake if you ignore the laws of nature and sell our water to other places who fail to heed nature's warning signs.

Not all Racine presenters were against the diversion, and when speaking about the downstream impact of the diversion proposal, some of the supporters of the diversion focused on the anticipated positive impacts it would have on the ecology surrounding the Root River as well as the potential positive economic impact of the diversion.

Summary

During the meeting of the Regional Body, the Waukesha proposal was subsequently modified as members of that decision-making body also had questions regarding what communities were included with the request as well as the distinction between the current water service area and the future, proposed water service area. In the end, the Compact Council and Regional Body approved Waukesha's diversion request, but only after the service area was reduced (the new service area excludes the towns of Delafield and Genesee and most of the town of Waukesha) and the number of gallons of water diverted per day was also reduced (from 10.1 million gallons to 8.76).

Waukesha's water diversion request set a precedent for the application of the Great Lakes Compact in addressing diversion requests from communities in straddling counties. Waukesha's request also highlighted how important yet problematic the idea of "community" is for water use and water policy considerations. As a result, when one speaks on behalf of a "community," it is on the part of a hybrid collective that consists of both human and nonhuman agents.

ANALYSIS: RACINE WATER DIVERSION REQUEST

In 2018, the city of Racine submitted a request to the Wisconsin Department of Natural Resources to divert up to seven million gallons of water per day from Lake Michigan. The city of Racine itself is entirely within the Great Lakes water basin. However, the Racine Water Utility provides water service to the village of Mount Pleasant. Most of this village is also within the Great Lakes water basin, except for a relatively small corner of 787 acres which also happens to be the site of where the Taiwan-based corporation, Foxconn, will be building a new electronics manufacturing facility. Because part of Mount Pleasant is outside of the Great Lakes Basin, the request for water technically counts as a diversion request. However, it is a diversion request from a "straddling community" and therefore, underwent a different review and approval process from the Waukesha diversion request. Specifically, it was up to the state of Wisconsin government to decide on whether or not to grant the request, whereas for the Waukesha proposal, the decision-making body was ultimately an international body that represented multiple state, provincial, and national interests.

Aside from the fact that this was a Great Lakes water diversion request which, as seen from the controversy with Waukesha's request, garners a great deal of public attention, concern, and scrutiny, another controversial dimension to this request was the involvement of a non-US, foreign corporation. A key argument against the water diversion was that it was not being requested on behalf of a "community" or for the public good, but in fact, was being requested primarily to satisfy the needs of a corporate, private, for-profit, foreign entity.

Community as Corporation

The main criticism against Racine's water diversion request was that ultimately, it was intended to benefit a corporate entity rather than residents. As stated in an opinion piece by Peter Annin (2018), author of *The Great Lakes Water Wars*, "The Compact allows exceptions for straddling communities, but what about straddling corporations?"

In terms of ventriloquism, opponents in essence claimed that the diversion request was not actually speaking on behalf of residents of the village of Mount Pleasant, but rather, was speaking on behalf of Foxconn and its water needs. During the public hearing, some speakers even altered the description of the diversion to describing it as the "Foxconn diversion application" rather than Racine diversion application:

> Clearly, we are concerned that [the] Foxconn diversion application is for industrial use and has not adequately provided detail about the amount and the

> quality of the water that will be returned through the City of Racine and Lake Michigan. Foxconn has not implemented sufficient conservation practices in its other industrial locations.

In the preceding remark, as noted, the speaker attributes the application to Foxconn, highlighting how the main purpose for the water will be to satisfy a "private, foreign industrial company's water needs" (stated earlier by speaker). Other presenters also emphasized how the diversion is to address the needs of a "private entity." Thus, the water diversion request is not speaking on behalf of the water needs of the residents of the village of Mount Pleasant. Rather, the diversion request is representing the needs of Foxconn.

Downstream Impacts of Water for the (Corporate) Communal Good

Supporters of the diversion request did not necessarily contradict opponents. Rather, they focused on the primarily economic impact that the diversion would have. As stated by the Chair of the Mount Pleasant Community Development Authority:

> We have been presented with a great opportunity to expand development of Mount Pleasant along the I-94 corridor. The time is right for considering how we can provide clean, reliable, safe water to the businesses that will locate in the area along with the thousands of workers they will employ.

What the chair emphasized is how the diversion will have a positive impact on businesses that in the future ("will") be in the area and for future employees of those businesses. In this respect, there is a similarity to Waukesha's proposal in anticipating a future, currently not existing, need that the diversion would address.

In contrast to the economic downstream (positive) impact, opponents to the diversion were concerned about the potential environmental impact of having a large, electronics manufacturing facility in the area. As expressed by a member of the League of Women Voters of Wisconsin (a group that opposed the diversion):

> The future public need for water in the village of Mount Pleasant may be compromised by this sizeable diversion to a nonpublic entity. This request for a diversion on behalf of water for a nonpublic entity threatens future public water use especially if the entity increases its footprint and need for water. Foxconn, the proposed end user of most of the diverted water intends to send as much as 60 to 70 percent of withdrawn water a day to Lake Michigan with no enforceable, RIGOROUS requirements for measuring contaminants in return waters. The waters returning to the lake must be treated to assure that toxins used in large screen panel production do not enter Lake Michigan.

> Among these toxins is a cancer causing chemical substance density also included is a host of toxic heavy metals.

This speaker as well as others were concerned about the downstream impact of Foxconn's manufacturing plant that would be the main beneficiary of the water diversion. Because of the need for a return flow, the water being return to Lake Michigan would have been processed at some point vis-à-vis the Foxconn facility. However, as part of brokering a deal to entice Foxconn to Wisconsin, the government of the state of Wisconsin waived Foxconn from several state environmental laws (Johnson 2017).

Summary

Legally, Racine's water diversion request fits within the parameters of the Great Lakes Compact. However, opponents to the Racine diversion request emphasized that the request would not even have been made if it were not for the imminent presence of Foxconn. This crucial difference between the Waukesha and Racine water diversion requests is highlighted by the fact that a hydrologist who supported the Waukesha diversion opposed the Racine diversion because of the crucial distinction between water for the communal good versus water for the corporate good (Gordon, 2018). The state of Wisconsin eventually did approve Racine's water diversion request, however, a number of legal challenges have been mounted against the request.

CONCLUSION

For the water diversion requests, "community" is a crucial issue. Legally, the Great Lakes Compact is set up to protect the Great Lakes while also recognizing the needs of communities for access to fresh, safe water. The Compact regulates which communities can request water and presents various definitions and conceptualizations of community. In application, however, this analysis identified the following issues and challenges for defining community with respect to water diversion policy and its implementation:

- There is broad agreement on the importance of the Great Lakes and the Great Lakes Compact.
- Community may be based on legal definitions or stakeholder interests, and they do not necessarily result in the same grouping.
- Water diversions affect multiple communities that include those requesting water and those nearby who are effected environmentally and, potentially, economically by the request.

- Rhetorically, "community" may be used as a broad term to gloss over the distinctions between multiple groups such as private citizens and businesses.

A ventriloquial analysis is particularly well-suited for examining the differences and nuances in how "community" was defined by different stakeholders. Ultimately, the ventriloquial perspective interrogates who is speaking on behalf of whom and/or what. A key disagreement across the water diversion requests was on the application of the Compact and how "community" is constituted by the water diversion requestors. Opponents of the diversions highlight the role of economic interests (e.g., businesses generally in the Waukesha case; Foxconn in the Racine case) in driving the diversion requests. In the case of Waukesha, because of the role of an international oversight agency, the diversion parameters were reshaped in ways that addressed critics' concerns. In the case of Racine, the diversion request was not subject to international oversight.

The ventriloquial approach highlights the multiplicity of human and nonhuman entities that are represented through a water diversion request. Ultimately, while the Compact provides some guidance, the language within the Compact lacks guidance in other areas, leaving it open to states to fill in various gaps. From a ventriloquial perspective, the downstream impact of water diversions are also multifaceted in affecting neighboring communities and economic development.

REFERENCES

Annin, Peter. 2006. *Great Lakes Water Wars.* Washington, DC: Island Press.

Annin, Peter. 2018. "Foxconn: A New Chapter in the Great Lakes Water Wars." *Milwaukee Journal Sentinel,* April 13, 2018. https://www.jsonline.com/story/opinion/contributors/2018/04/13/foxconn-rekindles-great-lakes-water-wars/511566002/.

Asen, Robert. 2017. "Neoliberalism, the Public Sphere, and a Public Good." *Quarterly Journal of Speech* 103, no. 4: 329–49.

Bencherki, Nicolas, Frédérik Matte, and François Cooren. 2020. *Authority and Power in Social Interaction: Methods and Analysis.* New York: Routledge.

Bergeron, Caroline D., and François Cooren. 2012. "The Collective Framing of Crisis Management: A Ventriloqual Analysis of Emergency Operations Centres." *Journal of Contingencies and Crisis Management* 20, no. 3: 120–37.

Castor, Theresa. 2016. "The Materiality of Discourse: Relational Positioning in a Fresh Water Controversy." *Communication Research and Practice* 2, no. 3: 334–50.

Castor, Theresa. 2018. "Sustainability and Textual Extensions of Institutional Discourse: Testing the Great Lakes Compact." *Critical Sociology* 44, no. 2: 341–56.

Clifton, Jonathan. 2017. "Leaders as Ventriloquists: Leader Identity and Influencing the Communicative Construction of the Organisation." *Leadership* 13, no. 3: 301–19.

Cooren, François. 2001. "Translation and Articulation in the Organization of Coalitions: The Great Whale River Case." *Communication Theory* 11, no. 2: 178–200.

Cooren, François. 2010. *Action and Agency in Dialogue: Passion, Incarnation and Ventriloquism.* Amsterdam, Netherlands: John Benjamins.

Egan, Dan. 2017. *The Death and Life of the Great Lakes.* New York: W. W. Norton & Company.

EPA. n.d. "Facts and Figures about the Great Lakes." Environmental Protection Agency. https://www.epa.gov/greatlakes/facts-and-figures-about-great-lakes.

Fairhurst, Gail T., and François Cooren. 2009. "Leadership as the Hybrid Production of Presence(s)." *Leadership* 5, no. 4: 469–90.

Gordon, Scott. 2018. "When Great Lakes Water Is 'Public' And When It Isn't: Foxconn's Lake Michigan Bid Raises Questions about Interpreting A Young Law." *WisContext,* April 19, 2018. https://www.wiscontext.org/when-great-lakes-water-public-and-when-it-isnt.

Johnson, Shawn. 2017. "Wisconsin Foxconn Deal Waives Environmental Regulations." *Wisconsin Public Radio,* July 28, 2017. https://www.wpr.org/wisconsin-foxconn-deal-waives-environmental-regulations.

Latour, Bruno. 2005. *Reassembling the Social: An Introduction to Actor-Network-Theory.* Oxford: Oxford University Press.

NOAA. 2019. "Great Lakes." National Oceanic and Atmospheric Administration. https://coast.noaa.gov/states/fast-facts/great-lakes.html.

Schoeneborn, Dennis, Steffen Blaschke, François Cooren, Robert D. McPhee, David Seidl, and James R. Taylor. 2014. "The Three Schools of CCO Thinking: Interactive Dialogue and Systematic Comparison." *Management Communication Quarterly* 28, no. 2: 285–316.

Soules, Marshall. 2007. "Jürgen Habermas and the Public Sphere." *Media Studies.ca.* http://www.media-studies.ca/articles/habermas.htm.

Tarlock, Dan. 2006. "Five Views of the Great Lakes and Why They Might Matter." *Minnesota Journal of International Law* 15: 21–41.

Vasilyeva, Alena, Jessica S. Robles, Jean A. Saludadez, Christian Schwägerl, and Theresa R. Castor. (2020). "The Varieties of (More or Less) Formal Authority." In *Authority and Power in Social Interaction: Methods and Analysis*, edited by Nicolas Bencherki, Frédérik Matte, and François Cooren, 37–55. New York: Routledge.

Chapter Six

Kansas and the Ogallala Aquifer

Greenwashing Attempts to Balance Water Conservation with Free Market Principles

Jordan Christiansen

Out of sight but not out of mind is the Ogallala Aquifer. It lies beneath the surface of eight states in a semiarid region of the North American Midwest known as the High Plains, with relatively little rainfall and few other water resources (Ashworth 2006). It's a dry and transitionary place between the prairies to the east and the mountains and deserts of the west. For many parts of the region, the Ogallala Aquifer, a reservoir for millions of years to rainfall and river runoff, is the only reliable water source. And, in recent decades, the groundwater from the Ogallala Aquifer has transformed the region into an oasis of agricultural viability (Opie 2000). The Kansas Water Office highlights how "the Ogallala portion of the High Plains aquifer is the primary source of water in western Kansas for all uses and is heavily developed, primarily for irrigation" (Kansas Water Authority 2015). The High Plains region has come to be called "America's breadbasket" (Parker 2016), and the economy supported by the aquifer drives a $20 billion industry that grows nearly one-fifth of the country's crops and livestock (Kansas Water Office 2014). This demand means that five trillion gallons of water are drawn from the Ogallala annually (Ashworth 2006, 10). Mass irrigation has resulted in nearly sixty million acre-feet of water being removed from the aquifer, with half of that occurring since 1980 alone (Ashworth 2006, 11). At such rates, the aquifer is being drained faster than it can naturally be replenished. In fact, if things stay the same, there are some areas that have less than twenty-five to fifty years of water remaining (Butler et al. 2013).

Of the states impacted by the aquifer's depletion, Kansas is particularly threatened. Kansas sits on top of 18 percent of the aquifer, and much of that lies beneath the western half of the state. However, "by 2010, 30 percent of the aquifer's water had been tapped," and an "additional 39 percent of the aquifer's reserve is projected to be used by 2060" (Butler et al. 2013). Concurrently, "If current irrigation trends continue, 69 percent of the groundwater stored in the High Plains Aquifer of Kansas will be depleted in 50 years" (Tammen 2013). Such gulping in Kansas continues, though, almost exclusively to support the agriculture industry. The Docking Institute of Public Affairs at Fort Hays State University found the economic impact of Ogallala to be $188.5 million per year (Torell, Libbin, and Miller 1990). Additionally, estimates have the depletion of the aquifer as an economic loss of $110 million each year between 1996 and 2005 (Fenichel et al. 2016). Water from the Ogallala Aquifer supports agriculture in Kansas, and its depletion poses threats of catastrophic economic losses.

The severity of this crisis does not mean that reducing consumption is impossible. "Current research indicates that even though modest reductions in groundwater usage could greatly reduce the rates of water-level decline" however, "full recovery of the aquifer to predevelopment levels would likely take decades to thousands of years" (Hanks 2018; Wilson 2019). Communities in Kansas are working hard to conserve the water, simply because they have to do so. For instance, Hays, Kansas—a community that relies entirely on the aquifer for its water—reduced its annual water consumption from 1.2 billion to 636 million gallons between 1979 and 2014 (Connor 2017). Hays was able to achieve this result by providing incentives to residents for reducing their household water use. The incentives involve rebates for installing high-efficiency appliances and implementing water-smart landscaping. The community of Hays shows how the aquifer crisis directly impacts all who call the High Plains home, and demonstrates how these communities are working toward sustainable water conservation.

From when Kansas Governor Sam Brownback took office in 2011, through his departure in 2018, preserving the Ogallala Aquifer was "one of his most talked about priorities" (Dillon 2018). A goal of his administration was to develop and implement a fifty-year plan for the management of Kansas's water, known as the "Kansas Water Vision." The vision's mission statement affirms a commitment to "Provide Kansans with the framework, policy and tools, developed in concert with stakeholders, to manage, secure and protect a reliable, long-term water supply while balancing conservation with economic growth" (Kansas Water Office 2014). From this mission statement and the policies put forth by Brownback's administration, one can see rhetorical attempts to negotiate water conservation with the demands of maintaining a free market economy. As Brownback stated during his 2017 Water Tour, "Preserving water resources is vital to the success of our state.

Without water, there is no future." To ensure Kansas's future, Brownback supported the transition from "use it or lose it" water policies to new policies of "first in time, first in line." Additionally, in 2012, Brownback's administration established Local Enhanced Management Areas (LEMAs), voluntary water conservation programs for farmers who wish to collectively reduce their water usage. Such efforts also contributed to the organization of the Ogallala Water Summit, an annual conference where universities, governments, farmers, ranchers, and other consumers come together to generate ideas for conserving the aquifer (Melgares 2018). With these changes, Sam Brownback certainly encouraged some modification of water consumption practices.

However, those modifications were not ultimately mandated by the state. They were instead set up as *voluntary* options for farmers and, thus far, have yet to significantly reduce water consumption rates and the inequitable distribution of water rights. In fact, "The very mention of stricter limits, even when they are locally imposed, provokes difficult and uncomfortable dialogue among irrigators" (Dillon 2018). Many farmers respond to water regulations by arguing, "It's my water right, and I'm going to do what I want with it" (11). The freedom to draw the maximum amount of water to ensure their own economic success has left many unwilling to reduce their consumption. Those who don't have or have limited access to the aquifer are left reaching for a straw to drink. Despite this, farmers across the state acknowledge the depletion of the aquifer is a pressing crisis (Buchanan 2018; Brown 2018). The apparent lack of action toward more aggressive conservation and more equitable water rights seems to be a result of a complicated double-bind wherein farmers are caught amid tensions between water conservation and neoliberal free market principles. Amid these tensions, farmers can cut back water use voluntarily but face the probability of lower productivity and, thus, less economic income. Conversely, farmers can consume water as usual and maintain their economic livelihood—at least for the short-term. Unpacking these tensions between water conservation and free market principles demonstrates how people in Kansas understand the need to conserve the only reliable water source they have; yet, "there is a curious disconnect between what they understand and what they do" (Ashworth 2006, 58). In short, Kansas farmers "have treated declines in the Ogallala Aquifer about the way they treat the weather: they talk about it, but they don't do much" (Buchanan 2018).

Another component of the Ogallala Aquifer crisis is that current water policies perpetuate the inequitable distribution of water rights. In Kansas, the water system is set up in ways that "privilege some users over others" (Dillon 2018, 4). In attempts to enact LEMA programs in their communities, farmers often report "getting pushback from neighboring irrigators who either don't want to limit themselves or don't like the terms of how they're being limited"

(Dillon 2018, 4). Additionally, Brownback's "first in time, first in line" policy allocates water rights to whoever can prove they started using the water first—meaning those who arrived later face usage restrictions depending on when they show up. In short, "water is allocated based on whoever started using it first, with later users—who have junior water rights—having to stand in line behind senior water right holders" (Dillon 2018, 4). This contributes to having senior water rights holders "consume about 22 percent of all groundwater" in the state (5). Ultimately, inequitable water rights and individual unwillingness create issues of water overconsumption and injustice, perpetuating water conflicts in the state.

For these reasons, the attempted negotiation of the double-bind between conserving the Ogallala Aquifer and prolonging Kansas's economic viability will be the focus of this chapter. Specifically, the chapter discusses how Brownback's confrontation of the Aquifer's depletion demonstrated rhetorical strategies that addressed the crisis while attempting to balance water conservation and free market principles. Therefore, the chapter expands upon the continued need for both water conservation and water justice in Kansas and elsewhere, highlighting the necessity of holding institutions accountable for supporting "equitable access to water" (62).

In order to unpack how Sam Brownback confronts this environmental crisis (Cox 2007), it is necessary to approach Brownback's rhetoric from the perspective of greenwashing. Greenwashing can be defined as "an attempt to influence the perceptions of consumers, media, politicians, and the public" to believe a group or individual is environmentally friendly or is engaging in environmentally friendly actions (Cox 2013, 286). Prior interdisciplinary greenwashing scholarship typically has focused on how corporations use greenwashing to deceive consumers into believing they have a green image and/or engage in environmentally friendly behaviors or products. Typically, greenwashing has been approached as ideological deception. This chapter will demonstrate how Brownback's greenwashing attempted to negotiate the tensions between water conservation and the viability of Kansas's agricultural economy. I posit what makes Brownback's greenwashing unique is that it appears to not rely solely on ideological deception. Rather, Brownback's greenwashing can be seen as *both* potential deception *and* possibly legitimate attempts toward water conservation. Therefore, I ask: How do Brownback's greenwashing strategies work to negotiate the double-bind surrounding the Ogallala Aquifer crisis and issues of water justice in Kansas while maintaining commitments to free market principles?

I argue that Brownback's justifications for conserving the Ogallala Aquifer utilized greenwashing strategies that attempted to maintain the conflicting practices of water conservation and upholding the viability of the Kansas economy. These strategies relied less on washing away the negative effects of current water usage, and instead, greenwashed Brownback and Kansas farmers as simultaneously environmentally responsible and economically viable. When faced with the environmental crisis of the Ogallala Aquifer's depletion, Brownback used appeals to economic growth and freedom to attempt to solve the crisis. Such strategic greenwashing allowed Brownback—and synecdochally the Kansas farmer—to rhetorically recognize that caring for the aquifer is caring for Kansas. In this way, such greenwashing sought not to deceive constituents, but rather, engage in rhetorical strategies that attempted to foster water conservation within contemporary neoliberal society and craft a green image. This green image put forward relied not on a rhetoric of *reducing* water consumption or economic prosperity, but rather, *maintaining* the future viability of both. Therefore, Brownback's greenwashing strategies can be read as both potential ideological deception and possible genuine attempts at water conservation.

This analysis will expand upon previous understandings of greenwashing, showing how in a contemporary environmental crisis, greenwashing strategies can possibly foster genuine attempts to conserve the environment in a world dominated by free market principles. In this way, this chapter will forward an understanding of greenwashing as multidimensional. Even though Brownback's greenwashing techniques may not have been intentionally deceptive or technically manipulative, because they were grounded in free market principles, they perpetuated structures of inequitable water rights in Kansas. In what follows, I provide further discussion regarding current theoretical understandings of greenwashing while advocating for approaching greenwashing beyond motive and deception. I then provide an analysis of Brownback's greenwashing strategies, demonstrating how they sought both water conservation and economic viability. Last, I conclude by arguing that even though Brownback's greenwashing techniques did work to address the crisis, because they were wedded to neoliberalism, they failed to resolve the lived tensions farmers face in balancing water conservation and economic viability.

GREENWASHING AND FREE MARKET PRINCIPLES

Greenwashing has its popular origins in the mid-1980s (Watson 2016) and, in the decades since then, has become a key concept within the discourses of environmental activism and environmental scholarship. Scheider et al. (2016) note that of all the strategies used by corporations to craft a green public image, "greenwashing has received the most attention" (15). Greenwashing is typically described as exemplified by people, corporations, or groups engaging in strategies that mislead others about the environmentally friendly or "green" qualities of their products or selves. Pezzullo (2003) defines greenwashing as not just "'greening' the appearances of products and commodity consumption, but also the deliberate disavowal of environmental effects" (346). Mason and Mason (2012) understand greenwashing as "tactics that mislead consumers regarding the environmental practices of a company" (481). Delmas and Burbano (2011) define greenwashing "as the intersection of two firm behaviors: poor environmental performance and positive communication about environmental performance" (65). What these definitions of greenwashing illustrate is how greenwashing "allows corporations to manipulate an image of environmental, social, and cultural responsiveness" (Munshi and Kurian 2005, 513). Overall, greenwashing has collectively come to be known as misleading claims about environmental performance to craft a pro-environmental public image.

There has been a dearth of interdisciplinary research exploring greenwashing. The research that does exist has focused mainly on greenwashing's functions, drivers, and effects in advertising and marketing campaigns (Miller 2016; Milstein and Dickinson 2012; Plec and Pettenger 2012; Budinksy and Bryant 2013; De Jong, Harkink, and Barth 2017; Cherry 2014; Lyon and Montgomery 2015; Bowen 2014). This scholarship has identified two trends of greenwashing: first, the examination of green products promoted as "ecologically friendly alternatives" when in reality they are not; and, second, analyzing corporate greenwashing strategies to construct an image of being environmentally responsible, when that may not be the case (Plec and Pettenger 2012, 460). Brett Bricker (2014) has highlighted how antienvironmentalist individuals and groups sometimes "deceptively name themselves as green sounding" (641). As such, when discussing greenwashing, "scholars often engage in ideological criticism focused on debunking and demystifying environmentally focused corporate communication as some form of rhetorical or ideological greenwashing" (Scheider et al. 2016, 15). This research has sought to explore the ideological dynamics of greenwashing; to expose corporations' intentions to deceive consumers that their companies or products are environmentally friendly.

The ideological dynamics explored by prior greenwashing scholarship have been closely tied with the belief and prioritization of free market principles, otherwise referred to as neoliberalism. Neoliberalism can be understood as "a discourse and set of practices that privilege market rationality, and individual freedom and responsibility above all else" (Scheider et al. 2016, 3). As Cox (2013) noted, behind much of the greenwashing rhetoric "is the belief in a free market" (285). Neoliberalism can be thought of in three ways: first, as a set of practices that alter relationships between markets and the state; second, as ideological beliefs for how to structure society; and third, as a set of discourses for organizing reality (Scheider et al. 2016, 10). Valuing economic success and market freedom above all else, neoliberalism is "the ideal central organizing principle of society" (10). Neoliberalism's prioritization of the free market and economic prosperity is positioned as being at odds with conserving the environment given that environmental conservation places restrictions on the free market to extract resources for profit. Therefore, greenwashing rhetoric is thoroughly intertwined with neoliberalism; and this is evident in Brownback's attempt to conserve the Ogallala Aquifer.

Clearly, prior greenwashing scholarship has worked to expose ideological manipulations by a speaker in order to put forth an image of being green. However, I propose that greenwashing is not always so one-dimensional. De Jong, Harkink, and Barth (2017) note, "In practice, an organization's environmental performance might not be clearly good or bad" (80). Lyon and Montgomery (2015) also argue, "Greenwash can range from slight exaggeration to full fabrication," and the research around greenwashing "needs to begin to think in terms of 'varieties of greenwash'" (225–26).

Thus, while greenwashing typically has been approached as ideological deception, for the critic, proving deception is difficult. Instead, what a critic can do is look for possible textual clues or strategies of greenwashing independent of motive. In this way, a critic can remain close to a text without having to prove deception. In approaching greenwashing this way, a critic still might find manipulation occurring, but, in the case of Brownback, one may find commitments to neoliberalism so strong that it is difficult to transition an economy toward water conservation.

In the Ogallala Aquifer crisis, wherein true replenishment of the aquifer to predevelopment levels is not a viable option (Evans 2017), Brownback used neoliberal appeals to attempt to solve the aquifer's depletion, making him seem both environmentally and economically friendly. As climate change continues to be a pressing concern in the public (Milman 2019), the likelihood for such use of greenwashing strategies to negotiate the relationship between climate crises and the free market could become more common. Kevin DeLuca (2001) once argued we must continue to question the dichotomy between "good nature-lovers and bad corporations" (645) in ways "that opens environmentalism to unexpected yet promising alliances with justice activists, unions, and corporations" (633). Answering this call, I forward our understandings of greenwashing by proposing to approach greenwashing as a collection of textual strategies independent of motive; in doing so, greenwashing may prove to be ideologically deceptive *or* simply an attempt to move toward greener environmental practices while negotiating the environment's relationship with the free market for, ultimately, the sustainability of both.

Understanding greenwashing as a spectrum of rhetorical strategies provides the critic with a few advantages. First, the critic is no longer required to prove that a politician or corporation is knowingly deceiving their audience. The burden of proof for the critic or environmental activist is lowered. Instead, this allows one to stay close to the text by looking for possible strategies of greenwashing independent of accusing deception. Second, in this sense, greenwashing could possibly be good or bad, or more likely, fall somewhere in between. In approaching greenwashing this way, a critic still might get to the conclusion that manipulation is occurring, but what Brownback's greenwashing demonstrates is strategies that position Brownback and Kansas as environmentally responsible without denying the negative effects of the current agricultural status quo. Therefore, greenwashing can be potentially understood as not always a "sellout but a tactical negotiation" between two seemingly disparate priorities, water conservation and agricultural viability. In this way, Brownback's greenwashing tries to walk a fine line to try and convince farmers that sustainable water use is attainable.

I will analyze the rhetorical strategies in Brownback's greenwashing, attempting to ascertain how Brownback worked toward water conservation while maintaining Kansas's agricultural economy. To do so, this case study examines Brownback's rhetoric made public through sources such as the *Hays Post*, as well as Brownback's official public statements to constituents during his 2017 Water Tour and to the U.S. Senate Committee of Agriculture, Nutrition and Forestry. The greenwashing strategies analyzed span the majority of Brownback's tenure as governor of Kansas, ranging between 2011 and 2017. From these instances, the analysis demonstrates that Brownback's greenwashing engaged in three strategic themes: (1) acknowledgment of the aquifer's depletion; (2) changing the present to maintain the future; and (3) Kansas is agriculture, and agriculture is Kansas.

Acknowledgment of the Ogallala's Depletion

Sam Brownback's first greenwashing strategy acknowledged the depletion of the Ogallala Aquifer as an environmental crisis. This greenwashing strategy allowed Brownback to seem concerned about both the environment and the economy. To do so, Brownback specifically identified the agricultural water practices that have contributed to the aquifer's depletion. For instance, early in his first term as governor, Brownback (2011) stated, "High-capacity well pumping, an increase in irrigated crop production and periodic droughts have all contributed to long-term declining water levels." One can see Brownback's recognition of the crisis when Brownback mentioned "long-term declining water levels," and how these declines are symptoms of specific farming practices such as "high-capacity well pumping." Operating as a sort of synecdochic structure, Brownback's rhetorical recognition of the crisis positioned his administration as one that "supported a proenvironmental agenda" (Bricker 2014, 638). This allowed Brownback to not only appear environmentally responsible, but also to demonstrate concern about the Kansas economy.

Another instance was when Brownback (2011) asserted, in the same official statement, "Without Ogallala resources, agriculture and all its related business could not be sustained, manufacturing could not continue, recreational opportunities would diminish and the communities in the area would cease to exist." Here, Brownback states a concern for maintaining both "Ogallala resources" and "all its related business." Specifically, this greenwashing provided a message that advocated for conserving the aquifer in order to support the needs of the agricultural economy, in order to avoid situations where "business could not be sustained" and "manufacturing could not continue."

In this first greenwashing strategy, one can see rhetorical attempts to maintain both economic prosperity and water conservation. Such rhetorical attempts acknowledge an environmental crisis already well-known by Kansas farmers. As one farmer put it, "We are trying to do something that conserves and extends the life of the aquifer for the land we own, and at the same time, conserve our profitability" (Dillon 2018, 11). Like the greenwashing from Brownback, this farmer is trying to balance conserving and extending "the life of the aquifer" in order to maintain economic "profitability," demonstrating a greenwashing strategy that attempts to balance the priorities of water conservation with the demands of the economy.

Overall, this first strategic greenwashing's acknowledgment of the aquifer crisis allowed Brownback to appear simultaneously environmentally responsible and economically viable. In stating, "Water is a finite resource" and without it "we will no longer be able to meet our state's current needs," Brownback recognized the environmental crisis surrounding the Ogallala, and such acknowledgment allowed Brownback to potentially generate discussion and action to attempt to conserve the aquifer and maintain the economic status quo. In this way, such greenwashing mirrored the material demands faced by Kansas farmers, trying to find a balance between water conservation and profit.

Change Present to Maintain Future Viability

Brownback's second greenwashing strategy centered around the issue of time. This strategy conveyed a message to conserve water today in order to preserve the future of the aquifer and Kansas. This greenwashing strategy argued that in order to maintain present water levels and economic viability, Kansas and its farmers must make changes in order to extend the economic status quo into the future. In other words, Kansans must reduce current water usage by voluntarily taking up solutions provided by the state. This allowed Brownback to appear as caring for both the aquifer and the economy. As Brownback mentioned in 2017, "Preserving water resources is vital to the success of our state. Without water, there is no future" (*Hays Post* 2017). To preserve for the future required a shift from conservation to sustainability. As Brownback (2017b) mentioned, "We are announcing a new and achievable vision for much of the Ogallala Aquifer region in Kansas. We are moving from a 'conserve and extend' vision to sustainability."

To initiate this shift, Sam Brownback's time-orientated greenwashing relied on messages to conserve the present in order to sustain the future. Brownback argued, "It is our legacy to future generations to take this information and put it to productive use in the conservation of this precious resource. We want to be remembered as the generation who took the proactive steps and not the generations who didn't" (*Hays Post* 2017). Brownback highlighted the need to conserve in the present in order to ensure prosperity for the future. Brownback does so by stating, "It is our legacy to future generations" to productively seek the "conservation of this precious resource" in order "to be remembered as the generation who took the proactive steps." Additionally, this greenwashed the responsibility to conserve the aquifer to those who rely on the aquifer the most: Kansas farmers. Such appeals to ensure the future viability of the aquifer strove to motivate Kansas farmers to leave behind a positive legacy for their families and for Kansas. Brownback stated, "In twenty, thirty, forty years or more, we will be judged by what we did, or did not do." He proposed a legacy for Kansas farmers based on what they did, responsibly using the water resource and maintaining the state's economic viability, in place of a legacy based on what they did not do, for leaving behind a drained aquifer and depleted economy.

In order to leave behind a positive legacy, this time-orientated greenwashing argued the solutions are presently out there. In other words, Brownback argued that farmers must voluntarily engage in water conservation by taking up the solutions currently provided by the state. This greenwashing positioned Brownback and the state as having addressed the crisis and provided solutions. It is simply up to the farmers to implement these solutions and use less water. This greenwashing allowed Brownback to convey an image of environmental responsibility and argued that sustainable use of the aquifer is attainable; the farmers simply had to do it themselves.

Brownback mentioned, "Kansans understand the severity of this issue and are working together to develop state solutions to address the challenges of the declining Ogallala." For instance, one solution is to switch to "new and modified programs . . . to achieve significant and permanent water conservation in areas of severe decline in the High Plains aquifer." During his tenure as governor, such "new and modified programs" were provided in the form of LEMAs. These programs relied heavily on the neoliberal ideal that locally driven actions have the largest potential for success, rather than state-enforced regulation. They also operated on the premise that market-driven water conservation would allow for the maintenance of the current economy into the future. In short, these solutions largely depended on water users to do the work themselves in imposing restrictions on water consumption. For instance, Brownback mentioned:

> With water management tools like Local Enhanced Management Areas, to Water Conservation Areas, to Water Banking, with the repeal of 'use it or lose it' water laws, coupled with new irrigation techniques like deep soil moisture probes, drop nozzle and mobile drip irrigation and sophisticated irrigation water management, it is possible to reduce our demands on the Aquifer to a level of sustainability and still grow our crops to feed the world.

Part of Brownback's Water Vision, these voluntary programs in combination with innovations in irrigation technology "have worked to create policies and programs which reward good stewardship of the state's water resources" (Brownback 2017a). As such, these programs demonstrated Brownback's greenwashing attempts to deal with the aquifer crisis while still emphasizing free market principles. These state programs also support the values of neoliberalism as they are completely voluntary, and, as such, greenwashed Brownback and Kansas as environmentally friendly through maintaining the present status quo of the free market into the future.

At first glance, Brownback's LEMAs appear to have the potential to extend the life of the aquifer. In 2012, a group of farmers located in Sheridan County agreed to collectively average an annual reduction of their water usage by 20 percent each year for five years. This agreement is the state's first and only LEMA. As of today, there are signs the reductions started by the LEMA are working to slowly help replenish the aquifer. As Brownback stated in 2017(a), "The data reveals that the voluntary efforts happening as a part of the fifty-year Water Vision are being rewarded. The Ogallala is replenishing itself faster than we previously know . . . sustainable use of the Ogallala aquifer is attainable." One Kansas farmer also noted how "extraordinary technological gains will make environmentally sustainable agriculture not just possible in Kansas, but the norm" (Brown 2018). What the small success of the Sheridan LEMA shows is that water conservation in Kansas is possible; however, there is also reason to be hesitant about the LEMA program's potential.

First, to date, there is only one LEMA in existence, and the Sheridan LEMA only covers sixty-three thousand of the millions of acres of farmland that drinks from the aquifer (Buchanan 2018; Marso 2015). More LEMAs need to be established in order to begin replenishing the aquifer at a large enough scale to realistically refill the tank. Second, such limited uptake is the result of the double-bind between water conservation and economic viability that Kansas farmers find themselves in. Many are unwilling to reduce the short-term economic profitability of their farmland when the long-term sustainability of the aquifer is so unclear. This is especially the case when farmers are competing against each other for access to the water. As one farmer mentioned, "It is very hard to look over the fence and see someone not complying when you are doing your best" (Dillon 2018, 19). When left with a choice between reducing water usage or maintaining economic viability, the majority of Kansas farmers choose the latter.

Overall, this time-orientated greenwashing highlighted how engaging in water conservation can have positive implications for the future of both Kansas and agriculture. Such greenwashing simultaneously argued that sustaining the aquifer now will allow the water to be maintained for the future of Kansas, and positioned Brownback as caring for both the aquifer and for Kansas. This greenwashing strategy presented messages that suggested the solutions for addressing the crisis are already "out there" and that, ultimately, it is up to the individual farmer to decide to reduce their water usage. These solutions allowed Brownback to present an image of being environmentally responsible. At the same time, solutions like the LEMAs, allow Brownback to support neoliberal values since they rely on voluntary action by Kansas farmers without mandating their implementation by the state. Consequently, such free-market conservation prioritizes neoliberal principles that perpetuate structures of inequitable water rights and continued water conflicts, creating a need for further water justice to create a truly sustainable water future.

Kansas Is Agriculture, Agriculture Is Kansas

A third greenwashing strategy from Brownback positioned Kansas and the agricultural economy as mutually constitutive. Agriculture and its economy are firmly grounded in the creation of Kansas and is a major segment of Kansan identity. As Brownback mentioned in 2011, "Kansas has a strong agricultural tradition that predates its statehood." Here such greenwashing tethered an "agricultural tradition" and story to the very origins of Kansas, even predating "its statehood." Tying the agricultural economy to Kansas, and its potential success and failures, allowed Brownback to seem concerned both about the environment and about Kansas. Without agriculture, Kansas does not exist, and there is no agriculture without the Ogallala Aquifer. As indicated earlier, the water from the Ogallala Aquifer supports Kansas's largest economic contributor. Brownback (2011) stressed, "Agriculture is our state's largest industry," and "The Ogallala aquifer is vital to the Kansas economy. It is the main source of water for all uses in the Western third of the state." However, as he himself also mentioned, the water from the aquifer is not unlimited, and, therefore, he implied Kansans need to maintain this resource to prevent catastrophic economic losses. He argued:

> Agriculture is key to the economic vitality of Kansas and water is key to agricultural production. In Western Kansas, we have a tremendous, but not unlimited resource in the Ogallala aquifer. The vitality of the western portion of the state depends heavily on agricultural production and business supported by this water resource. (2011)

Evident in this greenwashing is the desire to conserve the aquifer in order to maintain the current free market economy. What this greenwashing rhetoric indicates is Brownback attempting to conserve the environment while maintaining a commitment to the principles of free-market agriculture and neoliberalism.

By attempting to conserve the aquifer through appeals to the free market, Brownback attempted to foster environmental conservation; however, such conservation still places the needs of the economy over sustaining the aquifer. For example, Brownback stated, "Water and the Kansas economy are directly linked. Water is a finite resource and without further planning and action we will no longer be able to meet our state's current needs, let alone growth" (*Hays Post* 2013). In this strategy, Kansas and agriculture are one and the same. Yet, Kansas needs to conserve this "finite resource" in order to continue to meet the state's "current needs, let alone growth." Therefore, Brownback's efforts to conserve the aquifer are also efforts to ensure the future success of Kansas while maintaining a commitment to the economic status quo. For instance, Brownback (2011) argued, "It is possible to reduce our demands on the Aquifer to a level of sustainability, and still grow our crops to feed the world." In this greenwashing strategy, one sees a commitment to the water when Brownback mentions, "It is possible to reduce our demands on the Aquifer," but also a commitment to the free market. In this way, the greenwashing strategy was able to maintain Kansas's agricultural identity and economic viability while also advocating for reduction in the state's water usage rates.

Overall, by arguing that the aquifer supports the biggest contributor to Kansas's economy, and that in order to maintain the free market we need to conserve the Ogallala Aquifer, this greenwashing presented an environmentally friendly image in order to preserve neoliberal ideals.

CONCLUSION

In this chapter, I have argued that Sam Brownback's justifications for addressing the Ogallala Aquifer crisis utilized greenwashing strategies in order to maintain both current levels of water consumption and the current agricultural economy. Instead of exclusively engaging in ideological deception by hiding the negative effects of current water usage, Brownback argued the aquifer crisis is being adequately handled, and, in the process, positioned himself and the rest of Kansas as both environmentally responsible and economically prosperous. As such, Brownback's green image relied on a rhetoric of maintenance for keeping the status quo, not a rhetoric for reducing water consumption or the free market. This rhetoric of maintenance also argued for maintaining current agricultural practices and legal structures that perpetuate inequitable water rights in Kansas. Such an insight allows one to understand Brownback's greenwashing not only as rhetorical deception but, also, as commitments to neoliberalism so strong that Brownback's pro-environmental solutions fall short in adequately addressing the aquifer crisis and the material tensions faced daily by Kansas farmers. In the end, Brownback's greenwashing upholds the values of neoliberal society, placing short-term economic prosperity over long-term water conservation.

This conclusion illustrates how future investigations into greenwashing can approach such strategies as multidimensional, advancing that greenwashing exists on a spectrum. Brownback's greenwashing attempted to craft a green image by acknowledging the water crisis, attempting to solve the crisis through market-driven conservation in order to maintain the present status quo—showing that greenwashing can be manipulative environmental communication *and* the manifestation of strategies designed to address an environmental crisis while still prioritizing economic freedom. Such greenwashing shows critics the need "to begin to think in terms of 'varieties of greenwash,'" and potentially begin to find ways to foster productive environmental conservation within a world dominated by free market principles (Lyon and Montgomery 2015, 225–26).

It is possible to see Brownback's greenwashing positively, initiating policies and fostering dialogue about water conservation during a contemporary environmental crisis. Indeed, in multiple ways, Brownback was able to initiate processes and conversations toward conserving the Ogallala Aquifer. On the other hand, though, Brownback's greenwashing operated through free market principles, wherein such solutions remained voluntary and limited in their application. Ultimately, his solutions placed the needs of the free market economy *over* the needs of attainable water conservation. In this way, Brownback's greenwashing failed not because it was deceptive but, rather, because the solutions were too wedded to neoliberalism.

Lastly, in terms of water justice, such rhetorical maintenance of the neoliberal status quo resulted in the perpetuation of larger patterns of water overconsumption and structures of inequitable water rights. This is because Brownback's greenwashing strategies failed to overcome the double-bind faced daily by farmers. While many farmers see the Sheridan LEMA making progress toward sustainability, many are unwilling to take up conservation solutions. As one Kansas farmer stated, "It's a step in the right direction; however it is a matter of who can get along with whom. . . . It is very hard to look over the fence and see someone not complying when you are doing your best. There is a fairness issue" (Dillon 2018, 19). The "first in time, first in line" policies left behind by Brownback perpetuate inequitable distribution of water rights in Kansas, where senior water-rights holders have more access to water than those who come later. Thus, Brownback's greenwashing failed to negotiate the daily tensions between water conservation and the economic viability faced by farmers, falling short in resolving inequitable water rights in Kansas.

Showing how as people attempt to move toward creating and living in a more environmentally sustainable and equitable world, environmental protection and water justice will have to try and negotiate themselves in conjunction with the environmental and economic needs of farming and feeding humanity—because, simply, food for life cannot be grown without water.

REFERENCES

Ashworth, William. 2006. *Ogallala Blue: Water and Life on the High Plains*. New York: W. W. Norton & Company.

Bowen, Frances. 2014. *After Greenwashing: Symbolic Corporate Environmentalism and Society*. Cambridge: Cambridge University Press.

Bricker, Brett Jacob. 2014. "Feigning Environmentalism: Antienvironmental Organizations, Strategic Naming, and Definitional Argument." *Western Journal of Communication* 78, no. 5: 636–52.

Brown, Corie. 2018. "Kansas Farmers Are Running Out of Water. Can They Pull Off an Environmental Miracle in Charles Koch's Home State?" *The New Food Economy*, December 20, 2018. https://newfoodeconomy.org/kansas-ogallala-aquifer-high-plains-decline-charles-koch/.

Brownback, Sam. 2011. "Statement for the Record." Senate Committee on Agriculture, Nutrition and Forestry, United States Senate, August 25, 2011. https://www.agriculture.senate.gov/imo/media/doc/82511Testimony_Brownback.pdf.

Brownback, Sam. 2017a. "Governor Sam Brownback Holds Water Tour: Sustainable Yield of the Ogallala Aquifer Is Attainable." Posted July 18, 2017. https://governor.kansas.gov/governor-sam-brownback-holds-water-tour-sustainable-yield-of-the-ogallala-aquifer-is-attainable/.

Brownback, Sam. 2017b. "Governor Sam Brownback Issues Statement on Attainability of Sustainable Yield from the Ogallala Aquifer." Posted on July 19, 2017. https://governor.kansas.gov/governor-sam-brownback-issues-statement-on-attainability-of-sustainable-yield-from-the-ogallala-aquifer/.

Buchanan, Rex. 2018. "Can the Ogallala Aquifer Be Saved? LEMAs May Help." *Kansas Public Radio,* April 12, 2018. http://kansaspublicradio.org/kpr-news/can-ogallala-aquifer-be-saved-lemas-may-help.

Budinsky, Jennifer, and Susan Bryant. 2013. "'It's Not Easy Being Green': The Greenwashing of Environmental Discourses in Advertising." *Canadian Journal of Communication* 38, no. 2: 207–26.

Butler, Jr., J. J., R. L. Stotler, D. O. Whittemore, and E. C. Reboulet. 2013. "Interpretation of Water-Level Changes in the High Plains Aquifer in Western Kansas." *Groundwater* 51, no. 2: 180–90.

Cherry, Miriam A. 2014. "The Law and Economics of Corporate Social Responsibility and Greenwashing." *UC Davis Business Law Journal* 14, no. 2: 281–303.

Connor, Kaley. 2017. "Hays Highlights Water Conservation Efforts." *The Hays Daily News*, October 15, 2017. https://www.hdnews.net/news/20171015/hays-highlights-water-conservation-efforts.

Cox, Robert. 2007. "Nature's 'Crisis Disciplines': Does Environmental Communication Have an Ethical Duty?" *Environmental Communication: A Journal of Nature and Culture* 1, no. 1: 5–20.

Cox, Robert. 2013. *Environmental Communication and the Public Sphere*, 3rd ed. Thousand Oaks, CA: Sage Publications.

De Jong, Menno D. T., Karen M. Harkink, and Susanne Barth. 2017. "Making Green Stuff? Effects of Corporate Greenwashing on Consumers." *Journal of Business and Technical Communication* 32, no. 1: 77–112.

Delmas, Magali A., and Vanessa Cuerel Burbano. 2011. "The Drivers of Greenwashing." *California Management Review* 54, no. 1: 64–87.

DeLuca, Kevin Michael. 2001. "Trains in the Wilderness: The Corporate Roots of Environmentalism." *Rhetoric & Public Affairs* 4, no. 4: 633–52.

Dillon, Karen. 2018. "In Dealing with the Ogallala Aquifer, Western Kansas Is Running Out of Water and Time." *Kansas Leadership Center*, August 13, 2018. https://klcjournal.com/in-dealing-with-the-ogallala-aquifer-western-kansas-is-running-out-of-water-and-time/.

Evans, Cathy. 2017. "Kansas Geological Survey Shows Stabilized Water Levels in Aquifer Possible." *Kansas Geological Survey*, July 31, 2017. http://www.kgs.ku.edu/General/News/2017/stabilize.htm.

Fenichel, Eli P., Joshua K. Abbott, Jude Bayham, Whitney Boone, Erin M. K. Haacker, and Lisa Pfeiffer. 2016. "Measuring the Value of Groundwater and Other Forms of Natural Capital." *PNAS* 113, no. 9: 2382–87.

Hanks, Kathy. 2018. "Heading Out to Measure Aquifer Levels, Crew Expects to See Some Decline from Previous Year." *Lawrence Journal-World*, December 28, 2018. https://www2.ljworld.com/news/ku/2018/dec/28/heading-out-to-measure-aquifer-levels-crew-expects-to-see-some-decline-from-previous-year/.

Hays Post. 2013. "Brownback's Calls for Action on Water Issues." *Hays Post*, October 24, 2013. https://www.hayspost.com/2013/10/24/brownbacks-calls-for-action-on-water-in-kansas/.

Hays Post. 2017. "Governor Issues Statement on Attainability of Sustainable Yield from Ogallala Aquifer." *Hays Post*, July 20, 2017. https://www.hayspost.com/2017/07/20/governor-issues-statement-on-attainability-of-sustainable-yield-from-ogallala-aquifer/.

Kansas Water Authority. 2015. *Kansas Water Authority 2015 Annual Report to the Governor and Legislature*. http://www.kwo.org/Kansas-Water-Authority.html.

Kansas Water Office. 2014. *Kansas Water Plan*. http://www.kwo.org/Water-Plan.html#2014%20Kansas%20Water%20Plan.

Lyon, Thomas P., and A. Wren Montgomery. 2015. "The Means and End of Greenwash," *Organization & Environment* 28, no. 2: 223–49.

Marso, Andy. 2015. "Water Plan Tackles Critical Issue of Supply and Demand: State Aims to Balance Economic Needs with Sustainability Goal." *Kansas Health Institute*, January 12, 2015. https://www.khi.org/news/article/water-plan-tackles-critical-issue-of-supply-and-demand.

Mason, Marianne, and Robert D. Mason. 2012. "Communicating a Green Corporate Perspective: Ideological Persuasion in the Corporate Environmental Report." *Journal of Business and Technical Communication* 26, no. 4: 479–506.

Melgares, Pat. 2018. "More Than 200 Come Together to Discuss Ogallala Aquifer." *K-State Research and Extension*, April 11, 2018. https://www.ksre.k-state.edu/news/stories/2018/04/ogallala-aquifer-summit.html.

Miller, Toby. 2016. "Greenwashed Sports and Environmental Activism: Formula 1 and FIFA." *Environmental Communication* 10, no. 6: 719–33.

Milman, Oliver. 2019. "Americans' Climate Change Concerns Surge to Record Levels, Poll Shows." *The Guardian*, January 22, 2019. https://www.theguardian.com/environment/2019/jan/22/climate-change-concern-americans-poll/.

Milstein, Tema, and Elizabeth Dickinson. 2012. "Gynocentric Greenwashing: The Discursive Gendering of Nature." *Communication, Culture & Critique* 5, no. 4: 510–32.

Munshi, Debashish, and Priya Kurian. 2005. "Imperializing Spin Cycles: A Postcolonial Look at Public Relations, Greenwashing, and the Separation of Publics." *Public Relations Review* 31, no. 4: 513–20.

Opie, John. 2000. *Ogallala, Water for a Dry Land*. Lincoln: University of Nebraska Press.

Parker, Laura. 2016. "What Happens to the U.S. Midwest When the Water's Gone?" *National Geographic*, August 2016. http://www.nationalgeographic.com/magazine/2016/08/vanishing-midwest-ogallala-aquifer-drought/.

Pezzullo, Phaedra C. 2003. "Resisting 'National Breast Cancer Awareness Month': The Rhetoric of Counterpublics and Their Cultural Performances." *Quarterly Journal of Speech*, 89, no. 4: 345–65.

Plec, Emily, and Mary Pettenger. 2012. "Greenwashing Consumption: The Didactic Framing of ExxonMobil's Energy Solutions." *Environmental Communication* 6, no. 4: 459–76.

Schneider, Jen, Steve Schwarze, Peter K. Bsumek, and Jennifer Peeples. 2016. *Under Pressure: Coal Industry Rhetoric and Neoliberalism*. London: Palgrave Macmillan.

Tammen, Greg. 2013. "Study Forecasts Future Water Levels of Crucial Agricultural Aquifer." *K-State News*, August 16, 2013. https://www.k-state.edu/media/newsreleases/aug13/groundwater82613.html.

Torell, Allen, James D. Libbin, and Michael D. Miller. 1990. "The Market Value of Water in the Ogallala Aquifer." *Land Economics* 66, no. 2: 163–75.

Watson, Bruce. 2016. "The Troubling Evolution of Corporate Greenwashing. *The Guardian*, August 20, 2016. https://www.theguardian.com/sustainable-business/2016/aug/20/greenwashing-environmentalism-lies-companies.

Wilson, Brownie. 2019. "Groundwater Levels Slightly Up or Near Steady in Western, Central Kansas." *KU Today*, April 15, 2019. https://news.ku.edu/2019/04/15/groundwater-levels-slightly-or-near-steady-western-central-kansas.

Chapter Seven

Naturalizing Environmental Injustice

How Privileged Residents Make Sense of Detroit's Water Shutoffs

Kelsey Mesmer, Mostafa Aniss, and Rahul Mitra

Problems of water security abound all over the world, proliferating in such forms as drinking water contamination, national security disputes, and depleting reserves—often exacerbated by rapid climate change and lack of adequate investment in water infrastructure (or their responsible management) (Dinar and Dinar 2017). Such instances almost always result in environmental injustice, affecting low-income people of color and other marginalized groups disproportionately. Existing social tensions often prevent long-term sustainable solutions to water security, such as upgrading infrastructure or access systems. In the United States, urban-suburban conflicts particularly perpetuate vicious cycles of environmental injustice, as in Detroit, Michigan, which continues to reel under massive infrastructure costs for a common water system that also services its far wealthier suburbs (Recchie et al. 2019).

Using Metro Detroit as a case study, our chapter addresses communicative meaning-making of water access and scarcity among both urban and suburban residents to explore the constitution of privilege and difference that works to "other" low-income, often black, people in the city. Although Detroit has evidenced signs of economic revitalization after years of decline and population flight to the suburbs, this renewal has largely been concentrated in the greater downtown area, prompting charges of negligence to the outer neighborhoods peopled by communities of color (Howell and Feldman 2017). While Detroit was in bankruptcy and under the administration of a state-appointed emergency manager, its newly restructured water authority started shutting off water connections to thousands of residential customers (Howell et al. 2019). In 2014 alone, the first year of mass shutoffs, more than 33,600 accounts were disconnected (Thibodeau 2015), and shutoffs have continued every year since, amounting to more than 142,000 disconnected households, according to grassroots activists (personal correspondence, We The People of Detroit). This resulted in turmoil, including a spike in water-borne illnesses (e.g., Hepatitis A) in the Metro area, an increase in tax foreclosures as delinquent water bills were rolled over onto already high property taxes, and a declaration by the United Nations that the water shutoffs were an unprecedented human rights travesty (Wylie-Kellermann 2017).

Despite this media and grassroots furor, suburban residents and leaders—who share the common water infrastructure built by the Detroit Water and Sewerage Department (DWSD) in the 1950s—remained ambivalent or even outright hostile toward the plight of fellow citizens who were denied access to clean, affordable water barely a few miles away (Sabourin 2016). And despite 80 percent of Detroiters being African-American, the city's first white mayor in decades, Mike Duggan, coasted toward reelection on the votes of middle-class residents who did not perceive themselves to be affected by the continuing water shutoffs. At stake here are thus underexplored tensions of privilege related to race, class, and geographical location that shape community members' sense-making of access to drinking water. Particularly relevant is the ongoing "naturalization" of culture (even as the natural environment influences culture) theorized by Donal Carbaugh (1996), along with the "othering" of both nature and residents deemed less worthy of access to a vital natural resource like water (Carbaugh and Boromisza-Habashi 2011; Carbaugh and Cerulli 2013; Milstein 2011). Anand (2012) calls this erasure of citizenship rights to a marginalized urban community "abjection," whereas Kornberg (2016) cites the "territorial stigma" that Detroit residents have historically shouldered as contributing to this ongoing othering. Thus, this chapter puts into conversation theoretical frameworks from environmental communication, social justice law, and anthropology to unpack how relatively privileged residents of Metro Detroit constitute their "relations-in-place" (Milstein et al. 2011) with both urban dwellers denied water and the access of water itself. The interview data we examine stem from a larger oral history project, which gathers narratives not only from residents, but also community and institutional stakeholders on water insecurity in Detroit.

In the following sections, we first outline the theoretical frameworks that guide us, pertaining to the naturalization of culture, othering of nature, and water-based abjection of marginalized urban residents. Next, we describe the body of data collected and analytical method used for this chapter. Then, we present our key findings on how water access in general, and Detroit's shutoffs in particular, were perceived by urban and suburban residents who were largely unaffected by the disconnections. We close with some takeaways for environmental justice scholars, practitioners, and policy makers.

COMMUNICATION, NATURE, AND OTHERING

Even as communication scholarship has long foregrounded the human actors designing and exchanging messages for accomplishment of mutual goals, several theorists of environmental communication have pushed for a more complex view that attends to the ongoing intersections among human and ecological elements (e.g., Cantrill and Oravec 1996). Broadly speaking, such research highlights the dialectical, even tensional, relationships between social culture and its natural environment. Antecedents of this line of thinking are varied, ranging from Donna Haraway's (1989) interdisciplinary work on "natureculture" as an amalgamation that disrupts overly simplistic debates of "nature versus culture," to eco-materialist strains in literary studies that emphasize the natural properties and agency of environmental contexts and elements (such as water) (e.g., Strang 2005), to even more recent adaptations of actor network theory that seek to understand climate change as a holistic action encompassing multiple environmental agents in addition to human beings (Latour 2018).

Our chapter is particularly guided by the work of Donal Carbaugh and his colleagues on how culture is inherently "naturalized" by environmental contexts and elements, and by critical perspectives on the "othering" of these elements theorized by Tema Milstein and others. By emphasizing the naturalization of culture, Carbaugh (1996; Carbaugh and Boromisza-Habashi 2011) makes the point that the way we do things, or the values, norms, standards, and discourses that emerge and hold sway in different cultures, is inevitably shaped by the environmental conditions at stake. At the same time, social culture—and the communicative practices it engenders—is formative, in that it produces particular ways of representing and talking about these environmental forces, so that communication re-presents them, as it were, in an ongoing loop of nature and culture. Carbaugh and Cerulli (2013) focus on the role of cultural discourses that shape social understandings of nature, and highlight five discursive hubs to explore them—identity, action, feeling, relations, and dwelling. They urge special attention to how local communities use discursive practices in mundane interactions to help make sense of who they are, what they are engaging in, the emotions these practices generate, and how these themes are connected to broader connections between social groups and the larger ecology.

Milstein's work builds on this framework in two important ways. First, attuned to the fragmented and tensional ways sense-making occurs, she unpacks the dialectics that commonly occur in Western public discourse when it comes to the environment (Milstein 2009, 2011). Three foundational tensions are highlighted: mastery over nature vis-à-vis harmony with nature, othering nature and environmental agents vis-à-vis connecting with them through diverse relationships, and exploitation of the environment vis-à-vis idealizing its scope and power. In Western public discourse, Milstein asserts primacy is accorded to mastery, othering and exploitation, marginalizing discursive strands of harmony, connecting and idealism—thereby "othering" nature as passive, lifeless, nonsentient, or merely instrumental. For instance, entire disciplinary fields such as natural resource management reduce environmental entities to instrumental "resources" (like oil or water) that can be extracted and exploited for human use. Required to disrupt this ongoing process of othering, then, is a shift in cultural discourse primacy from separation and exploitation toward greater ecological interdependence and collective (rather than personal) responsibility, geared toward identifying *with* nature (Milstein 2011).

Second, crucial to our study is the notion of "relations-in-place," which emphasizes how social relationships are always formed *in situ*, within and because of particular sites, locations, and ecologies, and are thus meaningless if we do not consider their environmental contexts. Milstein et al. (2011) used community-based participatory research to understand local meaning-making of self-identity, childhood memories, and current social problems in New Mexico, finding that place figured enormously in these cultural discourses. Place held both historical and material significance, forming participants' memories and experiences through the availability of food, for instance, or their ancestral connection to farming land.

Despite helping rethink the communicative intersections of nature and culture, this body of work largely ignores how privilege, place, and social histories intersect to create volatile power inequities through everyday discourses that often elide the gravest of environmental injustices. Water, in particular, is both material and political, notes Soyini Madison (2007) in her performative unpacking of the political economies and hierarchies that constitute both "consumers" and "users" of water, while reducing it to a passive commodity constituted entirely by political treaties and the free market. Thus, our project extends this line of thinking to considering how race, class, and geographical location intersect with the material facets of water access to shape power relations and privileges. Moreover, by considering the issue of water scarcity in urban contexts, we highlight the institutional structures—at state, local, and utility management levels—that underlie these cultural discourses and perpetuate power inequities. Specifically, we emphasize how water rights are used to marginalize low-income groups and people of color, amid long-running urban-versus-suburban conflict.

URBAN ABJECTION AND WATER ACCESS IN DETROIT

Anthropologists of water have long noted how it constitutes and shapes social worlds, influencing broader human values, norms and processes of equity, creation and maintenance of organizational structures, political discourses and conflicts, and knowledge systems that confer legitimacy (or not) on social groups (Lauer and Aswani 2010; Long et al. 2013; Orlove and Caton 2010). Across both rural and urban contexts, the conferral of water rights helps constitute the legitimacy and agency of social groups, determining which groups are powerful and which occupy subaltern status. Critical scholars are especially interested in the contestation and challenge of status quo structures and devices that confer such subalternity, through local community organizing, alternative interpretations of policy texts, and coalition-building with unlikely partners, for instance.

The interconnections among institutional structures, water access and social privilege are aptly demonstrated in Paerregaard et al.'s (2016) examination of Peru's new water law based on "integrated water resources management" that sought to include all citizens in managing the country's water—but also inadvertently created hierarchies of privilege and power depending on location, class, and community history. Moreover, they note that casting water as a resource to be managed foregrounds the privatization and commercialization of the environment, and altogether erases alternative ways of understanding water and social relationships vis-à-vis water, such as dwelling-with water, indigenous worship, or even political action. In addition to rules and laws, scholars have pointed out the disciplining nature of water infrastructure, so that mundane pipes, plumbing, water lines, and other forms of infrastructure shape broader social consciousness about water access and relations among communities that might have differential access to such infrastructure (Radonic and Kelly-Richards 2015). Deitz and Meehan (2019) connect what they term "plumbing poverty"—or the systematic erosion of water infrastructure, specifically among low-income communities, leading to a lack of safe, clean, and adequate access to water for residents—specifically with environmental injustice, when it comes to household water insecurity in the United States. Inadequate state and federal investment in water infrastructure mean that low-income urban areas like Detroit end up bearing the brunt of heavy sewage overflows and recurrent flooding, leading to residents internalizing a sense of precariousness because of the lack of local resources to address the issues (Sampson et al. 2019).

In line with previous work on water rights and water citizenship, we argue that this systemic othering of both nature and marginalized urban communities might be termed "abjection," aligned with dominant neoliberal institutions, laws, and market structures. For Anand (2012), abjection is "a social and political process through which particular populations are pushed beyond the biopolitical care of the state or other institutions, even as they remain central to the constitution of such social (or political) collectives" (488). While his original work traced the abjection of minority Muslim and low-income communities in the suburbs of Indian metropolis Mumbai, the United States's history of suburban sprawl and white flight reverses the direction of abjection, so that it is inner-city urban groups that are "pushed down, or cast out of social and political systems they could once access and claim" (489). The notion of abjection, however, ought also to be extended to include the ongoing naturalization of culture and othering of nature, discussed earlier. Indeed, the original term remains fairly focused on social privilege and relegates both infrastructure and water to instruments wielded by this privilege, rather than unpacking the complex cultural discourses and ecological sense-making at stake. One important contribution might be in demonstrating how abjecting discourses depend on privileging mastery, othering, and exploitation of nature; another might well be showing how they deny privileged social groups' relations-in-place, both with marginalized communities and with nature.

For their part, community organizers and activists mobilizing around water access in Detroit are keenly aware of this intersection of social privilege, place, and water rights. Activist pastor Bill Wylie-Kellermann's (2017) book is proudly titled *Where the Water Goes Around*, the English translation for the Anishinaabe name for the Detroit ("de troit," or "the strait" in French) region, in a clear nod to the region's colonial history which has been indelibly shaped by its water bodies. In contrast to scholars far removed from the scene (and community) who documented the community organizing against Detroit's mass water shutoffs, Howell et al. (2019) highlight the need to challenge conventional meanings of water access, environment, and community for long-term, meaningful change (also see Howell and Feldman 2017). Recognizing that water rights have been used as a weapon against the poor and marginalized, Howell et al. (2019) call for myriad coalitions, multimedia products, and educational protests to shift public consciousness around the human right to water to something more akin to Milstein's ecological interdependence and collective responsibility. Kornberg's (2016) historical genealogy of "territorial stigma" demonstrates that Detroit's household water insecurity is due to, at least in part, decades-long institutional racism that targeted majority-black neighborhoods with high bills and unreliable supply, along with targeted propaganda by white suburban leaders who claimed Detroit's leadership was corrupt and incompetent while coveting the city's sophisticated water distribution system worth billions of dollars. The formation of the Great Lakes Water Authority (GLWA), in the aftermath of the city's bankruptcy—due in no small part because Flint went off the Detroit water system, which led to its own crisis with lead poisoning—achieved this control of suburban interests over the city's system, and resulted in an operational culture that privileged privatization and further commodification of water (Sabourin 2016). While this institutional history has been documented by policy scholars, such as the Haas Center at the University of California Berkeley (Recchie et al. 2019), local groups like We The People of Detroit (2016) have also engaged in citizen research to map out geographically how water shutoffs and unduly high monthly bills disproportionately affect low-income communities of color (even in affluent neighborhoods) who pay retail rates for household water, while suburban customers pay wholesale to GLWA.

Drawing on the literature reviewed thus far, our guiding research question may thus be posed: How is water privilege discursively reified (in metro Detroit), leading to the abjection of residents struggling with household water insecurity?

METHOD

The data examined in this chapter is part of a larger publicly available oral history project, titled "Detroit Water Stories" (see www.detroitwaterstories.wordpress.com), for which the team has gathered narratives of water scarcity, shutoffs, flooding and other infrastructure failures, and contamination fears from residents, activists, and institutional stakeholders.

Participants

Data for this chapter was collected over two academic semesters, fall 2017 and fall 2018, with student teams. Four of the student researchers were undergraduates, whereas the others were MA and PhD students enrolled in qualitative research methodology seminars. Both undergraduate and graduate students were trained in research design, interviewing, and qualitative data analysis by the third author. The student researchers were instructed to reach out through their personal and professional networks to interview metro Detroit residents—people living in the three counties of Wayne, Oakland, and Macomb counties specifically—who had either experienced problems related to water access or knew of people who had. Several students also visited public spaces and events, such as the Detroit riverfront and Belle Isle (an island park in the Detroit River, near downtown), to approach random passersby and talk to them about their experiences with water access and security.

In total, 46 Metro Detroit residents were interviewed for this chapter—of whom 25 were Detroit residents and 21 were suburban residents who lived in the tri-county area. (Detroit is located in Wayne county, and shares borders with Oakland and Macomb). Twenty-four were women, 18 were men, and 3 did not disclose their gender identity. Participants' age ranged from 18 to 77, with the average being 39 years. Seventeen participants were black, 20 were white, 1 Asian, 2 mixed (white and black), 1 saw themselves as "Other," and 3 participants declined to state their ethnicity—the majority of suburban residents were white, while the majority of Detroit residents were black, reflecting the demographic makeup of the Metro Detroit area and history. Our participants were also well-educated and earned a household income well over the median Detroit household income of $26,249. Seventeen participants had a bachelor's degree, 13 had a postgraduate degree, 6 had an associate's degree, while 2 had completed "some college," 5 had a high school diploma or GED, and 3 did not answer the education question. Only 8 participants noted their salary to be at or below the $25,000–$50,000 range, with 11 participants in the next range of $50,000–$75,0000, 4 in the $75,000–$100,000 range, and 5 enjoying household income in excess of $100,000; 13 participants declined to share their household income.

Procedures

A semistructured interview protocol was developed, with the help of the student researchers, to ensure that participants would be able to explore their thoughts, ideas, emotions, and experiences at length during the interviews. The protocol began with questions about participants' communities and neighborhoods, to trace their connections to their lived sites, after which they were asked to describe a memorable event where water played an important role and to reflect on how they thought access to water affected their community in different ways. Participants were also asked if they or anyone they knew had experienced problems accessing clean and safe water, and to describe both what happened and how it impacted their lives. They were asked to reflect on where they thought their community's water supply originated, how it was managed and distributed, and the role of both institutional stakeholders and community members in protecting this water supply. Finally, participants were asked if they were familiar with the ongoing water shutoffs to thousands of Detroit residents, and asked to explain how they knew of them (if they did), what they thought of the situation, and what might be done to address the problems at hand. Interviews concluded with a general question that allowed participants to leave with parting comments or closing thoughts.

Throughout the interview process, the tone was deliberately kept informal and conversational (Brinkmann and Kvale 2015). Knowing fully well that class, race, and suburban privilege might mean that several participants would be unaware of or not very sympathetic to the plight of poor Detroiters suffering from water shutoffs but still wanting to engage these perspectives, the interviewers were trained to ask questions in calm even tones, and even provide explanations or missing gaps in knowledge if asked by participants. In fact, explaining what the water shutoffs were and how long they had been going on became a routine part of our role as interviewers. The majority of our suburban interview subjects were not aware that the shutoffs had been going on at all, and thus, interviewers developed short explanatory scripts so that participants could at least speculate on the causes for, effects of, and possible solutions to the shutoffs. Furthermore, interviewers frequently had to using leading questions to get participants to answer questions in a relevant manner, as even some participants' understanding of "water access" was marred by their privilege. For example, water access was frequently thought of by suburban participants as one's proximity to lakes and rivers, with water use and access being synonymous with recreation. This is juxtaposed to the focal point of our study—thousands of Detroit residents who at one point or another did not have access to running water in their own homes. The interviews ranged from 20 to 60 minutes, depending on the convenience of the participants. All interviews were audio-recorded and transcribed professionally. Since participants were assured of confidentiality and anonymity, we used pseudonyms to report the data.

Data Analysis

We approached our data from a critical-interpretive perspective, in which our role in the analysis was not only "'making sense' of scene from the participants' point of view—examining not only behaviors but intentions and emotions" (Tracy 2013, 48)—but also pointing out power imbalances rooted in "historical considerations of power and class" (48) in and around Detroit. In this way, we looked not only at our situated data, but drew on our understandings of the multiyear water crisis in Detroit, race relations in metro Detroit, and urban abjection to provide context to the data, illuminating participants' lack-of-knowledge—or sometimes their belief in misinformation—and calling out racism and classism rooted in privilege.

Data was analyzed using Charmaz's (2006) constant comparison method. Using ATLAS.ti software, the first and second author went line-by-line through the transcripts engaging in open coding. This happened separately at first, then the two authors met to compare their codebooks and discuss their coding procedures, and then separated again and came back together again to compare until they were satisfied that their codebooks matched. During this process, special attention was paid to relationships and paradoxes of power and privilege based in race, location, and income. We also engaged in memoing, both in the form of notated asides during the coding process, and in the form of long-form blog posts, shared on our project website. We engaged in dialogue among ourselves and with community leaders (residents, activists, and experts of water law and environmental justice) to reflect on and make sense of our data. These processes allowed us to "figure out the fundamental stories in the data" (Tracy 2013, 196) and see how codes related to one another. Sample open codes included *ignorance of water crisis*, *don't work hard to pay for others*, and *water is a human right*. In our second round of focused coding, these open codes were grouped into six axial codes: *nature of water*, *personal responsibility*, *role of government*, *civic solutions*, *knowledge of water issues*, and *community bubbles*. After distinguishing between the macro- and microlevel discourses represented by these axial codes, these became the main themes in our findings, discussed next.

FINDINGS

Our findings demonstrated four key discourse frames used by participants in both urban and suburban contexts to make sense of their access to water. These ranged from emphasizing residents' personal responsibility to manage water supplies and pay their utility bills on time, to shifting across paradoxical roles of the government as both problematic actor and solution-provider, to underlining the need for civic action and solutions, to relying on relatively narrow community boundaries for denying the real scope of the water crisis or mischaracterizing that crisis in a way that can lead to abjection.

While we describe each of these frames below, we also trace how they demonstrate competing tensions between macrolevel cultural discourses on the meaning of water as a natural entity or element. Three such discourses become apparent in the metro Detroit context—*water as life-giving* (and thus, potentially dangerous if denied), *water as key right* (and thus, leading to social entitlement to adequate supplies of water), and *water as commodity* (and thus, an object that can be both bought and owned by human actors). Each of these cultural discourses is tensional. Consider, for instance, the discourse of water as life-giving: while it foregrounds the benevolent properties and powers of water, participants frequently highlighted the flip side of not having adequate access to water, which resulted in severe malnutrition, disease, and even death. Similarly, if water was considered an inalienable right for all beings—an understanding that stemmed from the recognition that it was crucial for life, but was then resituated in terms of social institutions of law and democracy—some participants also noted that there were several residents who tended to treat it as an absolute entitlement and thus abused this right. This line of reasoning then often resulted in marginalizing or "abjectifyng" people whose water was shut off as being unworthy of this right in the first place. Finally, there is the third cultural discourse of water as a commodity or tangible resource, which lends itself to the very human institution of the marketplace, making water subject to the laws, norms, rules, and other powers of this institution. Yet, value in the marketplace is always relative, so that successive instances of buying and selling inevitably lower and raise the value of a thing. Inevitably, then, as a commodity, there is inherent tension between the understanding of water as a private or public good or service to be bought or sold, or distributed among broader groups for general welfare, and between water and other commodities for sale in the marketplace.

As we show below, these cultural discourses are not discrete, but overlap in both reifying and resistive ways across each of the sense-making frames used. At stake, thus, is tracing how social privilege operates in complex and layered ways, shaped by (and shaping in turn) the biopolitical power of natural environments and elements.

Personal Responsibility

Arguably the most popular frame among both urban and suburban resident participants was personal responsibility, echoing neoliberal story lines of individuals creating their own financial situation, for better or worse. Participants adopting this frame placed the blame for Detroit's water shutoffs on individuals who should have made better choices and paid their bills on time. For example, Amelia, a suburban resident, explained that the water bill is a necessity and is the first thing that should be paid before going on to pay for other luxuries:

> If you don't pay your water bill, of course it's going to get shut off. . . . You know if you use the resource and they told you, you have to pay for it, pay for it. I never had a shutoff with my water. Not saying that I got so much wealth, I don't, but I know I would have to pay my bill. I can't go buy a Maserati when I know my water bill is due, okay?

This excerpt demonstrates how personal responsibility is emphasized both by abjectifying residents unable to pay their bills as feckless and spendthrift (and thus unworthy of empathy), and by othering water itself as an instrumental "resource" over which one has mastery. Not only is water foregrounded as a *commodity*, but Amelia also perceived those unable to pay as feeling entitled to their water access (which is the flip side of the *water as key right* cultural discourse). Similarly, William, a white suburban resident, believed paying (or not paying) the water bill was a choice vis-à-vis other competing but nonessential commodities such as cell phones.

Others adamantly believed that DWSD was rightfully disconnecting the water of irresponsible citizens; Hunter, a white suburban resident, said:

> To me, it sounds like the city is finally doing something right. If you don't pay your bills, you get what you get. If I don't pay my bills, if I don't pay the electricity bill, well, they come shut my lights off. If I don't pay my car

> payments, they take my car away. If I don't pay my mortgage, they'll come take me out of my home. That's just part of life. I don't know, don't have a family, don't buy a house if you can't afford to live there. If you can't afford to pay your water bills, which aren't very expensive, then quite frankly, you need to look at yourself and think about how did I get myself in this situation because we all have our own decisions to make. We all have our own choices.

The misinformation evident here is immense. Although he assumed Detroiters' water bills are similar to his own, the truth is that suburban customers pay wholesale rates for their water to GLWA compared to Detroiters who pay retail rates to DWSD (as a result of the restructuring agreement between DWSD and GLWA). Moreover, suburban customers are billed every three months, compared to Detroiters' monthly payments. Hunter reifies this misinformation by framing the problem as a matter of Detroiters not making the right choices (which abjectifies them, as with Amelia), and as "just part of life," which negates empathy for those less privileged than himself. The territorial stigma theorized by Kornberg (2016) is also evident here, when he notes that by aggressively shutting off water for nonpaying customers, "the city is finally doing something right" because "if you don't pay your bills, you get what you get." The personal responsibility frame evinced here also draws on the broader pull-yourself-up-by-the-bootstraps neoliberal ideology, averring that "we all have our own choices," so that *if only* affected Detroiters made the right choices, they would be fine. Such an understanding ignores any structural factors associated with the water shutoffs and reinforces the dominant media narrative of Detroit's poor, inner-city residents trying to milk the system. In another nod to the bootstraps ideology, Hunter continued, "Maybe they'll learn to kick it into gear and figure out how to pay their bills. Maybe it will teach them a little personal responsibility."

Reiterating that these false narratives were disseminated in the local media coverage, Bernie, a black Detroiter, noted, "[It's] not even seen on the news in-depth but just more or less in passing, 'Oh, people didn't pay their water bills, they're getting shutoff notices.' And then, 'Hey, we'll have a day you can come down and pay your bills, the lines were out of the door.'" In other words, the mainstream media coverage was casual and offhanded, without much consideration of the institutional causes or the human costs at stake, which influenced broader public understanding of the issue. In contrast, Bernie saw the effects of the water shutoffs on school-aged children firsthand, when she worked as a nurse for Detroit's public school system. Broadly speaking, then, the personal responsibility frame draws on cultural discourses of both *entitlement* (water as right) and *commodity*, while favoring the poles of mastery over water and othering of both water and low-income residents.

Governance as Problematic Solution

Participants also adopted a frame that saw the government as responsible for the water shutoffs, paradoxically both blaming the government and holding out hope for better management of its affairs—which served to absolve participants themselves of needing to care or contribute toward a specific solution. Some participants drew on decades-long rhetoric blaming city leadership as corrupt and incompetent (Kornberg 2016; Sabourin 2016), despite their underlying empathy for those affected by the shutoffs. For instance, Bernie said,

> I think that we obviously have a system that isn't working. It's one of those things where if you haven't experienced it, then you may not see how flawed the system is. I can't even tell you who put the system in place. But let's say that whoever those people are, they've never experienced water shutoffs, and they've never been in the situation to feel as though the water rates are expensive. This wouldn't be an issue for them, and then they may have a difficult time understanding why some people wouldn't be able to afford it.

Bernie zeroes in on systemic problems facing water access for some Detroit residents here, and recognizes the "flawed system" at stake, along with the lack of exposure to such problems that leads to the lack of empathy on the part of most people. In particular, she focuses on how most people in city leadership are unaffected themselves by the high costs of water access, and so are themselves out of touch with the problems facing their low-income constituents. Citing the wholesale-retail ratio of suburban-urban payment of water bills, Amelia (who used to live in Detroit before moving to the suburbs) noted ruefully of DWSD leadership, "If I am in control of something and my people as citizens that is within that county, their water bill should be lower than the people you have to run pipe fifty or five hundred miles away from your source." She was decidedly in the minority, however, when it came to participants recognizing the disparities between suburban and city residents' water bills.

Participants who saw flaws in the system believed water shutoffs should be determined only on a case-by-case basis if they were going to happen at all. Even many of the participants who adhered to the personal responsibility frame discussed earlier averred that exceptions should be made for households with children and senior citizens. Brock, a black suburban resident, said

> I do feel if you have fully functioning adults in the house, [it's] situational but if there's children in the house, absolutely not. There's no excuse, there's no way around it because the children don't get a say and they go without and that's just—I don't know, the kids, that's different. Limitations are always good.

Thus, despite his belief that "fully functioning adults" should be able to pay their water bills on time and avoid shutoffs, despite the structural odds, whatever they may be, he felt that households with children needed to get a pass because kids lacked the evidently all-powerful agency he imagined adults to possess. Others thought that water shutoffs should be held off when a senior citizen must choose between paying for their medication or their water bill, or when there was a person with a disability or illness in the home.

Many participants questioned the morality of the DWSD and GLWA for shutting off anyone's water, realizing, as Brock put it, "business and ethics are two different things." Crystal, a black Detroit resident who lived without water in her home for almost three months at one point, also wrestled with the complexity of the situation:

> The simple answer would be that the people who don't pay their bills on time is to blame, but you also got to take into consideration that financially, a lot of people are having trouble keeping up with payments because a lot of these water shut offs are happening in really low-income neighborhoods, neighborhoods filled with residents who don't have maybe a steady income or a livable wage to pay for everything. . . . The water company is just—I don't know, it seems like there should be a little bit more leeway for this sort of stuff or a little bit more compassion towards people who can't make up payments or anything. You would think somebody's sense of morality will kick in, but . . . it's just a corporation. You can't really expect them to show some form of compassion towards a bunch of people who don't give them the money that they want.

Excerpts such as this situate blame on the shoulders of the water utility and Detroit's city leaders who lack compassion for residents trapped in systemic poverty and struggling to make ends meet. Although both DWSD and GLWA are technically public entities, and not "corporations," they are squarely identified with profit-centered logics, so that "You can't really expect them to show some form of compassion towards a bunch of people who don't give them the money that they want." These corporatist logics also manifest in the way that shutoffs have been rolled out with little to no warning, and for seemingly minor offenses. For example, Tracy, a black Detroiter, recalls neighbors having their water shut off for being only a day late in paying their bill "without even so much as a reminder call or some type of grace period or anything. Knowing that water is something that's a necessity for human life, it seemed like it was a bit unethical and inhumane." Government leaders are thus framed as harsh and ruthless, in addition to being out of touch with their constituents.

Paradoxically, most participants acknowledged that the city's government was responsible for and should manage and regulate water access for residents because the water needs to be treated and water infrastructure needs to be maintained. They also averred that residents should pay something for their water to cover the costs of treatment, although this should be priced at a rate everyone can afford. Evelyn, a black suburban resident, for instance, said that water should belong to everyone, with civic leaders taking on the custodial role of "managing it and safeguarding it" for Metro Detroit residents; these leaders should essentially just be covering overhead costs, not making a profit off the water. This showcases a tension between all three cultural discourses of water as *life-giving*, as a *key right*, and as *commodity*. Despite recognizing water as necessary for life, our participants situated it squarely within social politico-legal-economic systems, which reduced it to a tangible resource for government distribution and as a market commodity. In electing both government and market as the twin social institutions in charge of "managing" water access, participants emphasized mastery, othering, and exploitation poles of the fundamental social-environmental tensions theorized by Milstein (2009). Moreover, in seeing government as both problem and solution, participants' sense-making frames allowed them to place faith in an external third party—local government, DWSD, other voters, and so forth—rather than see themselves as having any agency even if they viewed the water shutoffs as immoral, preserving their own social privilege.

General Civic Solutions

A small group of participants did articulate goals for civic action, which every citizen could take on to address the unfolding water crisis, but these steps were often so broad and general that they betrayed a lack of urgency and understanding of the severity of the situation. For instance, several participants believed it was one's civic duty to stay informed on such issues and to hold the government accountable when they see injustice. Mia, a white suburban resident, said it was important to ask questions of civic leaders at local board and city council meetings, and to be "more aware of water sources and who's managing it, and what people are paying." A few others suggested volunteering with local community groups and nonprofit organizations working on water insecurity issues in and around the city.

In some cases, civic action was reduced simplistically to having empathy for people affected by water shutoffs. Jane, a suburban resident, said, "There ain't no training in the world to teach you how to treat a human being, so at the end of the day that's just the main thing. Could you do this to your sister or your grandma?" In other cases, a very broad philosophy of "don't do bad stuff to the water" (Hunter) was espoused, so that residents should be mindful about what trash and toxins end up in the waterways. Some participants advocated general education for area residents, but did not specify how this might address water affordability or even be dispersed across to multiple townships and cities. Jane, for instance, urged "teaching education of that to not only the young ones but also our older folks . . . because we only have this one planet to live on." For the most part, then, although these suggestions seemed well-intentioned, participants' use of the civic solutions frame ended up reifying their own privilege, by focusing largely on personal behaviors and attitudes, while discounting both deeper systemic problems and the recognition of waterways as complex environmental agents. As with the preceding sense-making frame, all three cultural discourses of water as *life-giving*, inalienable *right*, and *commodity* are evident in this frame, yet human mastery and exploitation dominate overall.

Perhaps even more problematic were instances where participants endorsed a view entirely antithetical to adopting civic solutions, by averring that the underlying systemic problems were simply much too complex for reasonable solutions—so that there was no point in trying! Carol, a white Detroit resident, recognized that "a lot of factors" were to blame for the shutoffs, including "city problems . . . homeownership . . . and the city not having the resources to do good management," and ended up noting there was no upshot to her taking on any action at all to fix any of these problems. Echoing the decades-long "territorial stigma" (Kornberg 2016) against Detroiters, Oliver, a suburban resident, said, "I think Detroit in and of itself is a rundown community and people simply don't have jobs to support themselves." This anti-engagement frame essentially reinforced participants' sense of themselves as existing in their own community bubbles, safe and privileged enough not to be affected by the ongoing water crisis, and able to absolve themselves of any action to play whatsoever.

Safe Community Bubbles

Finally, participants employed sense-making frames that emphasized their geographical and environmental apartness from less privileged Detroiters who faced water scarcity and unaffordability—their relations-in-place, to borrow Milstein et al.'s (2011) terminology, but also these relations-*out-of*-place. Three community bubbles were evident, which utilized the cultural discourses of water in different ways. The first of these bubbles we termed NWIMBY, or "Nothing Wrong in My Backyard," because participants fell back on their financial or environmental privilege to see themselves safe and thus downplay the broader water crisis. Some participants who had, in fact, experienced life without access to household water, downplayed this incident because they were able to easily purchase water from the grocery or depend on friends and family for help—resources that are not easily available to low-income Detroiters struggling with water shutoffs and high water bills. Suburban Madison, for instance, recalled,

> I know *one time* our pipes froze. That was not fun for a day or two. . . . You had to be like really strategic about what you needed to do. I think we luckily just had cases of water in our garage. . . . So, we used that. . . . I was really lucky that it was only like a day, day and a half. We could just be out of the house and like we have bottles of water if we want to drink it. We could go out to eat and stuff, and didn't have to be too concerned with it. . . . I know we asked one of our neighbors to take a look. He was able to help us turn off all the pipes, so that they didn't cause more damage. That was one of the moments that we were like, oh yeah, we can go to our neighbors to ask for help with things.

Madison's memory of life without water was much more mundane than having your water disconnected because you were unable to pay your bill, and she fell back on the social and financial privilege at her disposal to use stocked-up water bottles and ask her neighbor for help. Even as she talked about the need to be "really strategic" in her use of limited water supplies, inherent in her story is the understanding this this was going to be a temporary problem, which would soon be resolved—which is not an expectation that Detroiters with shutoffs have.

Several participants also cited their experiences of recreational activities on the Great Lakes, and other small bodies of water, which underlined their geographical and social privilege. This was not wholly unexpected, since most Michigan residents, including many Detroiters, have long had an intimate relationship with the Great Lakes. For instance, Detroit resident Phil said, "I mean, I was growing up and into the outdoors and stuff. . . . So you know, we would just go up and we were on a lake. . . . [Those were] always fond memories." Similarly, Hazel who also lived in Detroit, noted how she often took water access as granted, recalling a big pool party for her daughter's birthday, when asked to recount an eventful memory related to water: "We filled the pool with water and all the little kids changed, splashed, jumped, rolled, and had a good time at her first birthday party." In sum, the NWIMBY frame allowed participants to see themselves as untouched by surrounding water crises, and hinged on cultural discourses that saw water access as both something they were *entitled to* (as a key right) and *commodity* they could easily afford.

Another community bubble frame that emerged was focused on differentiating participants' contexts from the Flint lead-poisoning tragedy, so that we labeled it, "Don't want to turn into Flint!" Several participants (especially suburban) noted the governmental apathy that characterized Flint, but in highlighting how their communities were different from—apart from—Flint, they also emphasized their territorial privilege that set themselves as better than residents of Flint. For example, Caleb mentioned that no cities' water issues were "as drastic as Flint," and Mia perceived Flint's water crises as particularly awful because there are "many deaths involved." Isaac said, "I'm perfectly happy with my water and the way it's been handled by my local government and by the regional government that runs the water department. . . . They haven't done anything stupid like the city council in Flint." Highlighting the stigma that accompanies such geographical othering, Jacob noted that he fielded queries from out-of-state friends whether he felt uncomfortable being on the same GLWA water system as Flint was: "It's funny, my out of state friend always asks me why I would drink tap water if the water is like Flint. I have to explain to them that it's not at all like Flint." This particular community bubble frame thus tapped into both the water as *life-giving* (and thus water contamination as dangerous!) and *key right* discourses, but twisted them such that participants' social privilege was preserved because they were entitled to better protection from water contamination, compared to less fortunate others (notably: Flint residents). Importantly, few Detroit participants used this frame; several like Connor actually noted how both Flint and Detroit residents were facing problems related to water infrastructure and institutional failure, so that they had much in common.

Nevertheless, a community bubble frame that was apparent among our Detroit participants emphasized "Detroit has the best water!" based on past federal and statewide data on DWSD's treatment practices. This frame worked somewhat like the previously noted NWIMBY, in that participants who averred this most strongly were generally the most likely to downplay concerns and fears from fellow residents about the quality of the system's water. An added possibility is that this frame allows more privileged Detroit residents to reduce concerns of household water security to only contamination, rather than consider the various other layers and problems at stake, such as aging infrastructure, affordability, and recurrent flooding (Deitz and Meehan 2019). For instance, Bernie said, "It is just always kind of been something ingrained in me, like, 'Oh, well, we live, you know, here. We have some of the best water in the world. We have the best.'" Nevertheless, she also continued later to offhandedly remark that "We drink a lot of bottled water now." Similarly, Hazel said, "In Detroit, we've always had clean drinking water because for years we had the best water in the country. People took it for granted." Nevertheless, some older suburban residents, like Amelia who had lived in the city for decades, evinced slight concern that "somehow" problems had arisen with the water supply—although they remained quite misinformed about the water shutoffs or its underlying systemic problems. She evinced, for instance, "When I was there [Detroit], it was the cleanest. We had the best water in Michigan. But I don't know what went wrong."

CONCLUSION

Our chapter unpacks how social privilege works in insidious ways to naturalize and reify environmental injustice. Specifically, using the context of Detroit's historic mass water shutoffs to low-income communities (primarily of color), we draw on interviews with both urban and suburban residents who were relatively unaffected by the shutoffs, to trace how they make sense of these events without feeling the need to be involved or intervene in any way. Our inquiry is not esoteric; rather, understanding the cultural discourses and sense-making frames used by privileged residents will enable grassroots activists and progressive policymakers to address constituents often opposed to the affirmative actions that environmental justice involves. Unearthing the discursive devices that allow such residents to see environmental injustice as "not my problem" or even as "inevitable" or "just desserts" will enable activists to design communication campaigns to better speak to residents' sense of naturalized privilege.

Our findings demonstrated the interplay of three underlying cultural discourses related to water—as life-giving, as key right, and as commodity—that circulate among participants' sense-making frames. Each of these cultural discourses demonstrate the biopolitical power inherent in natural environments, such that water becomes far more than a lifeless resource or element, but one capable of exerting power and influence over broader spaces, places, and human actors, even as it is also used, contained, restricted, and/or distributed by such actors. The sense-making frames applied by participants ranged from personal responsibility, to government as problematic solution, to overly general civic solutions, to emphasizing local community bubbles. While these frames overlapped and saw multiple iterations of the cultural discourses noted earlier, they broadly absolved key human actors of meaningful action—social or political—and uniformly sought to position water as an instrumental resource, valued almost entirely by the marketplace, and available only to people who could afford to pay for it.

To say that our participants were underinformed or misinformed would be accurate, but nevertheless miss the point. Although both urban and suburban participants lacked key details of the ongoing water affordability crisis in Detroit, underlying this misinformation was social privilege, shaped by political and economic ideologies as much as by ecological boundaries and naming conventions (Milstein 2011; Milstein et al. 2011). Naming water in terms of "resource" or "payed commodity" enabled participants to elide the life-giving necessities of water for less privileged residents; even those forced to come face-to-face with the environmental injustices at stake found it relatively easy to shrug off responsibility by blaming third parties such as government, out-of-touch city leaders, or even the residents denied water for failing to make the "right choices" in their lives. Our chapter thus shows how abjection of low-income residents, who were denied due process and status of citizenship akin to their own, occurs not merely through the rhetorical naming of "us" versus "them," but through the complex interplay of naming water, infrastructure, geographies, and agency (Anand 2012; Paerregaard et al. 2016). It is this confluence of agencies—both social and ecological—that partakes in the othering of some residents—low-income, inner-city black and brown residents mostly—as somehow less worthy or deserving of water, even as water itself is reduced to a private good for purchase.

Milstein (2009, 2011) suggests that one crucial way of restoring the balance from mastery-othering-exploitation of nature toward harmony-connecting-idealism of the environment is adopting the practice of "witnessing," rather than simply spectating. When one bears witness, as compared to merely observing or looking, there is deep resonance with one's ecology—an understanding, both visceral and intimate, that one is situated *within and by* this broader ecology, which prevents the callous categorization of "things" as apart from oneself. Bearing witness involves listening, identifying oneself with our larger environment(s), and embracing silence of one's voice to privilege the perspective of others—and especially *nonhuman* others in the ecology at large. These lessons could be cultivated in environmental justice programs that focus on ecological interdependence and collective responsibility, rather than the community bubbles and personal responsibility frames that our participants evinced. Such programs might push privileged residents to recognize how their lives are interwoven with those of lesser privileged citizens, and with the interconnected water supply and distribution system that sustains all life in the region. This training opportunity might also involve deeper reeducation about the power, agency, value, and vigor of water bodies and watersheds, far beyond justified by the buying and selling of the market, or even by politically expedient laws. Bearing witness to water access means recognizing that "one water" is not just a marketing tagline for GLWA, but a universal truth of social and ecological existence in SE Michigan.

Our findings also bring attention to how social privilege operates vis-à-vis place, especially when we note how participants fell back on narrowly defined community bubbles as their safe space, to avoid facing the trauma of the ongoing water crisis in Detroit. Just as communities can productively break through restraints by situating their relationships and discourses in-place (Carbaugh and Cerulli 2013; Milstein et al. 2011), our findings show how place can also eject or exile "other" beings from consideration of the values and norms they would otherwise hold dear. When privileged residents look within their small bubbles and see nothing wrong, or they see the resources at hand to stave off disaster for themselves, they tend to bury their heads deep in the sand and ignore the trauma of environmental injustice for large swathes of people who reside only a few miles away. Or, because they would rather not see themselves in the same predicament as others not too far away (Flint), they exert privilege to willfully blind themselves to the injustice occurring just beneath their nose (Detroit). Place, distance, and identification are thus crucial tropes ripe for further examination in future research on environmental privilege and power relations.

Finally, our research urges environmental communication scholars to recognize that while natural elements and agents are important to take note of in the constitution of everyday life, they are often mediated through the built environment, social institutions, and legal mandates that make the environment "real" for human beings. For collective action to be enabled, environmental justice programs must unpack how institutions and infrastructures shape social privilege vis-à-vis natural elements—even when elements are absolutely crucial to life, such as water. Institutions like the market and regional politics thus frame not merely who is the "true" citizen backed by state authority, but what these unequal powers and privileges mean for policies regarding, behaviors toward, and outcomes for environmental entities.

REFERENCES

Anand, Nikhil. 2012. "Municipal Disconnect: On Abject Water and Its Urban Infrastructures." *Ethnography* 13, no. 4: 487–509.

Brinkmann, Svend, and Steinar Kvale. 2015. *InterViews: Learning the Craft of Qualitative Research Interviewing*, 3rd ed. Thousand Oaks, CA: Sage.

Cantrill, James G, and Christine L. Oravec. 1996. *The Symbolic Earth: Discourse and Our Creation of the Environment.* Lexington: University Press of Kentucky.

Carbaugh, Donal. 1996. "Naturalizing Communication and Culture." In *The Symbolic Earth: Discourse and Our Creation of the Environment,* edited by James G. Cantrill and Christine L. Oravec, 38–57. Lexington: University Press of Kentucky.

Carbaugh, Donal, and David Boromisza-Habashi. 2011. "Discourse beyond Language: Cultural Rhetoric, Revelatory Insight, and Nature." In *The Rhetorical Emergence of Culture,* edited by Christian Meyer and Felix Girke, 101–18. New York: Berghahn Books.

Carbaugh, Donal, and Tovar Cerulli. 2013. "Cultural Discourses of Dwelling: Investigating Environmental Communication as a Place-Based Practice." *Environmental Communication* 7, no. 1: 4–23.

Charmaz, Kathy. 2006. *Constructing Grounded Theory: A Practical Guide through Qualitative Analysis*. Thousand Oaks, CA: Sage.

Deitz, Shiloh, and Katie Meehan. 2019. "Plumbing Poverty: Mapping Hot Spots of Racial and Geographic Inequality in U.S. Household Water Insecurity." *Annals of the American Association of Geographers* 109, no. 4: 1092–109.

Dinar, Shlomi, and Ariel Dinar. 2017. *International Water Scarcity and Variability: Managing Resource Use across Political Boundaries*. Berkeley: University of California Press.

Haraway, Donna J. 1989. *Primate Visions: Gender, Race, and Nature in the World of Modern Science*. New York: Routledge.

Howell, Sharon, and Richard Feldman. 2017. "Visions in Conflict: A City of Possibilities." In *Why Detroit Matters: Decline, Renewal, and Hope in a Divided City*, edited by Brian Doucet, 209–20. Chicago: Policy Press.

Howell, Sharon, Michael D. Doan, and Ali Harbin. 2019. "Detroit to Flint and Back Again: Solidarity Forever." *Critical Sociology* 45, no. 1: 63–83.

Kornberg, Dana. 2016. "The Structural Origins of Territorial Stigma: Water and Racial Politics in Metropolitan Detroit, 1950s–2010s." *International Journal of Urban and Regional Research* 40, no. 2: 263–83.

Latour, Bruno. 2018. *Down to Earth: Politics in the New Climate Regime*. Medford, MA: Polity.

Lauer, Matthew, and Shankar Aswani. 2010. "Indigenous Knowledge and Long-Term Ecological Change: Detection, Interpretation, and Responses to Changing Ecological Conditions in Pacific Island Communities." *Environmental Management* 45, no. 5: 985–97.

Long, Rachel, Elisha Renne, Thomas Robins, Mark Wilson, Kenneth Pelig-Ba, Mozhgon Rajaee, Allison Yee, Elizabeth Koomson, Codi Sharp, Jing Lu, and Niladri Basu. 2013. "Water Values in a Ghanaian Small-Scale Gold Mining Community." *Human Organization* 72, no. 3: 199–210.

Madison, D. Soyini. 2007. "Performing Ethnography: The Political Economy of Water." *Performance Research* 12, no. 3: 16–27.

Milstein, Tema. 2009. "'Somethin' Tells Me That It's All Happening at the Zoo': Discourse, Power, and Conservationism." *Environmental Communication* 3, no. 1: 25–48.

Milstein, Tema. 2011. "Nature Identification: The Power of Pointing and Naming." *Environmental Communication* 5, no. 1: 3–24.

Milstein, Tema, Claudia Anguiano, Jennifer Sandoval, Yea-Wen Chen, and Elizabeth Dickinson. 2011. "Communicating a 'New' Environmental Vernacular: A Sense of Relations-in-Place." *Communication Monographs* 78, no. 4: 486–510.

Orlove, Ben, and Steven C. Caton. 2010. "Water Sustainability: Anthropological Approaches and Prospects." *Annual Review of Anthropology* 39: 401–15.

Paerregaard, Karsten, Astrid Bredholt Stensrud, and Astrid Oberborbeck Andersen. 2016. "Water Citizenship: Negotiating Water Rights and Contesting Water Culture in the Peruvian Andes." *Latin American Research Review* 51, no. 1: 198–217.

Radonic, Lucero, and Sarah Kelly-Richards. 2015. "Pipes and Praxis: A Methodological Contribution to the Urban Political Ecology of Water." *Journal of Political Ecology* 22, no. 1: 357–465.

Recchie, Anna, Joseph J. Recchie, john a. powell, Lauren Lyons, Ponsella Hardaway, and Wendy Ake. 2019. *Water Equity and Security in Detroit's Water and Sewer District*. Berkeley: Haas Institute for a Fair and Inclusive Society at University of California Berkeley. www.haasinstitute.berkeley.edu/Detroitwaterequity.

Sabourin, Claire. 2016. "Responding to the Detroit Water Crisis: The Great Lakes Water Authority and the City of Detroit." *Journal of Law and Policy* 51: 305–27.

Sampson, Natalie R., Camel E. Price, Julia Kaseem, Jessica Doan, and Janine Hussein. 2019. "'We're Just Sitting Ducks': Recurrent Household Flooding as an Underreported Environmental Health Threat in Detroit's Changing Climate." *International Journal of Environmental Research and Public Health* 16, no. 1: 6.

Strang, Veronica. 2005. "Common Senses: Water, Sensory Experience and the Generation of Meaning." *Journal of Material Culture* 10, no. 1: 92–120.

Thibodeau, Ian. 2015. "After 33,607 Water Shut-Offs in 2014, Detroit to Focus on Commercial Accounts." *Mlive*, March 17, 2015. https://www.mlive.com/news/detroit/2015/03/businesses_next_up_for_detroit.html.

Tracy, Sarah J. 2013. *Qualitative Research Methods: Collecting Evidence, Crafting Analysis, Communicating Impact*. Hoboken, NJ: Wiley-Blackwell.

We The People of Detroit Community Research Collective. 2018. *Mapping the Water Crisis: The Dismantling of African-American Neighborhoods in Detroit*. Detroit: We The People.

Wylie-Kellermann, Bill. 2017. *Where the Water Goes Around: Beloved Detroit*. Eugene, OR: Cascade Books.

Chapter Eight

Reviving Sister Water

Hydro-Anthropomorphism, Catholic Social Justice, and Pope Francis's Eco-Rhetoric for the Care of Creation

Christopher J. Oldenburg

Since his election in 2013, Pope Francis has made environmental justice a dominant and permanent concern of his pontificate. In 2015, Francis became the first pope to devote an entire encyclical to ecology, thereby addressing a largely neglected theme of Catholic social teaching (Massaro 2015, 179; O'Brien and Shannon 2016, 589). *Laudato Si'* (*Praise Be to You*), with the subtitle *On Care of Our Common Home*, unequivocally repudiates the deleterious effects of anthropogenic climate change, the technocratic paradigm, and ecocide. In short, *Laudato Si'* is a call for an "ecological conversion" exhorting humanity to rethink, renew, and reconcile our relationship with the environment.

More recently, on September 1, 2018, Pope Francis delivered a papal message on the "World Day of Prayer for the Care of Creation." He focused his remarks specifically on the *topos* of water. According to a 2017 WHO/UNICEF report, 2.1 billion people are denied access to clean drinkable water. Moreover, half of the world's hospital beds are filled with people afflicted by waterborne illness, and 340,000 children under the age of five die each year from these diseases (WHO/UNICEF 2017). In light of these facts, it is clear that water injustice is an indissoluble rhetorical, moral, and humanitarian exigence. Addressing a global audience, Francis resoundingly reaffirmed that water access is a social justice issue. He averred, "Access to safe drinkable water is a basic and universal human right, since it is essential to human survival and, as such, is a condition for the exercise of other human rights." Water justice for Francis is quite literally a pro-life issue. It is *sine qua non* for ensuring the inalienable dignity of every human being. Francis (2016) explained, "Both everyday experience and scientific research show that the gravest effect of all attacks on the environment are suffered by the poorest" (48). Thus, there exists a social justice connection between the ecological concern for water and the poor.

The purpose of this chapter is to examine Pope Francis's eco-rhetorical and religious message on the environmental issue of water. Francis invites his global audience to ecological conversion by forging a confluence between social justice and water access. I argue that Pope Francis's rhetoric calls for the renewal of humanity's relationship with water and the poor. The evaluation of Pope Francis's speech employs a decidedly classical rhetorical approach. There exists a long and notable history of Catholic theologians drawing from the deep reservoir of the rhetorical tradition to advance religion and promote justice. In fact, many of the early Church fathers were proponents, practitioners, critics, and teachers of rhetoric (of course, St. Augustine is the most prominent, but there were several other; see d'Ambroisio 2014). The selection of a classical rhetorical armature for the analysis of Pope Francis's eco-rhetoric is both a genuflection to that venerable confluence as well as an edifying contemporary paradigm of it. Water is a classical rhetorical *topos* (see Rosenfeld 1989, 250; Rosenfeld 2006, 198–200; for water as the archetypal journey motif, see Osborn 2018, 95–134) and anthropomorphism a classical figure. Both concepts also possess dense theological signification. Francis's synthesis of both concepts reveals the inventive rhetorical strategy of what I call "hydro-anthropomorphism" employed to further "flesh" and "flush" out Francis's Catholic social justice model of "integral ecology."

Quintilian defined anthropomorphism or personification as "when we give voice to things which nature has not given voice" (Quintilian 1856, IX.ii.32). Therefore, applying Quintilian's definition in the context of Pope Francis's view of water justice, I define hydro-anthropomorphism as an eco-trope that functions to embody the presence, rights, and excluded expressions of both the biophysical Other (water) and the poor. It serves as a counter-corrective "voice" to the epistemological constitution of water as an abstraction, objectified by pure science, polluted, privatized, and commodified by economics and neoliberalism, geopoliticized by world leaders, and, as a basic human condition, denied to the poor. Thus, Francis views water as having an integral, interdependent, and divinely ordered relationship with humans. The logic of my analysis suggests that to humanize water allows us to see water as a *human right*.

The order of this chapter proceeds as follows: First, I briefly situate Francis's eco-rhetoric within the scholarship of rhetoric and environment. Second, Francis's view of integral ecology and its relation to Catholic social justice is articulated. Third, I provide a brief review of anthropomorphism and offer a rationale for Pope Francis's use. Finally, before concluding, I conduct a close rhetorical analysis of Francis's papal message, "The World Day of Prayer for the Care of Creation."

RHETORIC AND *ENVIRONMENTALITY*: POPE FRANCIS AS AN ECO-RHETORICIAN

Andrew McMurry (2017), quite rightly, concludes in the penultimate sentence of his opening paragraph in his chapter on "Rhetoric and Environment," that environment "is now rhetorically charged root to tip: its very nebulousness is proof that the biophysical Other will not be extricated, materially, or symbolically, from humankind itself" (733). It appears rhetoric's hegemonic anthropocentricism is clearly ensconced, and there is no denying that discourse about the environment profoundly influences human interaction with nature.

Reviewing the predominant precincts of scholarship in rhetoric and environment, McMurry describes two salient perspectives that address environmental exigencies: "the practico-political and the ethico-epistemic" (735–41). McMurry acknowledges that these perspectives are not mutually exclusive but do stake autonomous claims; study select discursive *topoi*, genres, and issues; and employ their own narratives and tropes. McMurry, who reaffirms both perspectives, characterizes the *practico-political* as the more realistic, "anchored in the world as we find it" (735), while the *ethico-epistemic* is more idealistic and involves the vital and "steers for the world that might be" (735). By McMurry's evaluation, the challenges and limitations of environmental rhetoric are endemic to both approaches and further exacerbate an incommensurate dichotomy between anthropocentric and ecological perspectives. Why?

According to McMurry, "While nothing human is alien to rhetoric, environment is not human" (740). And "the natural environment itself does not communicate at all; it can only stimulate the social systems to produce communication about it" (741). The fact remains that Eden cannot literally speak to Athens. This does not mean, either by indifference or insidious force, we should "silence springs," but the emergence of ecocriticism, deep ecology, political ecology, ecofeminism, and other more radically eco-centered approaches may overcorrect and thwart efforts at prudential balance (for more on these approaches see Garrard 2012; Clowney and Mosto 2009; Gaard and Murphy 1998). As Pope Francis (2016) cautioned, "A misguided anthropocentrism need not yield to 'biocentrism'[,] for that would entail adding yet another imbalance, failing to solve present problems and adding new ones" (118). As we will see, Francis's hydro-anthropomorphic rhetoric, casted in the form of a prayer, allies more with the "ethico-epistemic" than with the "practico-political" approach.

Pope Francis's view attempts to bridge the conflictual currents in the impasse between humanism and ecology that seem to have left us up a creek. As Paul Lynch (2017) has argued, Francis's more holistic approach endeavors to reconcile the first part of what Bruno Latour calls the "Modern Constitution," that fallacious and intransigent decree that human culture (society and religion) and science and nature must "remain absolutely distinct" (464–467). In terms of water justice, it is clear that Francis repudiates this "never the twains shall meet" imperative, which Jamie Linton (2010) has called the "Constitution of Modern Water" (181–90). Francis's "both/and" (humans *and* ecology, nature *and* society, facts *and* values) perspective promotes solidarity or communion "with," not dominion "over," creation. Restoring humanity's proper balance and relationship with God's creation is now a pivotal and integral task of Catholic social justice (Massaro 2015, 174–84; O'Brien and Shannon 2016, 589–92). Forging ahead, the goal is not that Athens speaks for Eden, but to view creation in agential relationship with our rhetorical culture, thereby striving for a harmonious balance. Such a perspective is what McMurry (2017) dubs *environmentality. Environmentality* is the idea that our rhetorical society is enveloped and "conditioned" by the ubiquity of nature and a nonhuman, living environment (742). McMurry's *environmentality* and Francis's integral ecology shares a tributary of similar thought. Recall, Francis (2018) argued that "water is a 'condition' for the exercise of other human rights." Our human rhetoricity is conditioned by an *enviromentality* in the same way water is the source from which other human rights flow. Water is perhaps the most pervasive, ostensible, and existential phenomenon of both *environmentality* and our humanity.

Laudato Si' articulates an "integral ecology" approach to environmental challenges. Integral ecology is a watershed in the history of papal encyclicals and Catholic social justice. Gary Chamberlain (2018) points out that Francis expands Pope Paul VI's 1967 encyclical *Populorum Progressio*, which espoused the concept of "integral development," the good of every man [*sic*] and the whole man [*sic*] to intersect with the natural environment and the whole Earth" (29). Such a perspective extends Catholic social thought beyond previous teaching on the environment from archaic stewardship and "wise use" paradigms to a more vital integrated approach. Integral ecology marks a decisive conversion from these previous Christian attitudes toward the natural environment (Massaro 2015, 175–81). Confuting the antiquated view that the Earth is God's gift to humans alone is a foundational premise of integral ecology and the first step in conversion. In fact, a social justice view of our current relationship with the environment must be viewed from a postlapsarian perspective. We have sinned against God's creation, and thus, social justice requires a reordering, a "setting things right" in terms of our relationship with the environment. Integral ecology advocates a new eco-ethical order that essentially believes "environment" should be seen as an interrelationship between nature and society. Integral ecology imagines a solidarity with other forms of creation in ecosystems by which humans are socially responsible for reconstituting an intimate relationship with the Earth. Francis's approach is a sacred version of American conservationist Aldo Leopold's (1966) ethical call for humans to see themselves as part of a "community of interdependent part" (204).

RHETORIC, WATER, AND ANTHROPOMORPHISM: ABERRATIONAL PERILS OR INTEGRAL PEARLS

Anthropologist Stewart Guthrie (1993), in his lofty book *Faces in the Clouds*, contends that anthropomorphism is endemic to human thought and action, and he floats the anthropological theory that religion is best understood as a systematic anthropomorphism (3). Relatedly, Guthrie notes that philosophers and rhetoricians acknowledge our consistent and comprehensive capacity to humanize our world. Natural scientists, historians of science, psychologists, and ethnographers caution against anthropomorphism's use and view it as aberrational (62–90).

Cassirer (1950) held that anthropomorphism was not knowledge, but "sheer obscurantism, nothing but an escape and a mask of our ignorance" (123). Philosophers and rhetoricians will be familiar with Fredrich Nietzsche's (1966) caustic indictment of truth as "a mobile army of metaphors, metonyms, and anthropomorphisms" (314); yet, ironically, Nietzsche also described this linguistic indeterminacy as metaphorically resting on "an unstable foundation of running water" (see Gare 1995, 58). In critiquing capitalism's commodification of nature, Neil Smith (1984) remarks how nature is reduced to Marx's use-value, value, exchange-value. He writes, "There can be no apology for anthropomorphism from this perspective: human society has put itself at the center of nature" (8). More recently, even advocates of political ecology like Bruno Latour (2004) are critical of anthropomorphism and have characterized it as recrudescent, premodern, and ruinous. Perhaps the term anthropomorphism carries too much "Cave" baggage. This is why, in an effort to discard all human DNA from the term, Latour prefered to see nonhumans as *actants*, acting agents, interveners (2004, 75).

Thus, anthropomorphism has been interpreted pejoratively, defined as a term for bathetic representations of nonhumans with human characteristics. It is further perceived as a vain, lazy, and flawed account of the world, an overall reductive mode that perpetuates anthropocentrism. However, Guthrie (1993) acknowledges the disrepute of anthropomorphism when used as merely a heuristic for understanding the world, but admits that "although [it] is mistaken it is not intrinsically absurd" (70). More recently, several scholars have sought to rethink anthropomorphism.

While rhetoricians have analyzed anthropomorphism in the political sphere (see, e.g., Gunn 2004), less studied are the rhetorical motives for its relational function in the environmental discourse of water justice. Perspectives critical of anthropomorphism notwithstanding, the rhetorical function of anthropomorphism does not always seek to objectify, derail political and environmental deliberations by hanging a millstone around the neck of the nonhuman and throwing it into the sea. Would not anthropomorphism's use depend on a rhetor's purpose? In point of fact, Latour (2010), changing his previously held views of anthropomorphism, argues for "cultivating a bit of anthropomorphism—the idea that human agency has some echoes in nonhuman nature to counter the narcissism of humans in charge of the world" (483). Moreover, ecocritic Greg Garrard (2012) notes that, while anthropomorphism may misrepresent and perpetuate an anthropocentric logic, there can be value in such use since anthropomorphism "describe[s] homologous human behaviors in species to which we are related" (155). Rejecting the aberrational view of anthropomorphism, Guthrie (1993) goes on to claim that seeing the environment as humanlike helps us establish a social relationship with it (5). Concomitantly, Jane Bennett (2010) has argued:

> Anthropomorphizing, the interpretation of what is not human or personal in terms of human or personal characteristics, is clearly a part of the story, but it is less clear how fatal it is. . . . In revealing similarities across categorical divides and lighting up structural parallels between material forms in "nature" and those in "culture," anthropomorphism can reveal isomorphisms. (98–99)

This seems to be an important observation for those concerned with an integral ecology approach. For Bennett, the anthropomorphic form ebbs and flows between human rhetoric and nature and divulges an eco-anthropohomophily, "a whole world of resonances and resemblances" (99) on which relationships and covenants can be renewed. Bennett's reconceptualization of anthropomorphism encourages a "betweenness," where one acknowledges, like McMurry's *environmentality*, that "I am no longer above, or outside a nonhuman 'environment'" (120). It is this relational aspect abiding in Pope Francis's rhetorical and theological use of anthropomorphism that will be my focus. For Pope Francis, environmental care and ecological conversion are only possible when this relationship is nurtured rhetorically, physically, and spiritually. Since water cannot speak, anthropomorphizing it is rhetorically nurturing because it argues for, and reinforces, the intimate and spiritual relationship humans have with water. Thus, anthropomorphizing water articulates a kind of *ecospeak* (an anthropomorphism of its own) that transcends the nature/society binary, thereby seeing water as having intrinsic value and being expressive of the Spirit responsible for creation of our common home.

As stated, water is a classical rhetorical *topos*. But more recently, even geographers and anthropologists recognize the constitutive relationship between rhetoric and water. Graeme Wynn's (2010) foreword to Jamie Linton's interrogatively titled book *What Is Water?* argues that "water becomes what it is in relation to other things and processes; it is what we make of it" (xv). Anthropomorphizing water is also exceedingly familiar to us. For example, we often speak of "fingers," "mouths," "faces," "heads," and "bodies" in referencing lakes, rivers, and oceans. Water is atavistic and familial; we speak of "the gene pool," and the Mississippi River, named by Algonkina-speaking Native Americans, can be translated as "Father of Waters." As we will see, St. Francis of Assisi, for whom Francis has taken as his papal moniker, famously anthropomorphizes "Sister water" in his Canticle of Creation poetry (Saint Sing 1985, 80). The difference between Francis of Assisi's and Francis of Rome's use may be the difference between poetry and rhetoric. Anthropomorphizing water, of course, is not a recent linguistic phenomenon. However, Francis's use of hydro-anthropomorphism is more rhetorically and ethically oriented than poetic or anthropological. His intention is to awaken his audience to an integral *environmentality* and spirituality of water by critiquing anthropocentric social systems that offer no representation to nonhuman creation or those poor human persons dying of thirst and waterborne diseases.

The opening of *Laudato Si'* recapitulates St. Francis of Assisi's encomium to "our Sister, Mother Earth." But, then, sharply shifts to lamenting the atrocities committed against our sister when he states:

> This sister now cries out to us because of the harm inflicted on her by our irresponsible abuse of the goods which God has endowed her. We have come to see ourselves as her lords and masters, entitled to plunder her at will. The violence present in our hearts, wounded by sin, is also reflected in the symptoms of sickness evident in the soil, in the water, in the air and in all forms of life.

We can interpret Francis's anthropomorphic account here as one in which a selfish and obtuse colonizer, afflicted with the venereal disease and toxic sins of power and profligacy assaults and wittingly infects our sister creation with a debilitating and chronic disorder. Francis clearly shares Guthrie's, Bennett's, and Latour's relational approach of seeing the human in nonhuman creation, but our current relationship, by all accounts, is clearly an abusive one. Francis's use of anthropomorphism is not romanticized for comfort or familiarity but presented to persuade others that our relationship with ecology is disordered, and we need be reminded that God's holy design includes preservation of and access to clean water. Chamberlain (2018) underscores this anthropomorphic and rhetorical reading of *Laudato Si'*'s opening with his own set of anaphoric rhetorical questions:

> If people felt water were a sister, would they let her be disfigured, kept unclean (contaminated, polluted)? Would they let her be bought and sold, traded (privatization)? Would they not care for her health and well-being? Such is Francis's view: sister water is not a metaphor but forms a real and intimate relationship with us humans. (163)

Chamberlain's criticism here reveals precisely why an eco-rhetoric of hydro-anthropomorphism is indeed an effective strategy in shifting prevailing, modern constitutional, and anthropocentric views of water.

What are the grounds for claiming our real and intimate relationship with water? Evoking Genesis 2:7, Francis argues that "we have forgotten that we ourselves are dust of the earth; our very bodies are made up of her elements, we breathe her air, and we receive life and refreshment from her waters" (2). What does it mean to be in relationship with water? Our relationship to water invites us to see creation as interconnected with humanity and as an element by which human life depends. Anthropomorphizing water is not done out of a desire for mere solace and companionship in the world. Rather it is used as an argument for seeing water's agency as not only necessary for our survival, but as the origin of a divinely designed universe, "God's spirit hovered over the water" (Genesis 1:2).

Hydro-anthropomorphism is designed to imbue an ethical and epistemological human agency to water. Water is one such creation whose relationship is not only analogous to humans but is literally in humans. Thus, anthropomorphizing water seems like a "natural" trope to employ if we are to see water as central to human and environmental justice. We must "throw cold water" on the view that water is merely a reified resource, detached from a social, political, and ethical milieu of human experience. Water is a right in and of herself, the agential wellspring from which the just association of all human and nonhuman relationships flow. I use the term "association" here as Latour (2004) does in which the meaning "social" is extended to make no distinction between subjects and objects, human and nonhuman (70–82). Likewise, Guthrie (1993) remarks on the powerful impact of anthropomorphism in stating "there is no certain line between the human and the nonhuman, we can better see that it is not unreasonable to look for features we are acquainted with in humans elsewhere as well" (89). Moreover, anthropomorphism does not equate to anthropocentrism. Like any rhetorical strategy, it depends on how and to what end it is being used.

ANALYSIS OF "WORLD DAY OF PRAYER FOR THE CARE OF CREATION"

Let us now turn to an examination of Francis's prayerful message. My main purpose in analyzing Francis's papal message is to demonstrate that he is a bravura eco-rhetorician, deeply concerned about reconciling humanity's relationship with creation and the implications it has for the social justice issue of water and the poor. The address is illustrative of Francis's commitment to "integral ecology," outlined above and in his encyclical, *Laudato Si'*. Moreover, Francis seeks to apply this vision to the water crisis specifically through the use of anthropomorphism and Catholicism's sacramental *doxa* of "living water," apotheosized in Jesus Christ's invitation to "come to me and drink." Francis invites his global audience, Christian or otherwise, to recognize the agency of water and the humanity of Christ in those who thirst.

As a prayerful and practical extension of *Laudato Si'*, and as an opportunity for Catholic and other Christian communities to renew their participation in caring for creation, Francis (2015a) proclaimed September 1 to be the annual observance of World Day of Prayer for the Care of Creation, delivering a papal message from the Vatican. In his brief, ten-paragraph prayerful missive, Francis began by thanking God for the marvelous gift of creation and implored God's mercy for the transgressions committed against creation. He called for a renewal of the relationship between humanity and creation. The highlighted topic of his "live streaming" message was water. Here, Francis called for its protection and just distribution. This was effectuated through an anthropomorphic rhetoric that constituted humanity's relationship to water as sacramental and a matter of social justice.

Francis greeted his global audience in the familial terms of "brothers and sisters." His audience is, of course, much broader than just the Catholic faithful, but also those "men and women of good will" committed to caring for and protecting creation. The salutation "brothers and sisters" adumbrates Francis's intertextual use of St. Francis of Assisi's anthropomorphic poetry of "Brother Sun" and "Sister Water" from his well-known Canticle. Additionally, familial terminology was intended to persuade his audience to see themselves as consanguineous with creation.

After some prayerful expression of gratitude to God for the gift of our common home, Francis acknowledged our collective failure in responsibly protecting the environment. With a contrite tone, Francis (2018) confessed:

> Rightly, there is a growing sense of need for a renewed and sound relationship between humanity and creation, and conviction that only an authentic and integral vision of humanity will permit us to take better care of our planet for the benefit of present and future generations. [Then, echoing *Laudato Si'*,] there can be no renewal of our relationship with nature without a renewal of

> humanity itself. There can be no ecology without an adequate anthropology. (118)

Francis, again, emphasizes the need for a comprehensive renewal of and intersubjective symbiosis among humans and creation. This passage clearly communicates Francis's "integral ecology" paradigm. Thus, Francis views the intrinsic value and sacredness of all living things and maintains that humans are *a part of* and not *apart from* nonhuman creation.

In the next paragraph, Francis (2018) injects the central theme of this year's "World Day of Prayer for the Care of Creation," the issue of water. Here, he primes the pump for the severity of the exigence and shares who has been hung out to dry by the water crisis. "It is a very simple and precious element, yet access to it is, sadly, for many people is difficult if not impossible." To reduce the epistemological distance, what Pope Francis has called the "globalization of indifference," (2013) as it relates to water, he employs antithesis to highlight and reinforce the gravity of disparity among water's significance and distribution. The divergent words "simple" and "impossible," when conjoined, engender an incongruity and invite auditors to wrestle with the tension of such antinomy. Simple things should not be impossible. This antithetical pressure is discharged when Francis (2018) declares, "Access to safe drinkable water is a basic and universal human right, since it is essential to human survival and, as such, is a condition for the exercise of other human rights."

The following paragraphs open the floodgates from which a deluge of hydro-anthropomorphisms cascades. "Water invites us to reflect on our origins," Francis states. To acknowledge that water "invites" is not only to anthropomorphize it, but also afford water rhetorical agency. This is an example of what Mucklebaur (2007) has characterized as "assignifying" rhetoric, which emphasizes force, action, or effect (239). In this particular passage, Francis is not signifying a specific meaning to water but underscoring its goading to "reflect." Water occasions us to "reflect," not through the anthropocentric gaze of Narcissus, but on humanity's *nidus*, which according to the Christian Genesis story marks humanity as the co-creation of earth and the breath of God. In other words, humans and water are literally consubstantial. And, physiologically, this is true. Francis reminds the audience "the human body is composed mostly of water." Water comprises from 75 percent body weight in infants to 55 percent in elderly and is essential for cellular homeostasis and life (Nicolaides 1998, 247). Accepting water's invitation to understand our co-creation is the first step in a kind integral ecological orientation, where the conditions for a more harmonious relationship with water and the environment can be made possible.

Moreover, rhetoric and water share a venerable socio-civic history. Conflating the ancient fragments of pre-Socratic philosopher Thales and sophist

Protagoras, one might adduce that "water (which comprises 75 percent of humans) is the measure of all things." Water's and civilization's primordial symbiosis is also recognized by Pope Francis. In what might be described as a "hymn to hydros," Francis (2018) remarks that "many civilizations throughout history arose near great rivers that marked their identity." Likewise Linton (2010) has argued, "Rivers were more than just influences or determinants of society; they were *part of* [original emphasis] society, figuring as full actors in the founding myths, the legends, and the rituals out of which social relations were spun" (93). Water, as Strang (2015) asserts, "is concerned with agency, the ability to 'make things happen.' In material terms, nothing can happen without water, and thus, its meanings as a generative, creative substance are intimately linked with . . . whole societies to reproduce, and to produce the things they need and desire" (53). Thus, Francis's example illustrates how rivers are the agential life-blood that imbues civilizations with their humanity, power, and vitality.

Francis's hydro-anthropomorphisms from here on become much more sacred and theologically dense. Referencing the Genesis story, Francis notes the "evocative image . . . of the spirit of the Creator [that] 'swept over the face of the waters.'" Fitzgerald (2012) notes that spirit can also be translated anthropomorphically as "breath," or in Hebrew, *ruach*, meaning "air" or "wind" (57). Considering these translations, a further theological and co-creational interrelationship among humanity and water is established. It reminds the audience that the same spirit that breathed into the first earth-man, Adam, is the same life-giving breath that created the waters.

Francis's attention to the anthropomorphic passage in Genesis "face of waters," may speak to a Levinasian nuanced ecological ethic. Levinas's (1989) definition of prayer is "repairing the ruins of creation" (233). Such a description aligns with Francis's social justice prayer for reconciling humanity's relationship with creation. Moreover, the strength of the parallel surfaces on the rhetorical use of "the face," which is a principal theme for Levinas. "Face and discourse are tied," Levinas (1985) states; "The face speaks" (87). In the face of social justice and the water crisis, what may the face of water evoke? It may evoke its status as the biophysical Other. And despite its foundational condition for humanity, the living face of water, from a Levinasian vantage, does not just speak, it commands. And its first words are: "Thou shalt not kill" (89).

For Francis, the face reflected in the biophysical Other is also the face of the poor. In common parlance, face is often used as a synecdoche for the whole human person. And to "lose face" is to metaphorically surrender dignity. Francis (2018) states, "Our world owes a great social debt towards the poor, who lack access to drinking water, because they are denied the right to a life consistent with their inalienable dignity." Thus, this debt may also apply to Levinas's acknowledgment that "the face of the Other is destitute; it is the poor for whom I can do all and to who I owe all." We could almost hear this unending and obligatory ethic of Levinas's being spoken to "Sister Water" and all those poor human sisters carrying water on heads for miles. We should respond to their labor and their call for the human right of clean, accessible water.

Next, Francis offers a critique of the economic appropriation of water. Our relationship with water becomes anthropocentric and odious when reified as an economic commodity guaranteed to us under the neoliberal God terms of "efficiency," "productivity," "individualism," and "privatization." Strang (2015) notes that hydrological metaphors inundate economic discourse. We talk of economic systems moving in "cycles and waves; they require injections and infusions. Wealth circulates and may (or may not) 'trickle down.' Markets are flooded; their indices can be buoyant or plunge. Economies and 'cash flows' are seen as a matter of 'liquidity'" (53). Recapitulating lines of argument from *Laudato Si'*, Francis (2016) repudiates such conceptual metaphors and advocates for the urgency of looking at water beyond capitalistic utility, "in which efficiency and productivity are entirely geared to our individual benefit" (159). He admonishes, "Every privatization of the natural good of water, at the expense of the human right to have access to this good, is unacceptable." Francis's position here is consistent with critical commentary on rapacious capitalism from his first apostolic exhortation, *Evangelii Gaudium* (2013; see also, Tornielli and Galeazzi 2015).

Toward the close of Francis's papal message, and by far the most theologically dense section, he invokes the "Living water" or "water of life" leitmotif familiar to Christians. "Living water" in a rudimentary sense simply means that water animates life processes from agricultural production to human hydration. Such an archetype saturates most societal contexts. In many cultures, "living water" maintains a salubrious and healing function. From the curative wonders of natural hot springs on which health resorts and spas are built, to the more faith-based, holy wells at Chartres, St. Winifred's, Glastonbury, and the miraculous spring at Lourdes, "living water" effectuates a crucial role in sustaining well-ordered health as well as spiritual healing (for more on the spirituality and sacramental nature of water, see Chamberlain 2018, 159–61). Strang (2015) asserts that "living water's" role in holistic human health, can be "readily transposed to consider religious and ecological health and 'order'" (46). When humanity's relationship with others as well as the environment, and specifically water, becomes disordered, polluted by the many sinful and sin-filled contaminants, purification is vital. For Christians, this takes the form of baptism, or what Francis calls the sacrament of "rebirth." Holy water cleanses the newly baptized of their original sin and becomes the source of "undying life." This eternal life, tasted first in the sacrament of baptism, can be consumed more fully in authentic unity with Jesus Christ. The "water of life" is a metaphor for salvation in Christ.

Next, Francis begins a litany of scriptural metonymies, metaphors, and anthropomorphisms comparing water, thirsting, drinking of Christ, and the Church's central mission, to a commission for ensuring the ecological and social justice of water.

> "If anyone thirsts, let them come to me and drink." To drink from Jesus means to encounter him personally, as the Lord, drawing from his words the meaning of life. May the words he spoke from the cross—"I thirst"—echo constantly in our hearts. The Lord continues to ask that his thirst be quenched, he thirsts for love. He asks us to give him to drink in all those who thirst in our own day, and to say to them, "I was thirsty, and you gave me to drink" To give to drink, in the global village, does not only entail gestures of charity, but also concrete choices and a constant commitment to ensure to all the primary good of water.

In Christ's invitation to anyone who "thirsts," "thirst" is used as a metonymy for the poor, those in desperate need (of the water). In what follows, Francis interlaces both the tropes of metonymy and anthropomorphism to further digest Christ's meaning. Anthropomorphic rhetoric is endemic to most religions, and Catholicism is no exception. In fact, it is ensconced as doctrine in the Catholic belief in the Incarnation. The most striking example of this is John 1:9, "the Word became flesh." The divine word, the logos, not only ordered the universe, but became incarnate in Jesus, the Son of God, in order to bring about salvation. Later, in 318 AD, Church father, Athanasius, referred to Christ as "God the word." Thus, the incarnation shares a relationship and an embodiment with creation. Athanasius (2012) tells us that the omnipresent divine word is "in every progress of nature" (17).

Further testament to rhetoric's and anthropomorphism's early place in Catholicism is instantiated when Athanasius cannot even write about Christ's ultimate act of human sacrifice and its impact without employing anthropomorphism himself. Christ's death made even the

> creation break silence. . . . All creation was confessing that he who was made manifest and suffered . . . was the Son of God and Saviour of all. For the sun hid his face, and the earth quaked, and the mountains were rent; all men were awed. Now these things showed that Christ on the cross was . . . God the Word [who] showed himself to men by his works. (19)

The spiritual logic follows that God, incarnated in Christ's human form and in every progress of nature, can also be transubstantiated as "life giving water." Therefore "living water" is also a metonymic extension or manifestation of the anthropomorphic doctrine of the incarnation expressed in the phrase "to drink from Jesus," who "promised a water capable of quenching human thirst forever" (John: 7:37, quoted in Francis 2018). Metaphorically, this is the water of eternal salvation anthropomorphized in encountering Christ.

One of the dominant themes of *Laudato Si'* is the inextricable nexus between the agony of the earth and the distress of the poor (see Dávila 2017). Referencing his namesake, St. Francis of Assisi, Pope Francis (2016) argues St. Francis is "an example par excellence [of the] inseparable . . . bond . . . between concern for nature, justice for the poor, commitment to society, and interior peace" (10). Anthropomorphizing the earth, in the opening of Francis's encyclical *Laudato Si'*, he writes "This is why the earth herself, burdened and laid waste, is among the most abandoned and maltreated of our poor." Echoing the scriptural passage Romans 8:22, Francis bemoans, "she [creation] 'groans in travail'" (Romans 8:22, quoted in Francis 2016, 3). An emphasis on the interconnection between ecology, water injustice, and poverty reinforces the *raison d'être* of Francis's pontificate: to reestablish "a poor Church for the poor" (see Oldenburg 2018). A long-standing priority of Catholic social thought is the preferential option for the poor and most vulnerable members of society. With Francis's anthropomorphic imagery, he now includes the Earth among the destitute, degraded, and downtrodden. As Francis (2015b) mentioned in his papal address to the "World Meeting of Popular Movements," recognizing the Other's struggle must now include nonhuman creation, the ability to see yourself and others in the face of the land.

Additionally, interconnections among Christ, the poor, and what Pope Francis has termed "water poverty" are elucidated in his World Day Address when he recapitulates Christ's words from the cross, "'I thirst,' echo constantly in our hearts." This is fitting, as the narrative of Christ is one that exhibits a life of lowly poverty. In his earlier Apostolic Exhortation, *Evangelii Gaudium*, citing the inclusion of the poor as one of the two pressing issues at this time in history, Francis (2013) declared, "Our faith in Christ, who became poor, and was always close to the poor and outcast, is the basis of our concern for the integral development of society's most neglected members" (49). In death, Christ humbly accepted crucifixion, emptying himself by becoming a slave. At the apogee of Christ's act of mortification, calling out "I thirst" is interpreted by Francis as not thirsting for water, but metaphorically for love, justice, and an end to structural sin. It's an imperative, a mission statement, to carry on his revolutionary social ministry, which today includes a loving and just treatment of our common creation. Catholics quench Christ's thirst for love when they advocate for water justice in their day and actively perform Christ's words reverberating in the Gospel of Mathew: "I was thirsty, and you gave me to drink." Another version of these ideas unites water and social justice in the Beatitude, "Blessed are those who thirst for righteousness, they shall have their fill" (Matthew 5:3–12).

Francis further explicates what "to give to drink" means in "the global village" and in doing so delineates a distinction between charity and justice. For Francis, charity entails a personal gesture, but the real issue of water justice, as a basic human right for all, is structural and requires everyone's care, help, and reordering of a society that must be committed to the preservation and moral distribution of our common, indispensable resource water.

It is in the closing line of Francis's sermon that his fluid embrace of water poverty and the link between environmental and social justice can be heard. He reiterates what he argued in *Laudato Si'*: "Every ecological approach needs to incorporate a social perspective that takes into account the fundamental rights of the poor and underprivileged" (2016, 93). The fragility of clean drinking water is not only an environmental issue but must consider those humans who thirst for this life-sustaining element.

CONCLUSION

This chapter set out to explicate the ecological discourse of Pope Francis's "World Day of Prayer for the Care of Creation" with a specific focus on water and its inextricable linkages to social justice and the poor. I have argued for Francis's status as an eco-rhetorician, whose discourse fits squarely within the current scholarship on rhetoric and environment. I have sought to demonstrate how his view of integral ecology is brought to life and advocates for a deeper relationship with water. This is accomplished primarily through the reinvention of the familiar, yet frequently dismissed, figure of anthropomorphism. Hydro-anthropomorphism, as a stylistic device, aims to animate water by granting it vivacity, agency, and a relationality in which humanity can be in solidarity with and work toward its protection and just distribution. As a trope, hydro-anthropomorphism is a form of *ecospeak* that acknowledges our *environmentality* and argues for a middle ground evinced in Francis's integral ecology approach.

Related specifically to Francis's papal mission, in employing hydro-anthropomorphic rhetoric, Francis taps anthropomorphism's central role in Christian *doxa* and doctrine and connects the humanity of water to the Catholic social teachings on the dignity of the human person and a preferential option for the poor. To protect water by characterizing it as both human and a human right advances Francis's chief call for ecological conversion as well as an effort to bridge the binaries of nature/culture, anthropocentric/ecological, science/religion. If humans are to reconcile with the Earth they must view it as having intrinsic value and set humanity in proper relationship with God's creation. This requires close attention to the rhetoric we employ. As Francis has warned, "If we no longer speak the language of fraternity and beauty in our relationship with the world," [if we eschew the stylistic and extra-rhetorical intimacy engendered by anthropomorphism and personification], "our attitude will be that of masters, consumers, ruthless exploiters, unable to set limits on their immediate needs" (2016, 11). As the global crisis of water scarcity, poverty, and justice exacerbates, eco-rhetoricians and spiritual leaders like Pope Francis will need to deploy the whole armamentarium of both the rhetorical tradition and the Catholic social tradition to stem the tide of continual violence inflicted on Sister water and the poor.

REFERENCES

Athanasius. 2012. *On the Incarnation*. Crestwood: St. Vladimir's Seminary Press.

Bennett, Jane. 2010. *Vibrant Matter: A Political Ecology of Things*. Durham, NC: Duke University Press.

Cassirer, Ernst. 1950. *The Problem of Knowledge: Philosophy, Science, and History since Hegel*. New Haven, CT: Yale University Press.

Chamberlain, Gary. 2018. *Because Water Is Life: Catholic Social Teaching Confronts Earth's Water Crises*. Winona, MN: Anselm Academic.

Clowney, David, and Patricia Mosto. 2009. *Earthcare: An Anthology in Environmental Ethics*. Lanham, MD: Rowman and Littlefield.

d'Ambroisio, Marcellino. 2014. *When the Church Was Young: Voices of the Early Fathers*. Cincinnati: Servant Books.

Dávila, Maria Teresa. 2017. "The Option for the Poor in *Laudato Si'*: Connecting Care of Creation with Care for the Poor." In *The Theological and Ecological Vision of Laudato Si': Everything Is Connected*, edited by Vincent J. Miller, 145–59. London: Bloomsbury.

FitzGerald, William. 2012. *Spiritual Modalities: Prayer as Rhetoric and Performance*. University Park: Pennsylvania State University Press.

Francis. 2013. *Evangelii Gaudium (The Joy of the Gospel)*. Washington, DC: United States Conference of Catholic Bishops.

Francis. 2015a. "For the Establishment of the World Day of Prayer for the Care of Creation." Vatican, September 1, 2015. http://w2.vatican.va/content/francesco/en/letters/2015/documents/papa-francesco_20150806_lettera-giornata-cura-creato.html.

Francis. 2015b. "Address of the Holy Father to the World Meeting of Popular Movements." Vatican, July 9, 2015. http://w2.vatican.va/content/francesco/en/speeches/2015/july/documents/papa-francesco_20150709_bolivia-movimenti-popolari.html.

Francis. 2016. *Laudato Si': On Care for Our Common Home*. Vatican City: Vatican Press.

Francis. 2018. "World Day of Prayer for the Care of Creation." Vatican, September 1, 2018. http://w2.vatican.va/content/francesco/en/messages/pont-messages/2018/documents/papa-francesco_20180901_messaggio-giornata-cura-creato.html.

Gaard, Greta Claire, and Patrick D. Murphy. 1998. *Ecofeminist Literary Criticism: Theory, Interpretation, Pedagogy*. Urbana: University of Illinois Press.

Gare, Arran. 1995. *Postmodernism and the Environmental Crisis*. New York: Routledge.

Garrard, Greg. 2012. *Ecocriticism*, 2nd ed. New York: Routledge.

Good News Bible: Catholic Study Edition. 1979. New York: Sadlier.

Gunn, Joshua. 2004. "The Rhetoric of Exorcism: George W. Bush and the Return to Political Demonology." *Western Journal of Communication* 68, no. 1: 1–23.

Guthrie, Stewart. 1993. *Faces in the Clouds: A New Theory of Religion*. New York: Oxford University Press.

Latour, Bruno. 2004. *Politics of Nature: How to Bring the Sciences into Democracy*. Cambridge, MA: Harvard University Press.

Latour, Bruno. 2010. "An Attempt at a Compositionist Manifesto." *New Literary History* 41, no. 3: 471–90.

Leopold, Aldo. 1966. *A Sand County Almanac*. New York: Ballantine Books.

Levinas, Emmanuel. 1985. *Ethics and Infinity: Conversations with Philippe Nemo*. Translated by Richard A. Cohen. Pittsburgh: Duquesne University Press.

Levinas, Emmanuel. 1989. "Prayer without Demand." In *The Levinas Reader*, edited by Sean Hand, 227–34. Oxford: Blackwell.

Linton, Jamie. 2010. *What Is Water? The History of a Modern Abstraction*. Vancouver: University of British Columbia Press.

Lynch, Paul. 2017. "On Care for Our Common Discourse: Pope Francis's Nonmodern Epideictic." *Rhetoric Society Quarterly* 47, no. 5: 463–82.

Massaro, Thomas. 2015. *Living Justice: Catholic Social Teaching in Action*. Lanham, MD: Rowman and Littlefield.

McMurry, Andrew. 2017. "Rhetoric and Environment." In *The Oxford Handbook of Rhetorical Studies*, edited by Michael John MacDonald, 733–44. New York: Oxford University Press.

Mucklebaur, John. 2007. "Rhetoric, Assignification, and the Other: A Response to Diane Davis." *Philosophy and Rhetoric* 40, no. 2: 238–47.

Nicolaides, S. 1998. "Physiology of Thirst." In *Hydration throughout Life*, edited by Maurice J. Arnaud, 3–8. Montrouge: John Libbey Eurotext.

Nietzsche, Friedrich. 1966. *Werke in Drei Bander*, vol. 3. Edited by Karl Schlechter. Munich: Carl Hanser.

O'Brien, David J. and Thomas A. Shannon. 2016. *Catholic Social Thought: Encyclicals and Documents from Pope Leo XIII to Pope Francis*, 3rd revised ed. Maryknoll, NY: Orbis Books.

Oldenburg, Christopher J. 2018. *The Rhetoric of Pope Francis: Critical Mercy and Conversion for the Twenty-First Century*. Lanham, MD: Lexington Books.

Osborn, Michael. 2018. *Michael Osborn on Metaphor and Style*. East Lansing: Michigan State University Press.

Quintilian. 1856. *Institutes of Oratory*. Translated by John Selby Watson. London: Henry G. Bonn.

Rosenfeld, Lawrence W. 1989. "Central Park and the Celebration of Civic Virtue." In *American Rhetoric: Context and Criticism*, edited by Thomas W. Benson, 221–66. Carbondale: Southern Illinois University Press.

Rosenfeld, Lawrence W. 2006. "After Walter Benjamin: The Paradise at the End of the Rainbow." In *Rhetorics of Display*, edited by Lawrence J. Prelli, 197–203. Columbia: University of South Carolina Press.

Saint Sing, Susan. 1985. *St. Francis, Poet of Creation: The Story of the Canticle of Brother Sun*. Chicago: Franciscan Herald Press.

Smith, Neil. 1984. *Uneven Development: Nature, Capital, and the Production of Space*. Oxford: Blackwell.

Strang, Veronica. 2015. *Water: Nature and Culture*. London: Reaktion Books.

Tornielli, Andrea, and Giacomo Galeazzi. 2015. *This Economy Kills: Pope Francis on Capitalism and Social Justice*. Translated by Demetrio S. Yocum. Toronto: Novalis.

WHO/UNICEF. 2017. "Water." *United Nations*. https://www.un.org/en/sections/issues-depth/water/.

Wynn, Graeme. 2010. Foreword to *What Is Water? The History of a Modern Abstraction*, by Jamie Linton, xi–xvi. Vancouver: University of British Columbia Press.

Section III

Crisis and Disaster

A central concern of environmental and water justice is that when disaster strikes, not all populations are affected equally, and that, more often than not, it is marginalized and vulnerable communities who bear the heaviest burdens. No recent example illustrates this inequity and injustice more clearly than the aftermaths of recent superstorms, floods, and unnatural disasters—water events whose deadly, destructive, and costly effects are exacerbated by human decisions; policy errors, constraints, and misjudgments; insufficient and aging infrastructure; structural systems of economics and oppression that keep poor communities and communities of color in the places most likely to be hit by devastating weather events; inequity in access to education and warnings; and unequal access to solutions and routes to safety in times of need. The aftermath of Hurricane Katrina in 2005 laid bare how weather events can be a matter of justice; after the hurricane hit and levees were breached, low-income and predominantly black parishes lacked basic food and freshwater resources and were profiled as looters in national media. The interclass and intercultural tensions exposed by the hurricane defined Katrina for many Americans (Bates and Swan 2007; Johnson and Rainey 2007). In 2017, when Hurricane Maria devastated Puerto Rico, Americans again saw poor and ethnic minority populations bearing the brunt of the disaster, with more than three thousand deaths and more than $90 billion in damage (see Lloréns et al. 2018).

But water crisis and disaster can be entirely human-made, as well. The modern environmentalist movement, for instance, was inspired in part by a water disaster near Cleveland, Ohio, when the Cuyahoga River, heavily polluted with industrial waste and dumping, caught fire in 1969. Media coverage of the event influenced the U.S. Congress in passing a series of environmental legislation bills, including the 1970 National Environmental Policy Act, which established the U.S. Environmental Protection Agency (EPA), and the 1972 Clean Water Act, which required hygienic maintenance of all U.S. rivers (Manke 2018).

In recent years, the contamination of municipal water in Flint, Michigan, has set a precedent for water justice awareness and activism. In 2014, as a cost-savings measure, the city adopted the Flint River as its water source, and residents almost immediately reported health issues and brown water running from their taps. Insufficient water treatment and lead leached from the aging water pipes, contaminating the water and exposing hundreds of thousands of families to elevated lead levels and Legionnaires' disease. After over a year of complaints, a federal emergency was declared in 2016 and Flint residents were told to avoid using their tap water indefinitely. Five years later, thousands of the pipes in question have still not been replaced (Clark 2018; Hanna-Attisha 2018; Pauli 2019).

In crises of this enormity, with stakes of literal life and death, it might seem that words and other rhetorics have little effect. But they do. In Flint, much of the change came from the persistence of local activists and online social media spread of personal stories and images and memetic campaigns to ensure national attention and awareness. In the digital media age, simply circulating images of a scarred body, a flooded village, or a glass of muddy tap water can garner public opinion and force public discussion and action toward change. In Kenya, for instance, in 2019, residents of the village of Marsabit confronted government officials about lax municipal water standards by serving them dirty water while the cameras rolled. Within twenty-four hours, news of the Marsabit water crisis was a topic of global concern (Walter 2019).

This section examines such rhetorical invention in four cases of crisis and disaster involving water and unequal distributions of resources and attention. Each chapter highlights a rhetorical concept useful in conceptualizing pain and loss from disaster and crisis across larger populations, and promoting public safety both before and after future disasters.

In chapter 9, James Coleman McGuffey examines the immersive environmental performance *Cry You One* (*CYO*), which invites audiences to reflect on intersections of interest and experience during an ongoing crisis of vanishing coastline in south Louisiana. Drawing on analysis of *CYO* secondary materials and participant observation at a *CYO* performance, he argues that *CYO* uses epideictic in order to refuse the erasure of what coastal Louisiana has lost, to amplify and intensify its audience's attunement to the vibrancy of the region's land and waterscapes, and to mobilize shared cultural and community resources in an effort to move forward. To establish this argument,

McGuffey coins the concept *copious dwelling*, an epideictic rhetoric performance invoking a sense of presence for its audiences. The chapter concludes with reflection on how performing copious dwelling may serve as a tactic in commemoration at/of other aquatic spaces.

In chapter 10, Christopher Scott Thomas turns to Flint and considers the *local and inventive community rhetorics* at play in a community striving for policy change, public recognition, accountability, and the basic right of access to safe drinking water years after the initial contamination caught public attention. In particular, he applies an ecofeminist lens to highlight the roles of Flint's women as mothers, grandmothers, and daughters, to campaign for attention, collaboration, and change. The central case study highlights a growing global trend of young persons leading the way in public rhetorics of environmental justice by tracing the public and social media appeals of Mari Copeny, an eleven-year-old Flint activist.

In chapter 11, Elizabeth Brunner looks to a remarkable crisis from 2013, when more than ten thousand bloated, dead pigs were found floating down the Huangpu River, or "Mother River of Shanghai." Shanghai residents used their computers and phones to voice concern over social media, and the public outcry prompted the government to release daily water-quality reports, but the government institutions ultimately deemed the water quality "normal" even as hundreds more carcasses arrived each day. Brunner examines the public response through *wild public networks*, wherein citizens using social media tools can contribute to social change, even in a political environment wherein censorship is rampant and surveillance omnipresent. To explore how the Huangpu incident provoked social media users to become activists, the chapter examines the incident itself, online reactions, government responses, and the political climate at the time that connected the incident to other food, air, and water safety concerns.

In chapter 12, Catherine J. Bruns's work serves as a transition from "Crisis and Disaster" to the book's final theme, "Plant and Animal Justice," by problematizing the utility of frames like "crisis" and "disaster" to begin with while considering satirical writer Rowan Jacobsen's 2016 "obituary" for the Great Barrier Reef. The "obituary" essay, recounting the long life and tragic death of "Australia's crown jewel," spread widely on social media as millions of readers expressed sadness at the passing of one of the world's most famous environmental icons. Yet scientists and environmentalists struck back at the "wildly irresponsible" commentary, arguing that Jacobsen's article failed to communicate that it was satirical, instead contributing to public misunderstanding of the reef's dire status and further hindering progress to protect it. The chapter positions the reef as a nonhuman environmental justice issue and argues that Jacobsen's obituary is a failed environmental justice intervention. Bruns seeks to understand how satire may be used to advocate for the justice of living environmental spectacles. She analyzes the obituary using a satirical framework and, in doing so, ultimately argues that if satire is to be a successful advocacy tool in the fight for environmental justice, the environmental communication discipline must develop a clear *standard of satire* that can be easily communicated to and correctly interpreted by audiences.

REFERENCES

Bates, Kristin Ann, and Richelle S. Swan. 2007. *Through the Eye of Katrina: Social Justice in the United States*. Durham, NC: Carolina Academic Press.

Clark, Anna. 2018. *The Poisoned City: Flint's Water and the American Urban Tragedy*. New York: Metropolitan Books.

Hanna-Attisha, Mona. 2018. *What the Eyes Don't See: A Story of Crisis, Resistance, and Hope in an American City*. New York: One World.

Johnson, Glenn S., and Shirley A. Rainey. 2007. "Hurricane Katrina: Public Health and Environmental Justice Issues Front and Centered." *Race, Gender & Class* 14, no. 1/2: 17–37.

Lloréns, Hilda, Ruth Santiago, Carlos G. Garcia-Quijano, and Catalina M. de Onís. 2018. "Hurricane Maria: Puerto Rico's Unnatural Disaster." *Social Justice: A Journal of Crime, Conflict, & World Order*, January 22, 2018. http://www.socialjusticejournal.org/hurricane-maria-puerto-ricos-unnatural-disaster/.

Manke, Kara. 2018. "Clean Water Act Dramatically Cut Pollution in U.S. Waterways." *Berkeley News*, October 8, 2018. https://news.berkeley.edu/2018/10/08/clean-water-act-dramatically-cut-pollution-in-u-s-waterways/.

Pauli, Benjamin J. 2019. *Flint Fights Back: Environmental Justice and Democracy in the Flint Water Crisis*. Cambridge, MA: MIT Press.

Walter, Jacob. 2019. "Kenya: Why Marsabit Residents Served Their Leaders with Dirty Bottled Water." *All Africa*, February 12, 2019. https://allafrica.com/stories/201902120766.html.

Chapter Nine

Copious Dwelling in a Sinking Landscape

James Coleman McGuffey

Coastal Louisiana is sinking. With every hour that passes, a wetland area the size of a football field disappears, lost to the sea due to natural and human-made causes (Rich 2014). On April 19, 2017, Governor John Bel Edwards declared a state of emergency for Louisiana's fading coasts, revealing that "the Louisiana coast is in a state of crisis that demands immediate and urgent action to avert further damage to one of our most vital resources" ("Gov. Edwards" 2017).

Wetlands are, simply defined, "areas where water covers the soil, or is present either at or near the surface of the soil all year or for varying periods of time during the year" ("What Is a Wetland?," n.d.). These water-saturated lands include various kinds of spaces, including bayous, woodlands, marshes, and swamps. Traditionally, wetlands provide a vital environmental service as a rich resource for its inhabitants, human and nonhuman, as a source for food, space for recreation, cultural resource, and storm barrier.

Today, the rising sea is swallowing such natural barriers as a result of natural causes, a lack of environmental protections, and human development (see Tibbets 2006; Marshall 2017). Notably, the impact of wetland loss continues to be painfully clear in a time when climate change is increasing the ferocity of hurricanes leading to flooding, ecological damage, the loss of communities, and death. This reality was morbidly on display in 2005 when Hurricane Katrina hit the region (Pezzullo 2009).

Moreover, climate change, the alteration of the region's waterways, and a significant toxic presence from the petrochemical industry have left Louisiana's population extremely vulnerable to an increasingly alien and unstable environment. Indeed, the region's inhabitants are facing what Lauren Berlant

(2007) refers to as "slow death" (see also, Fleetwood 2006 and Nixon 2013). Slow death, Berlant (2011) argues, is undergoing the "physical wearing out of a population in a way that points to its deterioration as a defining condition of its experience and historical existence" (95). Many of the industries promising economic security in the region are significant contributors to its instability. Couple this with the compounding uncertainties created by climate change and you have a place and a population mired in "deterioration" (95).

Our national indifference to the inhabitants of this region is a quintessential story of environmental injustice, wherein the people are reduced to expendable objects in the name of powerful interests. Nicole Fleetwood (2006), in her analysis of media coverage during Katrina, writes that the hurricane "exposed how systems of power invest in material and infrastructural technologies to maintain certain imbalances" (769). She argues that the inhabitants of coastal Louisiana have, through technological determinism, been produced as "disposable for the state" (770). The "deterioration" of these populations is not a matter of people simply left behind, but rather a product of deeply ingrained institutional discriminatory practices. Robert D. Bullard and Beverly Wright (2012) point out that the destruction of the vulnerable populations in this region "has its roots in an economic system characterized by economic exploitation, racial oppression, and devaluation of human life and the environment" (7). Technological progress is built upon the exploited bodies of people living in coastal Louisiana. This pattern of denigrated labor and unsustainable technological choices came into the media spotlight again in April 2010, when the BP Deepwater Horizon spill became the largest accidental oil spill in world history (Moss 2010).

Given these social and ecological conditions, it is perhaps no surprise that Louisiana is home to the first group of people acknowledged by the United States as its own "climate refugees," a possibility made reality when the inhabitants of Isle de Jean Charles were allocated $48 million to move their community as they struggle with the effects of climate change (Davenport and Robertson 2016). Their homes are sinking. Here, and throughout coastal Louisiana, land and water have come at odds with each other. Any security created by a harmony between land and water has been replaced by precariousness. Vandana Shiva (2008) argues that this precariousness created by water instability is a product of impairing and destroying water culture through "the most violent technologies for rerouting rivers, imprisoning rivers and drying out rivers . . . [which] disrupt the water cycle" (500). The disruption of the water cycle, Shiva argues, is not simply a product of technology, but the "current thinking that water is just another commodity on the planet, you don't have to give it any special respect" (500).

Previously, communication scholars have persuasively drawn on the notion of "presence" as a means of improving the state of environmental discourse and judgment (Pezzullo 2007; Peeples 2011). This critique emphasizes performances that aim to reveal the seedy underside of our current risk culture, exposing the tragedy and trauma that accompanies unchecked technological progress. Not only emphasizing crisis, Phaedra C. Pezzullo (2003) identifies how these tactics may "rhetorically invent more livable designs for life" (228).

The context of southern Louisiana's disappearing wetlands raises questions central to environmental communication dilemmas and a sense of presence: how can locals most persuasively share why they care about this specific place and the crises they perceive that they face? Which modes of address will aid them in overcoming the slow death and deterioration of the region? How can one challenge the acts of forgetting that perpetuate and proliferate environmental degradation and convince more people to feel a sense of connection to the fate of the region? Is it possible to move people who do not live in a place to feel a sense of connection to it anyway?

In this chapter, I focus on the environmental performance *Cry You One* (*CYO*). Like the wetlands themselves, this commemorative act is hybrid: "Part song, part story, part procession for our lost land, *Cry You One* celebrates the people and the cultures of South Louisiana while turning clear eyes on the crisis of our vanishing coast" ("Cry You One" 2016). Drawing primarily on my analysis of their secondary materials and my participant observation of one of their performances, I argue that *CYO* uses epideictic rhetoric to refuse the erasure of what coastal Louisiana has lost, to amplify and intensify its audience's attunement to the vibrancy of the region's land and waterscapes, and to mobilize shared cultural and community resources in an effort to move forward.

To establish this argument, I begin by defining "copious dwelling" as a type of epideictic performance that attempts to invoke a sense of presence for its audiences. From here, I will move my analysis to the copious dwelling enacted epideictically in *CYO*, in which I emphasize three temporal orientations to the landscapes visited in the performance. The first orientation to the landscape emphasizes the past through enunciations of lost. The second orientation emphasizes the many ways of experiencing the performative landscape in the present. The third orientation emphasizes the perseverance of coastal Louisiana cultures, and encourages imagining possibilities for moving forward in the future. I conclude this chapter by reflecting on how performing copious dwelling may serve as an inventional tactic of aquatic commemoration.

COPIOUS DWELLING AS AQUATIC COMMEMORATION

CYO is a response to the loss of land, culture, and communities caused by environmental destruction of coastal Louisiana. Moreover, it is an attempt to overcome the environmental impasse haunting the region and the world more broadly. This impasse, a product of disagreement and disavowal over the environmental impacts of a fossil fuel–driven economy, is producing devastating consequences for Louisiana. How is one to approach the loss of community—as a gathering of human and nonhuman elements—while striving to maintain the culture, which is a crucial feature of the gathering and does not agree on how to respond?

Often when communities face loss, they turn to the epideictic genre of rhetoric. Epideictic performances such as funerals and memorials provide an opportunity to deal with the perils of presence and the inherent exigencies of loss. As a genre with deep and complicated ties to presence and witnessing, epideictic provides resources for challenging the impasse bringing harm to the region's communities. The inventive power of epideictic rhetoric lies in what Ned O'Gorman (2005) identifies as "providing common images for people and shaping their sense of scale" (33). In drawing upon the registry of tropes regarding amplification and magnitude is the ability to render things present and visible before a community. Aquatic commemorations are the epideictic performances that draw upon amplification and magnitude to alter an audience's sense of scale in service of water culture. As such, these performances strive to provide a reflexive and immersed sense of being part of a water culture. Michael Sipiora (1991) ascribes these reflexive and immersive potentials to epideictic explaining, "the epideictic orator's anticipation of the future is not the foretelling of what will someday be present. It is instead the engendering of a recognition of openness to possibilities that are coming toward the audience out of the horizon" (242). To this end, I examine a particular type of amplification within the epideictic of *CYO* that works to construct an openness to future potentials.

CYO engages in aquatic commemoration by performing what I call "copious dwelling." With this term, I am thinking of the rhetorical concept of *copia* in relation to Heidegger's sense of *dwelling*. I link these two concepts because they shine light on the relationship between language and our orientation with and amidst "things"; further, in thinking about them together, I hope to help foster greater appreciation of their heuristic value. *Copious dwelling*, I argue, is a mode of epideictic performance that appreciates the complexity and entangled nature of things as a gathering.

Copia and Inventive Sedimentation

Beginning with the rhetorical tradition of *copia*, described by Quintilian (1921) and Erasmus (1963), I suggest that the rhetorical performance of *copia* can cultivate alternative attunements to the world. Too often *copia* is dismissed as an exercise in ornamentation. However, as Conley (1985) has pointed out, "to ignore *copia* is, implicitly, to place excessively narrow restrictions on 'what counts' as an argument, the discussion of which generally centers on 'logical' invention and leaves all else out" (96). A closer examination of *copia* unearths an orientation to how things—in the vibrant sense—are taken up and explored in discourse.

Copia is a rhetorical method of accumulating an abundance of expression. Quintilian (1921) informs us that to prepare students for oratory, they must first develop a "copious supply of words and matter" (Book 10, chapter 1, section 5). Copiousness succeeds "not by presenting audiences with an elaborate picture, but by bringing us into actual touch with the things themselves" (Book 10, chapter 1, section 16). Thus, *copia* enables rhetors to visit a sense of presence in a compelling way, not just as the act of goading audiences to feel present, as if they could touch things, but also as the condition of possibility of gathering inventional resources.

In *De Copia*, Erasmus (1963) further describes the rhetorical power of *copia* as no mere mode of ornamentation, but rather as fueling one's capacity for rhetorical invention. More recently, scholars have taken up this notion of *copia* to better understand its role in invention rather than style. In this same current, Conley (1985) reminds us that Cicero and Erasmus alike "realized that things are not very profitably reduced to univocal names. Things, whether or not they are in dispute, often are ambiguous and shifty" (102). Consider, for example, how Foucault (1972) emphasizes the messiness of identifying origins of discourse. *Copia* provides an interpretive resource required to "manage things and then 'make sense' of them" (Conley 1985, 102). Thus, we see *copia* deployed as a phenomenological tool for unlocking or uncovering, as well as an orientation for encountering and considering complex and entangled worldly things.

Thomas Sloane (1997) reiterates this point, arguing that *copia* "is not an easily graspable idea. Most modern scholars think of it as having something to do with style. After all, the word seems synonymous with 'abundance.' It does, but it refers to another kind of rhetorical abundance as well: abundance of thought. It therefore clearly has something to do with invention" (56). In other words, *copia* as a concept of abundance plays a significant role as a type of rhetorical sensibility for thinking about the world and how to move audiences to feel its tangibility.

Copia arguably lends itself to an openness to the great potentiality of things, as it challenges reductive and essentializing modes of discourse by inviting serious inquiry into the "variegated nature" of things. As Debra Hawhee (2017) in her study of *accumulatio* argues, "Variety, after all, brings energy to every sphere" (159). Hawhee adds that as an approach to living and to writing, *copia*, by way of accumulation "presumes the importance of matter and what matter can do, whether in a motley heap or in an organized collection" (137). Far from adorning things with symbolic panache, *copia* requires taking seriously its subject matter, exploring its potentialities and relationalities that lie beyond the horizon of its existence in the current moment.

Marcus Boon (2011) further explains that "the rhetorics of copia are the modes in which things are made to appear with no prior or more solid ground than knowledge of the practices that lead to their production or revelation. Again, this is not merely rhetoric as discourse or social construction, but rhetoric as a mode of co-participation with materiality and other forces in the fabrication of objects and events" (64–65). Therefore, *copia* decidedly resists foreclosing a thing's potentiality for being. As an orientation, *copia* is "inherently multiple, excessive, and abundant," feeding things so that they may grow (76). In this capacity, *copia* unlocks things from the shackles of instrumental logics that reduce them to object, because, in the receptivity for their abundant possibilities, alternative attunements to the thing and the world arise.

This potential of *copia* is particularly relevant to consider in discourses of crisis. Often, when crisis is invoked in discourse, solutions are posed and enacted that satiate the immediate exigence without consideration for their long-term implications. The call for immediate solutions supersedes other considerations. Pezzullo (2007) points out that in this "emergency time" decisions are made based upon limited analysis, "with little deliberation or prudence involved" (177). Moreover, as Naomi Klein (2007) argues, emergency discourses often become "orchestrated raids on the public sphere," narrowing potential solutions to meet certain criteria benefiting powerful vested interests (6). Consequently, discourse addressing crisis would benefit greatly from outlets for prudence, considering the vast world of things and how they interconnect. Next, I suggest this sensibility might be productively imagined through dwelling.

Channeling *Copia* into Dwelling

In expanding upon the possibilities of aquatic commemoration, the above description of *copia* reflects upon an orientation to how things are taken up and explored in discourse. However, it is also important to understand not only how things emerge and reside in discourse, but also to take seriously the role of being-in-the-world amidst-things—amidst-water. Consequently, I articulate *copia* as a form of the Heideggerian conception of "dwelling." A key principle to understanding Heidegger's sense of dwelling is to recognize, as Thomas Rickert (2013) puts it, that "dwelling indicates not a worldview but a principle of being" (223). I want to explore this notion of dwelling with *copia* in mind.

According to Heidegger (1971), dwelling means "to cherish and protect, to preserve and care for, specifically to till the soil, to cultivate the vine" (145). He adds, "The fundamental character of dwelling is the sparing and preserving" (147). Dwelling is an orientation, a description of what it means to be amidst-things in such a manner as to respect their potentials and possibilities. Most importantly, dwelling is an orientation predicated upon the conditions for which things may be revealed—their being made known. Similarly, dwelling emphasizes the importance of place in directing our everyday activities. Heidegger would argue that we are em-placed, where our everyday understanding of the world is cultivated through our everyday interactions with the places we exist. Paying attention to our modes attunement amplifies the importance of our surroundings in our being-amidst things.

Heidegger points to poetry—poetic dwelling—as a form of dwelling that cultivates an openness to things and their being-in-the-world. Poetry is an activity that reflects on humanity's sense of scale. In the work of poetry "is measure-taking, understood in the strict sense of the word by which man [*sic*] first receives the measure for the breadth of his [*sic*] being" (219). Dwelling is to encounter and witness things revealed, *poetic* dwelling avoids technological modes of being, which reduce the appearance of things to use value. Rather, it allows the freedom for things to reveal their own possibilities of being amidst a world of things. Graham Harman (2007) explains, "To dwell poetically on the earth means to stand amidst the presence of the hidden but hinting gods, which brings us into closer proximity to the essence of things" (116–17).

In exploring what it means to dwell *rhetorically*, Michael Hyde (1983) explains that it works to both maintain the presence of things in society, as well as to explore new possible presences. He concludes that "rhetoric achieves its 'purest form' when, abandoning its task to *preserve* something thought to be a truth, it attempts to disclose and thereby *create* that which remains concealed by this truth" (216). It is with this sentiment that I argue for thinking about copious dwelling as both a tactic and an orientation that enables poetic dwelling.

In this chapter, I am particularly interested in how copious dwelling promotes water culture by challenging dominant worldviews that encourage a desire to control, disrupt, and alienate ourselves from water. In its potential to direct audiences to immerse themselves within a water culture, it offers an opportunity for dwelling poetically, where one "is cooperating with the earth in the bringing forth of *truthful* things, things that bear on themselves the marks of what brought them forth" (Edwards 2002, 130). Performances of copious dwelling provide an opportunity for places—as gatherings of things—to show up abundantly, acknowledging past presences and allowing for alternative futures. Ultimately, it may short-circuit the teleology of technological instrumentalism, channeling our sense of potential over a desire for control.

Rhetorically, *copia* demands active engagement and concern with a great number or abundance of things. Copious dwelling is a cultivated discursive orientation attuning audiences to the vibrant dynamics of things through abundance such that it functions ambiently, as what Rickert (2013) calls an "activity that works through and with things as they show up in the world alongside us" (216). Since it requires rhetorical labor not only to feel connected in these many ways but also to not become overwhelmed by that sense of connection, we would do well to study more concrete exemplars and failures to ground and to nuance these theoretical arguments. With that in mind, I now turn to *Cry You One* as an exemplar of copious dwelling.

CELEBRATION AND CRISIS: AN ANALYSIS OF *CRY YOU ONE*

Launched in 2013, *Cry You One* (*CYO*) is "performance that journeys into the heart of our disappearing wetlands" ("Cry You One" 2016). The playwright claims the title was from a Cajun musician whom he had once heard introduce a song using that phrase (MacCash 2013). It also invokes a play on the famous 1953 song, "Cry Me a River" by Arthur Hamilton which has been remade by several artists and coined the idea of this colloquial response to someone who had hurt the person speaking/singing (Berlau 2010). Primarily, it is the result of a collaboration between Artspot Productions (http://www.artspotproductions.org/) and Mondo Bizarro (http://www.mondobizarro.org/), which are companies of multidisciplinary artists dedicated to inspiring and strengthening communities. Although relatively small compared to Broadway productions, they have received coverage in *Times-Picayune*, *Houston Chronicle*, and other notable press venues. Touring the country, *CYO* has visited places from Vermont to Arizona and Texas. At each location, *CYO* is a "site-responsive performance," enacted with core themes and acts, but also integrating what the performance group learns about the land where they are performing. In order to expand its reach as an act of commemorating the disappearing wetlands, *CYO* also serves as an online storytelling platform through "People Profiles" and "Stories" of "Land," "People," and "Culture" shared on social media, including their website: http://www.cryouone.com/.

The overall structure of *CYO* weaves together two primary story lines in two hours. The first follows the eco-tour group M.O.S.Q.U.I.T.O, which stands for "Mississippi Outlet Scientific Queries United In Tactful Observation." The acronym is telling: it invokes the old proverb that the smallest among us can make a difference ("If you think you are too small to make a difference, try spending the night in a tent with a mosquito"); it also foreshadows the science-based complexity of devastation and loss in the region through the clustered and compact nature of the acronym, while drawing on a local nonhuman icon of the region. In doing so, this narrative attempts to balance the overwhelming information about the challenges faced in southern Louisiana with a desire to foster a sense of agency in audiences. The second story line of *CYO* involves another local cultural icon: it is a jazz funeral procession for the wetlands, where mourning and celebration come together to pay respect to the region's past, and to begin imagining the process of moving forward. The audience is asked to sing and dance along with each other and the cast while traveling through instances of celebration and mourning in performative landscapes.

CYO not only is an interactive, multiact show, it also requires the audience to travel physically so that the play and the location interact, inflecting each other differently in each performance. Most obviously, these destinations are off the beaten path. For example, the Clear Creek festival grounds in Rockcastle County, Kentucky, are located in the foothills of the Appalachian Mountains. Getting to the space required climbing a somewhat treacherous gravel road that winds its way through thick woodlands, before reaching a clearing at the top of a large hill. Clear Creek itself lies at the bottom of the hill winding alongside the road leading to the performance site, visible until the ascent.

Although *CYO* proceeds without intermission or breaks in the performance, it is not fixed to a single space. Rather, *CYO* emphasizes travel in the form of leaving one's home. The opening sequence takes place in the same space as the audience's arrival. This sequence sets the stage for our impending journey into the performative landscape of the Louisiana wetlands, and our subsequent retreat from them. We begin our visit following the cast down a small gravel road through the woodlands, visiting sites that are part of the Clear Creek festival grounds. The path is narrow; however, there are small clearings on both sides of the road, which become stops along the way during the performance, where we become better acquainted with the wetlands as well as one another, and will later have to leave in evacuation. At the end of the narrow road is another clearing with a temporary tent-like structure featuring photographs of the deteriorating wetlands, as well as a more permanent small theater stage, which displays signs stating: "Mountains Burst into Tears," and "Coal Lies under Foot." This site would be the final stop of the performance, a place to reflect on the land we learned about, lost, and are invited to save. It also serves a stark reminder of a major exigence haunting the location of the performance: Appalachia. In order to bring together various threads of copious dwelling occurring throughout the performance, my analysis proceeds in four parts, beginning with the staging or "Arrival: When the Water Came" of the performance. Then, I organize the remaining parts through three temporal dimensions: "Looking Back: Waves In, Waves Out"; "Looking Around: The Ecotour"; and "Looking Forward: The Answer Is in Your Dreams."

Arrival: When the Water Came

Upon arrival, we check in at a temporary desk below a banner donning the aforementioned acronym, M.O.S.Q.U.I.T.O. Each audience member is assigned an animal or totem, which corresponds to a tour guide we will be grouped with during the performance. We discover each cast member plays the part of a tour guide and their totem. There are four groups in total: Alligators, Boars, Snakes, and Spiders. I am assigned to Snakes. While people are checking in, the cast members are informally introducing themselves to audience members in their M.O.S.Q.U.I.T.O tour guide characters. Once the audience assembles to the initial stage, the tour guides begin more formally introducing themselves with short monologues. In short: the Boar guide is Tom Dulac, acting as a member of the Deepwater Horizon Oil spill cleanup; the Snake guide is Sabine Jean-Louis, an environmental justice advocate playing herself; the Gator guide is Zelda Culotte, acting as a member of a local fishing village; and the Spider guide is Dr. Carol Carl, playing the role of a scientist (a botanical epidemiologist).

From the opening, cultural conflict between the M.O.S.Q.U.I.T.O. characters is apparent. A fifth man in a carnivale mask brings forth a suitcase full of cash asking what should be done to help the region. While a suitcase full of money alone cannot resolve the challenges faced in southern Louisiana, we are presented with an abundance of options. The characters disagree (sometimes strongly) about what they should do. Zelda and Dr. Carl argue over how to best restore the land in the region. Zelda suggests using recycled plastics to help rebuild disappearing islands along the coast. Dr. Carl suggests we find ways to replenish the flow of sediment into the Mississippi Delta. The others suggest using the money to relocate. Each character is not only competing for limited funds, but also limited time. Indeed, Dr. Carl literally times other characters with a stopwatch making sure they do not go over, embodying the ways scientific measurements can be insensitive to different communication styles and emotional appeals. (This push and pull of conflicting perspectives and competition for limited resources carries on as the tour guides interact throughout.)

After these opening introductions and competing solutions, the cast collectively tells a story about the travels of the river and how that has changed. This "river story" offers a tale of a Mississippi changed and a region dying. When not narrating the story, the others engage in song and dance, amplifying the feeling of the river's flow journeying to its Delta. They describe the emergence of the Mississippi from its headwaters in comparison to a mountain cat carrying her young to term:

> A drip and a drop.
> A trickle of icy water tumbling from Lake Itasca, Minnesota.
> Traversing ten states wending and winding its way.
> A mountain cat will birth her young before that water reaches the gulf,
> 90 days.

The background chorus sings in a manner comparable to steady flow: "I've been rollin' out to the high tide / I've been rollin' down to the low tide." Through metaphor and music, story and the chorus invite us to feel the rhythm of the river, its vast reach across land, and how long the water must travel across states, as well as up and back from coastlines. Both as small as a drip and as massive as a tide, the water's movement spans scale.

Expressing metaphors to help us imagine the fluidity of the River continues as a theme:

> a wanderer,
> a traveling singer,
> changing course 5 times in 5000 years.
> Overflowing its banks, flooding cornfields.

Here, personification embodies the river as unground and on the move, sometimes changing where it flows and, sometimes, changing how much the water rises. The performers continue, pointing out:

> Trailing pebbles and dirt downstream.
> The river lifted the land,
> held it by the hand like a lover rocking to and fro,
> and like a lover, let it go.
> That land made New Orleans.
> It made Bayou Teche.
> It made the words we now know.

Describing the river as responsible for the creation of New Orleans and Bayou Teche emphasizes the unheralded works of the Mississippi River as a powerful tool for invention. Moving seamlessly between personification of a lover, describing the creation of landforms, and imagining words "we now know" as part of the flow of the river, agency is imagined as including people, geography, and language.

The power of the river, however, is not without struggle. Acknowledging the gift of the river, they carry on into a building crescendo:

and we love that land.
We counted on it.
We built on its cabins, plantations, cities.
Cotton and Cane.
Back break and chain.
And when the water came.
AND WHEN THE WATER CAME!

In this final exclamation, the background chorus disrupts the flowing rhythm for sharp perturbations with the lyrics: "All the old places / All the old places / All the old places." We are invited to remember the loss, the trauma of Katrina as historical, as destructive, as heartbreaking as an ex-lover.

The story now narrates the river not as a traveler, but of an unruly force, "1844 harvest lost great sheaves of tobacco died. 1851 trees overtopped water too fast too wide. Too much money lost too much time. Congress said we have got to divide the land from the water." The story proceeds to begin describing the fate of the Mississippi River once Congress decided to "divide the land from the water." These lines highlight the moment when the United States took an historic departure from our previous relationship with the water. Indeed, we see the emergence of crisis discourse, inviting us to believe that what is necessary to save the region is to adopt radical change.

Tom Dulac describes the fate of the river as resting upon "two men, well two boys," who offer very different orientations to safeguarding against the river's tides. The first boy, "went to the bottom to feel the silt between his toes, make cutoffs, make jetties, let that river GO!" The second, "he said no. He had a plan and it started with an L and an E and a V E E, and that spelled money, and power, and pride." Here *CYO* is articulating the origin of Mississippi River management strategies, by focusing on the beginning of those perspectives. The first, rooted in knowing the river's contributions to the ecosystem, the second in knowing its contributions to the economy. The battle ensues, and "he won. The stronger son. The levees got built, the river got tamed made straight where it used to snake like a moccasin." We see the emergence of technological determinism emerging, where the river's fate shifts from considerations of the things it *is* doing to what it *can* do. A consideration made in favor of economic interest.

CYO concludes the river story with reflections on death, and the lessons that may be gathered from it:

You keep something confined and one of two things happens.
It fights or it dies, and lots of land died.
Everywhere you see graveyard trees,
knobby Cypress trees submerged in endless water.
And folks will say the river is angry, but rivers don't get angry.
They don't take revenge.

They don't contemplate what it is to be river.
They simply are.
Burbling and bubbling from somewhere unseen.
Busting through your door.
So tell me where is the man who told the river what the river is for.
He's gone to the tide. Buried by the sea.
Show me river what it means to be free, be free.
Can you hear it? Listen. Listen.

This final statement emphasizes that rivers often only become adversarial as a result of human action. When land dies, when rivers are "busting through your door," these are the results of risks taken in assuming command and control over a river's fate. These concluding thoughts are amplified by a chorus singing, "When the levee breaks, I will overtop/When the levee breaks, I will rise and fall." To understand what river (can) do, we must "listen."

The final major element of the introduction initiates the rest of the performance as a moment of what I would call aquatic commemoration, through a hymn sharing the title of the performance, *Cry You One*.

Out here,
you never know who's praying.
This air like a church invisible,
palmettos swaying their spindly fronds,
irises born hoping to be saved
among the many.
Some days it's just a heron
blinking way up there on a bald cypress branch
like the sky's white eyelid come down
a little closer.
This is a story of prayer.
This is what weeping looks like
in the bayou's full mirror.
This is the story of the winged-eyelid
come down
to cry you one.

Within this brief hymn are powerful elements of attunement, inaugurating the performance as a procession for the wetlands. The "air like a church invisible," establishes the upcoming journey as taking place on hallowed ground. Pointing out palmettos, irises, and a heron begins to point to the many nonhuman elements of the place that are vulnerable, and part of the loss. Finally, by explaining that this "is what weeping looks like in the bayou's full mirror," the audience is encouraged to keep an eye out for all those things to be celebrated and mourned, so that we may take the time to follow the performance's advice and "cry you one." The "Arrival" sets the stage for audiences to begin the journey that is *CYO*. It sets the tone for the occasion as epideictic, and it equips audiences with the tools to begin exploring the landscape wondrously.

Looking Back: Waves In, Waves Out

In the second act, Tom Dulac begins by describing the Mississippi River as an artery. Pulling up his sleeve and pointing to the veins in his arm he observes that as with those veins the Mississippi River is not a straight line, but it winds and twists with tributaries branching into it. He continues the metaphor by explaining that just as arteries bring nutrition and life to the entire body, the Mississippi brings the soil and sediment necessary for the sustained life of coastal Louisiana. Other cast members begin to chime in, noting that the artery of the Mississippi River has been weakened through transformation, struggling to bring the necessities that keep coastal Louisiana above water.

The cast continues using the human body to highlight coastal loss as Zelda, a member of a fishing community in the Gulf, notes that Louisiana has often been described as a boot-shaped state. She explains to the audience why this is not accurate any more, while showing how the foot has been disappearing by covering the foot of Tom Dulac's boot. These bodily metaphors, taken together, raise questions about what happens when certain elements of your body malfunction, or disappear altogether. After all, a weak artery or a missing foot are destined to change a body forever. These metaphors begin to highlight how a changing part of the world can have significant impacts to the rest of the world itself.

Music plays a significant role in drawing the audience into the copious dwelling of the *CYO* landscape. Frequently, the audience is invited to sing along as chorus, or to dance to the rhythm of drums and fiddle. Often, the music of *CYO* plays a crucial affective role in inhabiting the performative landscape, a potential for the spark of emotion emerging from the gathering of lyrics, of tune, of bodies together, and land as setting and emphasis. The audience is asked to sing alongside the cast in one particularly haunting line of lyrics, and to do so utilizing the musical form of a "round," where the different starting points for the melody are determined by totem. The melody follows, "Sinking, sinking, waves in, waves out / The sea, the wind, the waves / The land becomes sea." Described as a "Gregorian-chant-like refrain," we are invited to sing the song as we are about to encounter photographs of things lost in the disappearing wetlands of Louisiana (Bruin, n.d.). The rhythmic nature of the chant exposes that disappearances can occur slowly such that if you get caught up in the chorus of the everyday, you may not see what is happening outside of its song.

Near the end of the performance, Tom Dulac delivers a eulogy for the wetlands, which through copious description attempts to enunciate the loss felt by the region's deterioration. He attempts to account for all those things the wetlands represent, and those things that exceed representation:

> *Mon amour, tu te rappelles de toi-même dans un territoire sauvage*
> my love, I'll remember you a wild territory
> a tumble of languages ran through you like silted water
> lodged in your meander
> laid your bones upon our mud
> you stood in naked wonder at our sunsets, great tides
> of bird, of insects and plants, of estuarine species
> brackish waters teeming with untold life
> fish and shrimp and oysters' mysterious doings
> moving to and fro with each nocturne mystery
> I'll remember you a little boy
> ancient history
> no-one fathomed how much you'd have to give
> they would rip it from you, eventually
> bleed it from your downy vest.

With a declaration of love, Dulac begins to celebrate the "wild territory," for its abundance. The eulogy draws upon language that avoids encapsulating the fading presence of the wetland, instead pointing to "a tumble of languages," the "meander" of the wetlands, "waters teeming with untold life," and the "mysterious doings moving to and fro with each nocturne mystery." Each of these phrases points to a trace of something that exceeds our capacity for knowing and speaking it. This eulogy is an attempt to face an *aporia* of speaking for those things not present, by displaying language that points to the many ways that it is impossible to account for the loss of one's home. More importantly, it invites the audience to linger in this "wild territory," and to "fathom how much" the wetlands would "have to give." Looking back into the vibrancy of a dying landscape provides the audience with an informative cue for how to be better attuned to abundant happenings of the places they travel in/through.

Looking Around: The Ecotour

Upon checking in for the performance, audience members had been assigned separate groups allowing us to speak with our selected guides more closely. At this point, the audience of approximately fifty people split into those groups. I walk toward the Snakes.

The guide of the Snakes, Sabine Jean-Louis, identifies herself as an environmental justice advocate who had grown up in the Cancer Alley region of Louisiana. A black woman in her late twenties or early thirties, she attributes her passion for environmental justice to the hauntingly high cancer rates in the region that have claimed the lives of so many, including her father. Her demeanor is soft and caring, though she is often tasked with reminding the other characters that the burden of environmental pollution and disaster is disproportionately placed on minority communities. In all our Snake meetings, she takes on a friendly and caring role for her group.

After reintroducing herself, Sabine encourages us to offer quick introductions. We describe where we are from and what we enjoy about the places we live. I share with the group a couple of my "sanctuaries" outside of Bloomington, Indiana, where woodlands meet water. Others talk about the communities they live in, or how their Appalachian home is a great stage to learn and perform. Among the Snakes, there are approximately ten to fifteen of us, men and women, ranging in ages from young children to late fifties. Many in the group live nearby the performance site. Some are members of local performance groups. While listening to each other's introductions, we are also asked to reflect upon the lessons we have learned, not just from the land of our homes, but also the places we have been.

Following our introductions, Sabine begins to describe the toxic legacy of Louisiana, revealing the immensity of damages and death at the hands of the petro-chemical companies. The tone of our conversation is candid, but calm. Sabine informs us about the sickness and death happening along the "Cancer Alley" corridor; yet, she does so seemingly without anger. Her collected tone is one of reflection, and she asks us to reflect upon our own encounters with sickness, and how we deal with the enormity of circumstances beyond our own control. As we share our thoughts and experiences, she offers us fresh raspberries telling us they have many health benefits.

Meanwhile, I can hear other groups having different interactions. In the background of our discussion are riled voices of main characters frustrated by coastal loss in Louisiana; shouting occurs elsewhere. In other directions, I hear whispers of the science of climate change. In another direction, I notice distinctive movements, quick and active, as a guide moves his group around. It appears that all of us are attuned to some distraction—glancing around and hearing background sound—each group is aware of the fact that we are only hearing part of the conversation. As we know from the opening act, these four guides are not providing the same tour.

My attention returns to my group. As Sabine continues her conversation linking the toxins permeating the land and waterscapes of Louisiana, she asks for two volunteers. I and another audience member step up. We are asked to work together as a pair of lungs. Breathing in and out as other audience members watch us. Sabine informs us that free radicals have been introduced into our cells and tumors have begun forming. Breathing in and out becomes harder for us. At the same time, our guide describes the effects of the free radicals upon our cells. Our breathing begins to shorten and crack, worsening until there is only silence. The performance chooses to linger and dwell in the pain and suffering caused by cancer. Here, magnitude is addressed rhetorically not by seeking out macro-level statistics but rather by living in the polluted body of an individual. This interactive performance magnifies the chilling death resulting from the tiny toxic poltergeists—invisible—but always present. We bear witness and mourn the trauma of those who have suffered and those whose bodies that have already been marked to suffer. This moment of the performance renders present the toxic torment of the Louisiana landscape by figuratively amplifying the theme of sickness and death through the aggregation of trauma. It is haunting to embody these ailing lungs. Simply attempting to mimic distressed breathing for a few short moments is quite painful. I shudder to imagine living through this every day.

Sabine emphasizes the importance of awareness passing out a warm ginger tea to our group, informing us, as with the raspberries earlier, that "it is good for you because it contains antioxidants which combat cancer." The subtle gesture again points to the invisible presence of toxins throughout the land, water, and bodies of the Gulf Coast region. The ecotour with Sabine emphasizes that those things outside us are also within us, and we should be careful to recognize this fact. Meanwhile, a backdrop of activities not quite recognizable highlight that the horizon of the places we visit will always include events that elude our immediate perception.

Looking Forward: "The Answer Is in Your Dreams"

Once we have finished our tea, Sabine tells us each group will be offering a contribution to an altar for the disappearing wetlands and the rich, vital culture they support. The Snake group's contribution comes in the form of a whisper bottle. Using the bottles of tea distributed to us, she asks us to think about the lessons we have learned today, the lessons we have learned from our home, and the future that we hope to have, and then with a whisper offer those suggestions into the bottle. We will be placing our glass bottles onto the altar having filled them with a whisper advice in dealing with environmental degradation. Upon placing the bottle onto the altar, we are encouraged to shout out our advice for everyone to hear.

Other groups begin placing their contributions to the altar, such as ropes knotted together, and flowered plants gathered from the land. Some audience members shouting their suggestions such as "end our dependence on fossil fuels," or "develop sustainable agriculture." Others choose not to speak and dance to the music or stand. No one is compelled to offer more than they feel comfortable doing.

> As offerings are brought to the altar, the cast begins singing:
> You cannot own it, it belongs to itself . . .

> Crickets in the house of prayer
> Chirping hallelujahs there
> Up into the bowl of the sky.
> If we sing our songs unto the land
> And if we listen to this land,
> What would it say to us and why?

These lyrics are a capstone for lessons that *CYO* has brought before the audience; namely by returning to thinking about the places we live not as spaces to be conquered, but places to be cooperated with, *CYO* asks us to listen as we generate our own contributions, inviting us to feel kinship in a time of chaos.

A singular answer is not handed down by *CYO*. Instead, *CYO* invites polyvocal dialogue with the audience, with water, with land, with animals, and with all those with a voice. In celebrating the desire to move forward, the cast begins a chorus, and the audience soon follows along:

> The answer is in your dreams.
> The answer is in your dreams.
> Gotta problem that you can't solve,
> the answer is in your dreams.

CYO encourages us to be inspired by the places we live and the cultures we take part in from our many different perspectives. Now is the time, they suggest, to take those inspirations seriously, to draw upon all the resources of the places we dwell and to copiously take up the task of imagining a future that no longer suffers the devastating loss facing coastal Louisiana.

The performance ends with dusk at the horizon. The cast stands before the audience atop a temporary stage. One by one, each cast member transforms into a totem, beginning their dances of "bio-mimicry" as the beat of drums fill the Appalachian woodland. With the final beat, the audience—including myself—erupts into emotional applause. The cast, no longer in character, thanks the audience and invites all of us to take part in a gumbo dinner, where we may begin the work of reflecting upon the lessons of the last two hours.

CONCLUSION: EMBODIED, POLYVOCAL ABUNDANCE AS COPIOUS DWELLING

In this chapter, I have explored the epideictic rhetoric of *Cry You One*. While it may not be surprising that a live performance of southern Louisiana and the Mississippi River might mourn loss, what was striking to me was how the performance also encouraged polyvocality, embodiment, and a sense of what

I call "copious dwelling" to celebrate our history, capacity to act, and attunement to the many ways of living in the world.

In terms of polyvocality, the performance shared conflicts among four primary perspectives and encouraged audience participation to open dialogue even more. *CYO* privileges polyvocality because it emphasizes that the exigence of dying wetlands exceeds the abilities of any individual expertise to overcome. The infighting among the tour guides serves as a reminder of the sublime challenges the region faces. That no character has a monopoly of voice or answers acknowledges the limits of individual perspective and the necessity of community-driven efforts. Moreover, the many instances where the guides turn to the audience for our experiences places a premium on voice and reflection as techniques for moving forward in the face of extreme change.

In addition to the many voices shared and invited through *CYO*, the performance made its liveness matter. That is, the show emphasized our location and extemporaneously evolved to foster a sense of agency in those of us attending. *CYO* showcased presence, to be present. However, it was not enough to simply see and to know what is happening in coastal Louisiana, but also to feel and be a part of it. Hence, *CYO* relied greatly upon sharing—sharing culture, sharing stories, and even sharing food—in the hopes of building a bond between audience and cast to witness something larger. The "presence" showcased by *CYO* is, as Pezzullo (2007) explains, one that refers to "the *structure of feeling* or one's *affective* experience when certain elements—and, perhaps, more importantly, relationships and communities—in space and time appear more immediate to us, such that we can imagine their 'realness' or 'feasibility' in palpable and significant ways" (9). The felt here-and-now of *CYO*'s epideictic performance amplifies our very being-amidst worldly things.

Bringing together voice and embodiment as they did, *CYO* fostered a sense of copious dwelling by highlighting how communities—those of coastal Louisiana as well as the one forged in the performance—are gatherings *in situ* (or in place). The performance drew on an abundant number of connections, and invited us to feel the significance of how living in a place shapes us and exceeds us.

These communities are not possible without the "radiance" of things bearing their mark on the everyday. *CYO* pays tribute to the gathering by bringing an abundance of voices together, and the performance has us visiting the "radiance" of this coastal/Appalachian performative landscape. Here, in the liminal region of *CYO*, a new "imagined community" is formed (Anderson 1991). Cast and audience members share their experiences of environmental destruction created by humanity's reckless exploitation and use of fossil fuels.

Tom Dulac, a former oil-rigger turned cleanup worker, voices frustration over the number of industrial disasters occurring in the region, and he points out that just as coastland is disappearing at an alarming rate, so too are mountains in Appalachia due to mountaintop removal mining. Appalachia and Louisiana not only share water and soil, but also loss. In this two-hour performance, *Cry You One* bridges the gap between Appalachia and Coastal Louisiana. They address the dilemma of how to make a distant location feel more present by connecting it with the ground under our feet. If we can feel "here," perhaps, we will find the capacity to care about "there."

Through alternating moments of celebrating the history, power, ecological functions, and cultural importance of the Mississippi river, and mourning the manipulation and destruction of the river caused by humanity, *CYO* conjures before the audience a vibrant—yet dying—coastal Louisiana. By rhetorically invoking both landscapes as part of a greater watershed, where flows—ecological and cultural—become a defining quality of both places, the scope and scale of the audience's experience is saturated with elements near and far. This saturation carries with it a weight to explore and recognize our larger connectedness as beings-amidst one another and water.

Beyond this performance, copious dwelling is a temporally and spatially abundant mode of being that rhetorically cultivates a powerful mode of encountering things. Far from reducing the world to a storehouse of "useful" objects, copious dwelling reveals a world of things that may inspire. Things that, with acknowledged rhetoricity, can prove invaluable in seeing the world anew. In this regard, *copious dwelling* is a powerful mode of aquatic commemoration insofar as it privileges the importance of being "immersed" and "amidst" water, amplifying a "consciousness" of water (Shiva 2008, 500). Most importantly, it seeks to create a culture that respects and cooperates with water, instead of conquering and carving it up.

REFERENCES

Anderson, Benedict. 1991. *Imagined Communities: Reflections on the Origin and Spread of Nationalism*. London: Verso.

Berlant, Lauren. 2007. "Slow Death (Sovereignty, Obesity, Lateral Agency)." *Critical Inquiry* 33, no. 2: 754–80.

Berlant, Lauren. 2011. *Cruel Optimism*. Durham, NC: Duke University Press.

Berlau, John. 2010. "Crying a River for Years." *Wall Street Journal*, December 15, 2010. https://cei.org/content/crying-river-years.

Boon, Marcus. 2011. *In Praise of Copying*. Cambridge, MA: Harvard University Press.

Bruin, David. n.d. "Music and Movements: *Cry You One* at the International Festival of Arts and Ideas." *HowlRound.com*. http://howlround.com/music-and-movements-cry-you-one-at-the-international-festival-of-arts-ideas.

Bullard, Robert D., and Beverly Wright. 2012. *The Wrong Complexion for Protection: How the Government Response to Disasters Endangers African-American Communities*. New York: New York University Press.

Conley, Thomas M. 1985. "The Beauty of Lists: *Copia* and Argument." *Journal of the American Forensic Association* 22, no. 2: 96–103.
"Cry You One." 2016. *Cry You One*. http://www.cryyouone.com/.
Davenport, Coral, and Campbell Robertson. 2016. "Resettling the First American 'Climate Refugees.'" *New York Times*.
Edwards, James C. 2002. "Poetic Dwelling on the Earth as Mortals." In *Heidegger Reexamined: Art, Poetry, and Technology*, vol. 3, edited by Hubert L. Dreyfus and Mark A. Wrathall. New York: Routledge.
Erasmus, Desiderius. 1963. *On Copia of Words and Ideas*. Translated by Donald B. King and H. David Rix. Milwaukee: Marquette University Press.
Fleetwood, Nicole R. 2006. "Failing Narratives, Initiating Technologies: Hurricane Katrina and the Production of a Weather Media Event." *American Quarterly* 53, no. 3: 767–87.
Foucault, Michel. 1972. *The Archaeology of Knowledge*. Translated by Allan Sheridan. New York: Harper Row.
"Gov. Edwards Declares State of Emergency for Louisiana Coast." 2017. *Office of the Governor of Louisiana*, April 19, 2017. http://gov.louisiana.gov/assets/EmergencyProclamations/43-JBE-2017-Coastal-Louisiana.pdf.
Harman, Graham. 2007. *Heidegger Explained: From Phenomenon to Thing*. Chicago: Open Court.
Hawhee, Debra. 2017. *Rhetoric in Tooth and Claw: Animals, Language, Sensation*. Chicago: University of Chicago Press.
Heidegger, Martin. 1971. *Poetry, Language, Thought*. Translated by Albert Hoftstadter. New York: Harper Perennial.
Hyde, Michael J. 1983. "Rhetorically, Man Dwells: On the Making-Known Function of Discourse." *Communication* 7, no. 2: 201–20.
Klein, Naomi. 2007. *The Shock Doctrine: The Rise of Disaster Capitalism*. New York: Picador.
MacCash, Doug. 2013. "'Cry You One,' a Theatrical Wetland Tour, Premieres Saturday in St. Bernard." *Nola.com*, October 23, 2013. http://www.nola.com/arts/index.ssf/2013/10/cry_you_one_theatrical_wetland.html.
Marshall, Bob. 2017. "Losing Ground: Southeast Louisiana Is Disappearing, Quickly." *Scientific American*. https://www.scientificamerican.com/article/losing-ground-southeast-louisiana-is-disappearing-quickly/.
Moss, Laura. 2010. "The 13 Largest Oil Spills in History." *Mother Nature Network*, July 16, 2010. https://www.mnn.com/earth-matters/wilderness-resources/stories/the-13-largest-oil-spills-in-history.
Nixon, Rob. 2013. *Slow Violence and the Environmentalism of the Poor*. Cambridge, MA: Harvard University Press.
O'Gorman, Ned. 2005. "Aristotle's Phantasia in the Rhetoric: Lexis, Appearance, and the Epideictic Function of Discourse." *Philosophy and Rhetoric* 38, no. 1: 16–40.
Peeples, Jennifer. 2011. "Toxic Sublime: Imaging Contaminated Landscapes." *Environmental Communication: A Journal of Nature and Culture* 5, no. 4: 373–92.
Pezzullo, Phaedra C. 2003. "Touring 'Cancer Alley,' Louisiana: Performances of Community and Memory for Environmental Justice." *Text and Performance Quarterly* 23, no. 3: 226–52.
Pezzullo, Phaedra C. 2007. *Toxic Tourism: Rhetorics of Pollution, Travel, and Environmental Justice*. Tuscaloosa: University of Alabama Press.
Pezzullo, Phaedra C. 2009. "Tourists and/as Disasters: Rebuilding, Remembering, and Responsibility in New Orleans." *Tourist Studies* 9, no. 1: 23–41.
Quintilian. 1921. *The Institutio Oratoria of Quintilian*. London: W. Heinemann.
Rich, Nathaniel. 2014. "The Most Ambitious Environmental Lawsuit Ever." *The New York Times*, October 2, 2014. http://www.nytimes.com/interactive/2014/10/02/magazine/mag-oil-lawsuit.html?module=Search&mabReward=relbias%3Aw%2C{%222%22%3A%22RI%3A14%22}&_r=0;.
Rickert, Thomas. 2013. *Ambient Rhetoric: The Attunements of Rhetorical Being*. Pittsburgh: University of Pittsburgh Press.

Shiva, Vandana. 2008. "From Water Crisis to Water Culture" (Interview with Andy Opel). *Cultural Studies* 22, no. 3/4: 498–509.

Sipiora, Michael, P. 1991. "Heidegger and Epideictic Discourse: The Rhetorical Performance of Meditative Thinking," *Philosophy Today* 35, no. 3: 239–53.

Sloane, Thomas O. 1997. *On the Contrary: The Protocol of Traditional Rhetoric*. Washington, DC: Catholic University of America.

Tibbets, John. 2006. "Louisiana's Wetlands: A Lesson in Nature Appreciation." *Environmental Health Perspective* 114, no. 1: A40–A43.

"What Is a Wetland?" n.d. Environmental Protection Agency. https://www.epa.gov/wetlands/what-wetland.

Chapter Ten

It's All Child's Play

Flint's Water Crisis, Environmental Justice, and Little Miss Flint's Ephebic Rhetorics

Christopher Scott Thomas

Reports from residents of dirty, foul-smelling water started soon after April 2014, when, in an effort to save money, the state of Michigan switched Flint's water source from Detroit's water system to the Flint River. The decision worried many residents, who were concerned with using water from a tributary notorious for its filth (Carmody 2016). In a series of unfortunate events, the water crisis in Flint quickly became a national discussion that exposed the local and state government's role in poisoning residents of one of the country's poorest cities (Martinez 2017). In fall of 2014, an outbreak of coliform bacteria was detected in Flint's tap water to which city officials issued a boil-water advisory. In an effort to quell rising concerns that the bacteria was due to chemicals in the water, the Michigan Department of Environmental Quality blamed the outbreak on cold weather, aging pipes, and a population decline (Lin, Rutter, and Park 2016). Soon after, the General Motors plant in Flint stopped using municipal water and cited its usage eroded car parts (CNN 2017). It was not until January 2015 that the city of Flint publicly acknowledged the citizens' concerns and declared that the city's water contained high levels of trihalomethanes, a by-product of disinfectant used to treat the water. What ensued was a community-wide conversation among citizens, government officials, scientists, doctors, and activists over urban water management and fair and equitable access to clean water.

While policy discussions regarding urban water management are often highly "technical" and require a certain level of expertise to participate, residents of Flint employed whatever rhetorical resources were available to raise awareness of the toxic water and hold the local and state government accountable for their negligence. In addition to using traditional means of political engagement (attending meetings and hearings, lobbying for policy changes, utilizing government agencies), residents also protested at mayoral and senatorial offices, held public performances of resistance, and many joined community-based social and environmental organizations. Most notably, the citizens of Flint sought to raise awareness of the water crisis, hold government officials accountable for their negligence, and provide their community with access to clean water. Lee Anne Walters and Melissa Mays, for example, began *Water You Fighting For?*, a local organization of women who "gather to gain and disperse information to improve Flint's community" (Wateryoufightingfor.com). Flint's water crisis is a fruitful case study to explore the means with which residents transform from private citizens into political activists, engage in local struggles over urban water management, and fight for a voice in the decision-making process. Constrained by limited resources, time, and "expert" knowledge, these advocates for environmental sustainability and social justice put to use whatever resources were available to protect their health, their homes, their communities, and their environments.

Flint's water crisis is, by many scholars' and activists' definition (Bryant and Mohai 1992; Bullard 1993), an environmental justice issue because it concerns the inequitable distribution of environmental hazards and goods, as well as the unfair treatment and limited involvement of all people with respect to the development, implementation, and enforcement of environmental laws, regulations, and practices. And while much of the narrative surrounding the water crisis focuses on institutionalized discourses, such as government accountability and the city's failing infrastructure, the rhetorical labor of Flint's citizens reveals the constitutive forces that make possible environmental justice. Through various rhetorics of social change and protest, residents identified common ground in their concerns over clean water, built networks of dedicated community members and activists, constructed and advanced appeals for sustainability and justice, and inspired mutual critique in others. This chapter more closely explores the rhetorical agencies taken up by Flint's residents over the (mis)management of the city's water and the various articulations and performances that expand the very definitions of the "environment" and "justice."

Among this small but vocal group of community activists is Amariyanna Copeny—also known as Mari—who, in 2015, won a local pageant at the age of eight in the midst of the water crisis in Flint, Michigan. Copeny winning the title of "Little Miss Flint" was an exigence in the city's water crisis that offered the eight-year-old opportunities to raise awareness of water justice issues and speak on behalf of the children in the community. Though, it was a letter Copeny sent to then President Barack Obama calling attention to

Flint's water crisis that propelled her into regional and national spotlights. Whereas Copeny's circulation in public discourse largely conceives of her as evidence of the environmental injustice occurring in Flint, her grassroots organizing indicates there is much more to her rhetoric than purely victimhood; she encourages community engagement, brings national attention to urban water management, raises money to improve children's education and health, and chastises officials for their negligence and complicity in the city's water crisis. Copeny's activism in Flint's water crisis offers a chance to move beyond the notion that children are apolitical victims of environmental injustice and, instead, suggests that the traditional demands of institutional democratization are, in fact, child's play.

The following chapter is organized into three sections. First, I detail the rhetorical strategies and tactics of community-level environmental justice movements and the enabling possibilities for children and young adults to mobilize social and political change. While this chapter focuses predominantly on children activists, this section will also provide a general overview regarding environmental justice activism and grassroots campaigns. Second, I examine the performative repertoire of Little Miss Flint—specifically Copeny's use of youthful, ephebic knowledge and digital activism—to argue she reconstitutes herself into a rational speaking subject capable of influencing environmental justice efforts. And third, I conclude by discussing how the case study of Little Miss Flint offers scholars interested in water justice opportunities to test the boundary conditions of rhetorical studies by considering children—and, in this instance, a female child of color—as genuine and capable speakers of social change and protest.

ENVIRONMENTAL JUSTICE, COMMUNITY ENGAGEMENT, AND CHILDREN'S ACTIVISM

As a social movement and scholarly endeavor, environmental justice contends that people of color and low-income communities bear a disproportionate burden of environmental pollution and toxins when compared to white, affluent neighborhoods (GAO 1983; United Church of Christ 1987; Bryant and Mohai 1992; Bullard 1993). Activists and academics have spent considerable time identifying the various linkages between social justice concerns (e.g., housing, food, health) and environmental issues (e.g., toxins, waste, pollution). For much of the canonical research on environmental justice, "justice" is conceived of as environmental equity—the principle that environmental problems should not be disproportionately concentrated in socially, economically, or politically disempowered communities (Bryant and Mohai 1992; Bullard 1993). Studies that explored the geographical processes that produce such disparate impacts surfaced in the 1980s and provided evidence of environmental inequity with regard to the racial and class disparities in the distribution of environmental hazards and goods. The findings of these studies, such as those produced by the U.S. General Accounting Office and the United Church of Christ, were the first to document environmental inequity and enabled environmental justice activists to employ civil rights discourses to advocate for changes to federal environmental policies and procedures. Beyond inspiring future scholars to remain invested in environmental justice efforts, the research assisted in helping expand the very definition of the environment to include the places people work, play, live, and worship.

While the concept of environmental equity has been fundamental to developing environmental justice as a critical enterprise, its emphasis on distribution is called into question because of its causal claims of injustice (Been 1993; Been 1995). Evidence showing the proximity of toxic facilities to certain groups of people, Been (1993; 1995) argues, is unable to prove whether the placement of the facility was due to the disempowered community who lives there already or whether the placement of the facility lowered land value and, as such, drew in lower-income communities who are often people of color. Environmental equity operationalizes injustice as isolated (placement of toxic facility) and intentional (in an area that is already marginalized) acts of discrimination (explicit). Robert Bullard (1994) argues, conversely, that this "chicken-or-the-egg" question is not an important one to ask because it does not account for the nonlegal discrimination against people of color, or what he and others refer to as institutional racism (60–62). Laura Pulido (1996), in her critical review of environmental racism research, agrees and argues that distributive frameworks fail to account for all scales of injustice because they presume, for example, what racism is and how it is experienced and "that there exist[s] a single form of racism responsible for environmental discrimination" (143). As a result of their effort to operationalize and establish methodological rigor for research on environmental justice, scholars have ignored the "inherently complex" and "contradictory ideology" at play in a given environmental justice struggle (Pulido 1996, 148). This chapter moves beyond interrogating the causes of Flint's water crisis as it relates to racism and/or classism. Instead, this chapter examines the rhetorics of social change and protest that citizens employ as they make sense of the water crisis and ennoble an individual and community-wide sense of environmental justice.

Within a purely distributive framework, the structural and institutional forces that determine what and how certain environmental goods and hazards get allocated are minimized and, at times, overlooked as important contextual evidence of injustice. Iris Marion Young (1990) criticizes distributive frameworks because they "ignore the social structure and institutional context" that come to shape allocation patterns (15). Structures and practices, in addition to the rules and norms that guide them, are mediated by language, symbols, and social interactions that condition "people's ability to participate in determining their actions and their ability to develop and exercise their capabilities" (Young 1990, 22). Thus, to understand environmental justice holistically is to also consider the process by which speakers collaborate and articulate collective calls for social, political, and environmental change. As Phaedra Pezzullo (2001) reminds us, "It is not enough to gain a place at the environmental decision-making tables" because "[once] those spaces are made…there are substantive challenges that constrain citizens from having effects" (3). Environmental equity is a concept that focuses on the *outcome* of decision-making, ignoring that there are constraints activists must navigate in order to affect what and how certain goods and hazards are allocated in their communities. Flint's water crisis offers a case study to begin addressing this shortcoming and move beyond classification to, instead, detail "the nature of subsequent negotiations between citizens and institutions" in their battle over what should be considered the "environment" and "justice" (Pezzullo 2001, 3).

Speakers and groups of speakers must navigate a litany of constraints that impair their capabilities during community-level environmental justice disputes. As a situation "plagued with uncertainties, almost all of which work against citizen activism" (Peeples and DeLuca 2006, 60), it is an onerous uphill battle for citizens. Juliana Maantay (2002), in her analysis on industrial zoning laws, argues that citizens, while given opportunities to engage in zoning decisions, are quite diminished in their appeals for justice because of a perceived lack of "technical" knowledge. In response to these constraints, often marked by an urgency to protect themselves, their families, and their community, activists employ practices and performances of resistance that help gain visibility for their concerns, critique dominant structures, craft collective identities, and encourage community participation. Environmental justice rhetors, argue Peeples and DeLuca (2006), encourage audience members to "trust in the skills and knowledge they already inherently possess to fight the battle for their children" (74–75). And because citizens exist in such a "difficult rhetorical situation," advocates must also "present their case with such unquestionable support that their claims will be seen as 'true' even though they may not be backed by objective, scientifically verifiable 'fact'" (Peeples and DeLuca 2006, 75).

Activists use the knowledge gained from their own bodies and from the community to formulate structural and institutional criticisms as well as convince others to share similar reflections and critiques. For instance, Lee Anne Walters was not the only resident concerned by the city's water advisory in January 2015. Flint residents began attending city council meetings and complained of the water's orange-brown color and ghastly smell, showing photographs of skin rashes and testifying about the loss of their hair (Lurie 2016). Many residents, including Walters, brought milk jugs filled with their tap water as evidence of the environmental consequences and human injustices of unsanitary water (Smith 2015; McQuaid 2016). Flint residents utilized personal stories, anecdotes, and experiences as proof that the city's water was dangerous, countering governmental "expert" discourses that claimed otherwise. Body and local knowledge illustrate the capacity for activists to use such evidence to support claims of injustice and empower others with similar experiences to speak out.

Beyond truth being constructed from individual experience, environmental justice activists use collective truths gained through a sense of place; that is, truth is also constructed "through shared location and experiences" (Peeples and DeLuca 2006, 80). For example, a group of women protesters staged a "die-in" on the steps of the Flint Water Plant on May 16, 2017, in an effort to raise awareness of the surge in miscarriages and other women's health issues in the area. Many participants dressed in white painter's suits covered in red paint to present a visual representation of the damage caused to women's reproductive systems. "I shared my story of losing twins during the water crisis because it is essential that people know that this is not a short-term, painless ordeal that we are going through in Flint," said one concerned Flint mom (Duell 2016). Beyond the material markers that signified the women's health struggles caused by the city's toxic water, individual women's experiences collectivize when shared together and reveal a broader, more structural understanding of environmental injustice and inequity. As a result, women's body and local knowledge is renovated into a collaborative experience to expose environmental injustice occurring similarly across various homes in the community. Women's body knowledge (loss of children) and community truths (long-term struggle with unsanitary water) coconstitute an understanding of Flint's water crisis that relies less on scientific expertise but, rather, on women's available means of persuasion to construct the truth of the matter.

While rhetorical scholars (Peeples 2003; Peeples and DeLuca 2006; Sowards 2010; Murphy 2017; Thomas 2019) focus on the strategies and tactics of women—particularly mothers—in constructing truth beyond empirical, scientific means, similar agencies can be extended to other groups of people to better make sense of and address environmental justice issues. One particular group—children—utilize similar rhetorical tactics to engage in political systems and effectuate social and political change, even when they are often seen as apolitical victims of injustice. Kevill-Davies's (2018) analysis of the *Fighting the Red Menace* collectable card set finds that children are "capable of understanding the geopolitical world in which they live" and "may be seen as having the ability to identify and position themselves within ideological narratives" (239). Even with increasing recognition of children and childhood in rhetorical studies, the research focuses primarily on societal constructions of children and/or objects produced for children. In an effort to contribute to these growing conversations, this chapter explores the ephebic rhetorics—symbolic action enabled and constructed by, from, and for children and young adults—renovated and harnessed *by children* to engender environmental justice.

The body knowledge and community truths that compose much of environmental justice advocacy go well beyond its use by adults and include the rhetorical strategies and tactics utilized by children to raise issue awareness, challenge those who threaten their families and communities, and inspire mutual critiques and reflections in others. Thus, while children and young adults are limited in their exercising of resistance and empowerment due to a lack of access to the traditional means of political engagement, they may find sources of agency in those very constraints. Kang (2017) finds in her analysis on Korean youth protest movements that teenagers "circulated messages and enacted performances that captivated them, built affective networks and temporary alliances surrounding these objects of captivation, and respond to authority with irreverent subversion" (751). Through the assemblage of a "shared yet underarticulated fear of contaminated beef," Korean youth developed a "potent critique" of neoliberal policies "without adopting the language of institutional politics" (Kang 2017, 752). Thus, children's political sensibilities represent a significant departure from traditional rhetorical studies on social and political change and childhood and suggests that the youthful position of children is an enabling mechanism to articulate body knowledge and community truths.

FLINT'S WATER CRISIS AND LITTLE MISS FLINT

In early 2016, after Flint residents were encouraged to use bottled water, Mari Copeny began "hearing about the crisis" and "felt she wanted to speak up" (Suggs 2019). With her "keen sense of bringing attention to issues," Copeny entered into the trenches of environmental justice activism and over the past three years raised more than $350,000 toward Flint's recovery. She also collected and distributed more than 500,000 bottles of water, over 550 bicycles, and 15,000 backpacks filled with school supplies (Suggs 2019). However, Copeny's entrance to water justice activism occurred in March 2016, when she sent a letter to then President Barack Obama asking to visit with him during a community-wide trip to Washington, DC, where Michigan Governor Rick Snyder was testifying before Congress. What ensued was a life-changing meeting that would propel Copeny, as "Little Miss Flint," into a national spotlight and provide a national and international stage to raise awareness and help Flint's water crisis.

Child of the Copeny Family

Much of Little Miss Flint's style of activism is grounded in and informed by her positionality as a child. In her letter to President Obama, Copeny both acknowledges the constraints her age puts on her ability to effectuate political change as well as recognizes her significance in helping understand the severity of Flint's water crisis. She entreats the president:

> I know this is probably an odd request but I would love for a chance to meet you or your wife. My mom said chances are you will be too busy with more important things, but there is a lot of people coming on these buses and even just a meeting from you or your wife would really lift people's spirits. (*Los Angeles Times* 2016)

Copeny's sense of justice is organized by her role as a child—both in her individual family unit and the Flint community—requesting a meeting with the president while also understanding the difficulty in doing so given Obama's previous engagements. Beyond being just "one of the children . . . affected [*sic*] by this water" (*Los Angeles Times* 2016), Copeny crafts herself into an advocate for her community—joining others, presumably adults, in a cross-country trip to protest then Governor Snyder during his testimony to Congress over Flint's water crisis. And, these political capabilities are further empowered by Obama's response:

> You're right that Presidents are often busy, but the truth is, in America, there is no more important title than citizen. And I am so proud of you for using your voice to speak on behalf of the children of Flint. (Murray 2016)

Copeny is bestowed a greater authority to speak about the water crisis through Obama's presumed presidential authority. Moreso, his response simultaneously acknowledges Copeny's constrained positions as a child while also maintaining the significance of her civic performances.

Little Miss Flint articulates discourses of social change and protest that encompass a generalized ethic of care that includes the children of Flint who are also impacted by the water crisis. In particular, Copeny utilizes ephebic rhetorics that rely on both body knowledge and community truths to raise issue awareness, hold public officials accountable for their negligence, and inspire similar reflections and critiques in Flint residents. Her ability to speak on behalf of the children of Flint helps empower particular agencies that rely on body knowledge and community truths and permits Copeny access to the reigns of political engagement often removed from children. While much of Copeny's public persona is framed by media outlets, she also works to inform that public persona with an explicit focus on her childlike experiences and the impact toxic water has on children. When interviewed about her daily experiences with dirty water, Copeny explains that Flint's water crisis "has changed almost every aspect of our lives" (Richardson 2017). For Copeny, children are constantly confronted with a lack of access to clean water from not being able to "get a glass of water without dealing with bottled water" to being unable to "play in the water in the summer time" (Richardson 2017). Utilizing whatever available resources, Little Miss Flint adjudicates the devastation of the water crisis through her inability to enjoy childhood daily activities, such as playing in the water during summer. She is able to draw attention to the trickle-down effect of tainted municipal water—an issue inherently structured and informed by governmentality—as being relevant to the most vulnerable members of the Flint community: children.

A Child of the Flint Community

Copeny's role as a child—in particular, her relationship to a larger family unit and community—acts as an organizing base around which to advocate for environmental justice. Even with being "just a normal kid that loves to play with my American Girl dolls, read comics, draw and cheer," Copeny crafts herself into a spokesperson who speaks out on Flint's water crisis "because the water is still causing issues for so many people" (Battoclette 2018). The evoking of childhood activities, such as playing with dolls, is simultaneously linked to Copeny's political activism and reminds audiences that Flint's water crisis *is so bad* that it demands even children's attention and labor to resolve. However, Little Miss Flint is unphased by the potential constraints to her political activism: "I want people to know that kids can change the world, you don't have to be grown to do it" (Richardson 2017). Throughout much of Copeny's activism in Flint, she makes clear that children are a powerful force for social and political change. Her own actions are meant to show that "even a kid like me can make a difference" and make sure "people don't forget about the people living in my city" (Richardson 2017). Copeny enables other children to see themselves in her own reflections and critiques with the hope they "realize they are above the water crisis and . . . can live their dreams" (Richardson 2017). Beyond being the grounds for her entrance into public conversations and movements for clean water, Copeny constructs a sense of truth about Flint's water crisis around her body knowledge (physical signs of toxic water), community truths (the neighborhood's history of neglect and injustice), and her ephebic experiences (toxic water's impact on children's everyday lives).

For many mothers struggling for environmental justice, their knowledge and experiences of being caretakers, providers, and life-givers both propel them into activism (Krauss 1993; Di Chiro 1993; Kurtz 2007) and ground their strategies for social and political change (Peeples 2003; Peeples and DeLuca 2006; Thomas 2019). The "canary cry" of mothers raises awareness to the severity of an environmental injustice because it flags the substantial impact it has on children. In addition, motherhood is a constitutive force that provides women a maternal *ethos* to speak in the name of and on behalf of children as well as the broader community dealing with environmental toxins and disturbances (Peeples 2003; Murphy 2017; Thomas; 2019). And while rhetorical scholars have examined motherhood in relationship to environmental justice activism, limited research exists extending similar agencies of social and political change to younger adults and children. In one noticeable example, a video produced by Purpose and Footage Films to support the 2017 People's Climate March shows Copeny, also called "Little Miss Rambo" in the video, painting a yellow sun on her face, adding glittery pink detail to a protest poster, lacing up her neon pink tennis shoes, and tying a bandana around her forehead reading: RESIST (Hill 2017). Copeny clarifies to the audience that she is joining the march for her brother and sister, for the world, for America, and for drinkable water—statements that bring together local, regional, and international spheres under a common struggle for access to clean water. Here, Copeny's activism is militarized through ephebic discourses and symbols and renovated into a platform with which she is able to speak on behalf of her family. Flint's water crisis acts as an exigency with which to show that being a "good child" means performing activities that may seem contradictory to traditional understanding of what is required of children in and outside the home.

Copeny expands the scope of Flint's environmental justice struggle beyond her individual family unit to encompass the effect it has on the community's children. Describing herself as just a "Flint kid that wants to help all the other Flint Kids" (Lewis 2017), Copeny reconstitutes herself from member of her individual family to community representative speaking on behalf of neighborhood children. And while she speaks for "all the residents of Flint," Copeny sees her primary task as speaking "mostly for the kids" (Richardson 2017). For example, she launched two letter-writing campaigns focused on protecting and nurturing the children of Flint impacted by the water crisis. The first campaign sought to raise money for children in the neighborhood to see the popular film *Black Panther*, raising over $16,000 to rent out a 150-seat theater and purchasing more than six hundred tickets. "Kids need to see themselves as superheroes," Copeny explained to the *Washington Post*, "Black kids are seen as victims—and we're not" (Klein 2018). Enlarging the impact of the water crisis, Copeny reveals the lasting impact environmental injustice has on physical and mental health but also the significance of providing Black youth modes for resistance and empowerment. As a result, Flint's water crisis is framed less as an isolated government failure and, instead, is placed within a broader context of racial and economic inequity.

The second campaign, flagged by the #DearFlintKids hashtag, asked people from all over the nation to write a letter to a child in Flint, letting them know that someone somewhere cares about them. Comparing it to the feeling of receiving cards and letters on her own tenth birthday, Copeny "wanted every kid in Flint to share that feeling and to be reminded that the world sees them, supports them, and loves them" (Thomas 2018). A sense of self-worth and value is reaffirmed by the collaborative project, one that moves the water crisis beyond the city and works to bring people together over shared concerns for Flint's children. Instead of that sense of self-worth and value being "given" by adults, Copeny expands the boundaries of rhetorical agency to include children and their enabling mechanisms for social and political change in saying, "I want them to know that I love them, and I want them to know the world thinks they can do anything they put their minds to" (Thomas 2018). More than being victims of Flint's water crisis, Copeny facilitates a communal and collaborative mode of activism that includes the potential agencies of resistance and empowerment generated by ephebic knowledge and experience.

What is remarkable about Copeny's water justice activism is how she turns her position as a child into an *ethos* that speaks on par with politicians, policy makers, and scientists. While it is reasonable to dismiss Little Miss Flint due to her minimal life experience and lack of formal education, particularly on issues of urban water management, there is also an enabling mechanism embedded in Copeny's ephebic, childlike rhetorics that ennobles a legitimate perspective, clarity, and authority on the topic. In articulating the ways dirty water impacts the everyday lives of children, its effects on the health and well-being of home and family, and placing these experiences among the arc of injustice in Flint, Copeny constructs and advances arguments unlike many others. Her positionality as a child is unassuming and not threatening, a presumption that provides credibility to speak truth to Copeny's and others' experiences of environmental inequity.

Limitations to Little Miss Flint's Ephebic Rhetorics

While Copeny is lauded for her success in raising awareness of Flint's water crisis and continued labor in building up and empowering her community, there are limitations to her activism that are important to acknowledge. First, Copeny often faces dominant discourses of race, sex, and age to depoliticize her rhetorical labor. Most noticeably, Copeny is described physically, behaviorally, and pragmatically as "adorable," "cute," and "precious." Numerous headlines, newspaper descriptions, and interviewers remark on Copeny's style of activism in ways that overemphasize her gender-bound and childlike docility, and underscore the nuances of her political labor. Even those who seemingly are in support of Copeny are guilty of this move; changing Copeny's moniker to "Little Miss Rambo," for example, plays off the popular franchise starring Sylvester Stallone to comment on both her overtly feminine performance and her sense of tenacity when it comes to protecting Flint. While only a minor example of how ephebic rhetorics can constrain children's activism, they also reveal the ways gender and age intersect in ways that overlook how youth play a significant role in contributing to social and environmental movements.

In another example, Copeny met with Donald Trump in September 2016 to discuss Flint's continued lack of access to clean water. The then Republican presidential candidate spoke at the Bethel United Methodist Church in Flint, to a crowd including Copeny and her mother, explaining what "he can offer the city if he becomes elected" (McNeal 2016). However, the discussion of the city's water crisis was cut short by the church pastor as Trump attempted to deride Democratic presidential candidate Hillary Clinton (Garcia 2016). What circulated after the talk was not potential policy changes but, rather, a particular image of Copeny posing alongside Trump. The picture shows Trump bending forward—almost to match Copeny's smaller stature—and smiling while holding onto her wrist. Copeny stands underneath Trump's hunched frame with a haunted look: her mouth closed, eyes wide, eyebrows burrowed, and arms closely held against her body, as if to protect herself. While she wanted to talk about Flint, news outlets overemphasized Copeny's appeared "terrified," "uncomfortable," and "unhappy" appearance and demeanor (Garcia 2016). The image circulated as political commentary on Trump's failure to connect with citizens, and Copeny's response was reconstituted into a token representing the general public's concerns with a potential Trump presidency.

It would also be a mistake to not comment—even if briefly—on the racial discourses also at play when describing a young black female as "unhappy." The trope of the angry black female is one discussed by many scholars (e.g., Walley-Jean 2009; Madison 2009; Harris-Perry 2011), contending the discourses construct black women as hyperemotional, irrational, and uncivil. The employment of these discourses works to overemphasize the style of a speaker at the expense of reflecting on the content to their speech. In spite of these constraints, Copeny's sense of citizenship reveals alternative means of agency beyond being recognized by dominant actors, institutions, and structures of power and, instead, suggests that resistance and empowerment can be a constitutive process that occurs within the networks of identity and locality. Whereas many can argue that childhood can limit the inventional capabilities of citizenship, Copeny's rhetorical strategies highlight avenues of agency made possible by her reconstitution of her local knowledge and youthful perspective into a source of resistance and empowerment.

CHILDREN ACTIVISTS AND ENVIRONMENTAL JUSTICE IN RHETORICAL STUDIES

In this chapter, I have argued that Little Miss Flint represents a rhetorical strategy for water justice that utilizes ephebic rhetorics—those rhetorics crafted by and for children—to raise issue awareness, critique dominant institutions, and inspire similar reflections and critiques in others. In particular, Copeny utilizes body and local knowledge grounded in her experiences as a child to articulate Flint's water crisis beyond the faucet to include the everyday impacts of water to daily life and the life-changing and life-threatening conditions of dirty water. In addition to using her body knowledge and local knowledge to assert her own contribution to Flint's grassroots movement for clean water, Little Miss Flint constituted herself into an envoy on behalf of other children and the broader community. By drawing attention to the everyday impacts of dirty water on the health and well-being of children, Copeny shifts the narrative of Flint's water crisis away from depicting children as purely victims and, instead, regards them as significant activists in the broader struggle for environmental justice.

The case study of Little Miss Flint contributes significantly to academic conversation regarding environmental justice, rhetoric, and social change and protest. First, Copeny's grassroots activism helps to unearth the ephebic dimensions that are often overlooked due to children being seen predominantly as victims of environmental injustice. Most notably, and of particular interests for the current chapter, is the role age plays in how children and young adults understand and partake in grassroots movements and public deliberations over water management and sanitation and equal access to clean water. Similar to rhetorical theories about women, motherhood, and environmental justice, Copeny's struggle for clean water reveals how children are daily observers who notice environmental disturbances and hazards. Beyond being a material condition that may propel children into the trenches of activism, we can consider ephebic rhetorics as a source of empowerment and resistance for young adults to employ in crafting and advancing calls for social and political change. It is critical that rhetorical scholars contribute to environmental justice research by enlarging the scope of what "justice" looks like by also considering *who* is struggling for it. While it is true that children are, in fact, the largest group of human victims from environmental degradation and toxins, they may also be effective and powerful advocates to address those conditions.

Second, Copeny's environmental justice work helps rhetorical scholars recalibrate both the presence and role of children in grassroots movements for social and political change. Commonly, the rhetorical labor of young children is not the focus of our studies, in part because they are often the subjects being spoken about and advocated for. And, this is particularly true in environmental justice research wherein everyday citizens battle governments and corporations in the name of defenseless children poisoned by toxic waste, pollution, and hazardous chemicals. Yet, the case of Little Miss Flint presents an opportunity to see how children can speak for themselves and on behalf of others and their community. Positioning children activists as the agents of our analysis, rather than as *topoi* tests the boundary conditions of rhetorical studies in myriad ways regarding style, organization, tone, modes of identification, and strategy.

As the effects of global climate change become more prevalent and life-threatening, children activists are becoming key leaders in speaking for social and political change. In May 2019, 1.5 million young people in more than 125 countries walked out of schools, colleges, and universities in "the biggest day of global climate action ever" (Turns 2019). And while climate change becomes a politically divisive issue in the United States, young people have the "social freedom to say it like it is" and articulate it in ways "better than adults in power" (Turns 2019). Most notably, sixteen-year-old Greta Thunberg began skipping school to picket outside of Sweden's parliament to demand tougher policies to reduce greenhouse gas emissions. Emphasizing the need for a better future for children, Thunberg chastises politicians who "have known about climate change for decades" but have "willingly handed over their responsibility for our future to profiteers who search for quick cash" (Thunberg 2019). Young people, argues Volker Quaschning, are "honest, innocent in a way, and speak straight to the problems" that they "didn't create but will have to pay for" (Hockenos 2019). Thus, children and young adults may present a significant opportunity to raise awareness of the impacts of global climate change and do so in ways that hold politicians and governments accountable for the well-being and health of future generations.

As our current climate—political and natural—continues to involve more issues impacting children, there is an inevitable likelihood that rhetorical scholars will have to deal with the fact that citizenship is not only for adults but, also, child's play.

REFERENCES

Battoclette, Augusta. 2018. "Meet 'Little Miss Flint': The Girl Speaking Up for Her City." *Affinity Magazine*, June 4, 2018. http://culture.affinitymagazine.us/meet-little-miss-flint-the-girl-speaking-up-for-her-city/.

Been, Vicki. 1993. "What's Fairness Got to Do with It? Environmental Justice and the Siting of Locally Undesirable Land Uses." *Cornell Law Review* 10, no. 1: 1001–87.

Been, Vicki. 1995. "Analyzing Evidence of Environmental Justice." *Journal of Land Use and Environmental Law* 11, no. 1: 1–28.

Bryant, Bunyan, and Mohai, Paul. 1992. *Race and the Incidence of Environmental Hazards: A Time for Discourse*. Boulder, CO: Westview Press.

Bullard, Robert D. 1993. *Confronting Environmental Racism*. Boston: South End Press.

Bullard, Robert D. 1994. "A New 'Chicken-or-Egg' Debate: Which Came First—The Neighborhood, or the Toxic Dump?" *The Workbook* 19, no. 2: 60–62.

Carmody, Time. 2016. "How the Flint River Got So Toxic." *The Verge*, February 26, 2016. https://www.theverge.com/2016/2/26/11117022/flint-michigan-water-crisis-lead-pollution-history.

CNN Library. 2017 "Flint Water Crisis Fast Facts." *CNN*, November 28, 2017. http://www.cnn.com/2016/03/04/us/flint-water-crisis-fast-facts/index.html.

Di Chiro, Giovanna. 1993. "Defining Environmental Justice: Women's Voices and Grass-roots Politics." *Socialist Review* 22: 115–23.

Duell, Desiree. 2016. "Mothers, Women of Flint 'Die' in at City's Water Plant." *People's Tribune*, June 2016. http://peoplestribune.org/pt-news/2016/06/mothers-women-flint-die-citys-water-plant/.

Garcia, Feliks. 2016. "Little Miss Flint Meets Donald Trump, Appears Terrified." *The Independent*, September 15, 2016. https://www.independent.co.uk/news/world/americas/little-miss-flint-donald-trump-picture-water-crisis-a7310136.html.

Government Accountability Office. 1983. *Siting of Hazardous Waste Landfills and Their Correlation with Racial and Economic Status of Surrounding Communities*. Washington, DC: U.S. Government Accountability Office.

Harris-Perry, Melissa V. 2011. *Sister Citizen: Shame, Stereotypes, and Black Women in America*. New Haven, CT: Yale University Press.

Hill, Zahara. 2017. "The Fierce Little Miss Flint Wants You to Join the People's Climate March." *Huffington Post*, May 1, 2017. https://www.huffpost.com/entry/the-fierce-little-miss-flint-wants-you-to-join-the-peoples-climate-march_n_5900f341e4b0af6d718b0126.

Hockenos, Paul. 2019. "The Kids Are Taking Charge of Climate Change." *Foreign Policy*, April 10, 2019. https://foreignpolicy.com/2019/04/10/the-kids-are-taking-charge-of-climate-change/.

Kang, Jiyeon. 2017. "Internet Activism Transforming Street Politics: South Korea's 2008 'Mad Cow' Protests and New Democratic Sensibilities." *Media, Culture and Society* 39, no. 5: 750–61.

Kevill-Davies, Harriette. 2018. "Children Crusading against Communism: Mobilizing Boys as Citizen Soldiers in the Early Cold War State." *Rhetoric and Public Affairs* 21, no 2: 235–79.

Klein, Allison. 2018. "10-Year-Old 'Little Miss Flint' Helped Hundreds of Underprivileged Kids See 'Black Panther' over the Weekend." *The Washington Post*, February 20, 2018. https://www.washingtonpost.com/news/inspired-life/wp/2018/02/20/10-year-old-little-miss-flint-helped-hundreds-of-underprivileged-kids-see-black-panther-over-the-weekend/?noredirect=on&utm_term=.a8931e5a081d.

Krauss, Celene. 1993. "Women and Toxic Waste Protests: Race, Class, and Gender as Resources of Resistance." *Qualitative Sociology* 16, no. 3: 251–59.

Kurtz, Hilda E. 2007. "Gender and Environmental Justice in Louisiana: Blurring the Boundaries of Public and Private Spheres." *Gender, Place, and Culture* 14, no. 1: 403–17.

Lewis, Philip. 2017. "'Little Miss Flint' Helps 1,000 Local Students Get New School Supplies." *Huffington Post*, August 7. 2017. https://www.huffpost.com/entry/little-miss-flint-school-supply-drive_n_5987b2b3e4b0cb15b1bf3aab?section=us_black-voices.

Lin, Jeremy, Jean Rutter, and Haeyoun Park. 2016. "Events that Led to Flint's Water Crisis." *New York Times*, January 1, 2016. http://www.nytimes.com/interactive/2016/01/21/us/flint-lead-water-timeline.html?_r=0.

Los Angeles Times. 2016. "Read the Letter from 'Little Miss Flint' That Stirred Obama to Visit Flint." *Los Angeles Times*, May 4, 2016. https://www.latimes.com/nation/nationnow/la-na-read-little-miss-flint-letter-20160504-snap-htmlstory.html.

Lurie, Julia. 2016. "Meet the Mom Who Helped Expose Flint's Toxic Water Nightmare." *Mother Jones*, January 21, 2016. https://www.motherjones.com/politics/2016/01/mother-exposed-flint-lead-contamination-water-crisis/.

Maantay, Juliana. 2002. "Zoning Law, Health, and Environment Justice: What's the Connection?" *The Journal of Law, Medicine & Ethics* 30, no. 4: 572–93.

Madison, D. Soyini. 2009. "Crazy Patriotism and Angry (Post)Black Women." *Communication and Critical/Cultural Studies* 6, no. 3: 321–26.

Martinez, Michael. 2016. "Flint, Michigan: Did Race and Poverty Factor into Flint's Water Crisis?" *CNN*. January 28, 2016. http://www.cnn.com/2016/01/26/us/flint-michigan-water-crisis-race-poverty/.

McNeal, Stephanie. 2016. "People Are Laughing over This Girl's Different Reactions to Meeting Obama and Trump." *Buzzfeed News*, September 15, 2016. https://www.buzzfeednews.com/article/stephaniemcneal/little-miss-flint-trump.

McQuaid, John. 2016. "Without These Whistleblowers, We May Never Have Known the Full Extent of the Flint Water Crisis." *Smithsonian Magazine*, December 2016. https://www.smithsonianmag.com/innovation/whistleblowers-marc-edwards-and-leeanne-walters-winner-smithsonians-social-progress-ingenuity-award-180961125/.

Murphy, Mollie K. 2017. "What's in the World Is in the Womb: Converging Environmental and Reproductive Justice through Synecdoche." *Women's Studies in Communication* 40, no. 2: 155–71.

Murray, Lydia. 2016. "Obama to Visit Flint to Hear from Community." *The Michigan Daily*, April 27, 2016. https://www.michigandaily.com/section/news/obama-visit-flint.

Peeples, Jennifer. 2003. "Trashing South-Central: Place and Identity in a Community-Level Environmental Justice Dispute." *Southern Communication Journal* 82, no. 2: 80–93.

Peeples, Jennifer, and Kevin Michael DeLuca. 2006. "The Truth of the Matter: Motherhood, Community and Environmental Justice." *Women's Studies in Communication* 29, no. 1: 58–74.

Pezzullo, Phaedra C. 2001. "Performing Critical Interruptions: Stories, Rhetorical Invention, and the Environmental Justice Movement." *Western Journal of Communication* 65, no. 1: 1–25.

Pulido, Laura. 1996. "A Critical Review of Methodology of Environmental Racism Research." *Antipode* 28, no. 2: 142–59.

Richardson, Alyssa. 2017. "Little Miss Flint: The Face of a Community Coping with Crisis." *Spartan News Room*. June 22, 2017. https://news.jrn.msu.edu/2017/06/little-miss-flint-the-face-of-a-community-coping-with-crisis/.

Smith, Lindsey. 2015. "How People in Flint Were Stripped of a Basic Human Need: Safe Drinking Water." *Michigan Radio*, December 15, 2015. https://www.michiganradio.org/post/how-people-flint-were-stripped-basic-human-need-safe-drinking-water.

Sowards, Stacey. 2010. "Rhetorical Agency as *Hacienda Cara* and Differential Consciousness through the Lens of Gender, Race, Ethnicity, and Class: An Examination of Dolores Huerta's Rhetoric." *Communication Theory* 20, no. 2: 223–47.

Suggs, Ernie. 2019. "Mari Copeny: Activist, 11, Is Face, Voice of Flint Water Crisis." *Atlanta News Now*, February 28, 2019. https://www.ajc.com/lifestyles/mari-copeny-activist-face-voice-flint-water-crisis/ltBhSYnuS7ViDK6W8DHeeP/.

Thomas, Christopher S. 2018. "The Mothers of East Los Angeles: (Other)Mothering for Environmental Justice." *Southern Communication Journal* 83, no. 5: 293–309.

Thomas, Cynthia E. 2019. "World to Flint Kids: We Still See You." *Yes!* August 30, 2019. https://www.yesmagazine.org/people-power/world-to-flint-kids-we-still-see-you-20180830.

Thunberg, Greta. 2019. "Young People Have Led the Climate Strike. Now We Need Adults to Join Us Too." *The Guardian*, May 23, 2019. https://www.theguardian.com/commentisfree/2019/may/23/greta-thunberg-young-people-climate-strikes-20-september.

Turns, Anna. 2019. "Meet Generation Greta: Young Climate Activists around the World." *The Guardian*, June 28, 2019. https://www.theguardian.com/environment/2019/jun/28/generation-greta-young-climate-activists-around-world.

United Church of Christ. 1987. *Toxic Wastes and Race in the United States: A National Report on the Racial and Socio-economic Characteristics of Communities with Hazardous Waste Sites*. United Church of Christ Commission for Racial Justice.

Walley-Jean, J. Celeste. 2009. "Debunking the Myth of the 'Angry Black Woman': An Exploration of Anger in Young African American Women." *Black Women, Gender and Families* 3, no. 2: 68–86.

WaterYouFightingFor? 2017. *WaterYouFightingFor? Homepage*. Accessed July 1, 2019. http://www.wateryoufightingfor.com/.

Young, Iris Marion. 1990. *Justice and the Politics of Difference.* Princeton, NJ: Princeton University Press.

Chapter Eleven

Environmental Crises and Hydrosocial Networks

Using Online Discontent to Promote Water Justice in Shanghai

Elizabeth Brunner

On March 7, 2013, clusters of dead pigs were found floating down the Huangpu River, or "Mother River of Shanghai," which divides Shanghai into the East (Pudong) and West (Puxi) Banks. By March 20, more than ten thousand pigs had been fished out of the Huangpu and several thousand others retrieved from tributaries (Zhou 2013). Their appearance led to panic in the city, which is home to some of China's wealthiest residents. The people of Shanghai rely on the Huangpu for drinking water, the country relies on it for shipping goods further inland, and it serves as an important revenue-generator for the tourism industry as a seemingly continuous stream of tourists board cruise vessels that offer a scenic view of Shanghai's impressive and varied architecture from the river. The appearance of bloated pig carcasses caused networks of Shanghai's 23.5 million residents to voice their concerns over the Chinese social media platform, Weibo. The public outcry prompted the government to respond. They released daily water-quality reports as well as food-quality reports that checked for the appearance of meat salvaged from the carcasses in nearby restaurants. Each day, these government institutions deemed the water quality "normal" and declared that no tainted pork was found in the restaurants.

These checks did not, however, satisfy many local residents who were wary of official reports, especially given the years of recent food scandals combined with the thinly veiled excuses local government officials provided for the appearance of the dead pigs. For example, one official from Zhejiang's provincial agricultural department claimed cold temperatures were to blame for pigs' deaths despite weather reports that indicated local temperatures had not dropped below zero. The outrageousness of this excuse exacerbated growing distrust in officials, prompting a rush of questions as well as an outpouring of criticism again on Weibo, the microblogging site wherein users are able to follow other users, post, share images, comment, and interact via private messaging, often described as the Chinese equivalent to Twitter (Lee 2012; Olesen 2012; Yeh 2012).

Following the Weibo posts and an investigation that eventually tracked the pigs upstream to the small city of Jiaxing paints a complicated portrait of water justice (as well as lack thereof) wherein "nature, technology and society" are "mutually constitutive, forming 'hydrosocial networks'" among rivers, unlikely allies, and government officials (Zwarteveen and Boelens 2014, 144). These hydrosocial networks are extended and intensified via media networks that include apps, smartphones, internet forums, and text messaging. In the case of the Huangpu River pigs, upstream rural farmers disposed of dead livestock in the river, which may have gone unnoticed had they not floated through one of the country's largest metropolises, where many residents owned smartphones and were connected to and active on social media platforms. In 2013, 591 million people in China were connected online and those using Weibo, QQ, and other social media platforms were able to broadcast their message to an ever-growing audience in a country where mass media is strictly controlled (China Internet Watch 2014). The horrific images traveled online like wildfire as people shared, forwarded, liked, and reposted them, drawing attention to the issue.

The Huangpu pigs provide a productive case study for environmental justice research for two major reasons. First, environmental justice movements, which originated in the United States, often focus on Western case studies, which tend to focus on racism, gender, and economic inequalities (Murphy 2017; Bullard 2000; Cutter 1995). Focusing on China, which has very different racial, class, and political dynamics, helps to expand the concept of environmental and water justice to examine how "poverty and rural and urban differentiation"—considered the "main causes" of environmental (in)justice—unfold in authoritarian countries (Mah and Wang 2017, 265). Second, this case study contributes to existing research by taking a networked approach that combines the concept of *hydrosocial networks* with *wild public networks*. In so doing, this chapter contemplates the mutually influential flows of power, politics, media, and pollution, thereby unpacking the complex systems that, in some cases, contribute to a "slow violence" of accumulated injustices that left some jobless and forced others to move (Davies 2018; Mah and Wang 2019). A networked approach also acknowledges the multiple and varied outcomes of any given water justice initiative. In the case of the Huangpu pigs, rural farmers unwittingly created what would become a striking but unplanned image event by dumping dead pig carcasses into waterways (DeLuca 1999; He 2017). Wealthier citizens in Shanghai carrying smartphones captured photographic proof of the event and shared images of the bloated bodies on multiple social media platforms. This led to an investigation that traced the pigs back upstream to the small town of Jiaxing, which then made evident the ties between struggling pig farmers, black markets, high-level officials, corruption, and years of food scandals. The questions here thus become: *How do hydrosocial networks intertwine with wild public networks to communicate crisis and how did the crisis variously impact water justice efforts? Further, how was water justice enacted differently for urban versus rural populations?* To answer these questions, I examine the incident itself, online reactions, government responses, flows of pollution and information, and the political climate at the time that connected this incident to other food-, air-, and water-safety issues. Doing so enables me to chart the overlapping hydrosocial and wild public networks as well as the movement that created new connections among Chinese citizens.

I argue that hydrosocial networks intertwine with wild public networks to create a series of successes and failures that vary dramatically across urban and rural populations. I begin by mapping the contemporaneous political and social terrains as they were impacted through the introduction and rapid expansion of wild public networks alongside scandals that resulted in decreased trust in officials. I then delve more deeply into the Huangpu pig incident to map the multiple factors that coalesced over hydrosocial networks to cause it. After hydrosocial networks are mapped, I turn to the overlapping wild public networks that exposed the incident to the world and then think through the consequences and unevenness of water justice for rural and urban populations.

RIVERS, POLLUTION, AND HYDROSOCIAL NETWORKS

In recent decades, China has experienced decaying trust in officials alongside increased trust in social media, which has strengthened what I term in other writings *wild public networks* (Brunner 2017, 2019). Wild public networks supplant the Habermasian public sphere with a dynamic understanding of contemporary argument in a densely panmediated environment awash with images, video, gifs, and creative inventions meant to sidestep censors. Wild public networks acknowledge and highlight the more rhizomatic interactions fostered by social media networks that subvert traditional media hierarchies and make possible new forms of activism wherein people hold officials and each other accountable. In an authoritarian country like China—which is also highly connected on these media platforms—internet forums, bulletin board systems (BBSs), and social media apps serve as a space to share information and discuss issues in a constantly shifting space outside of state-controlled media. Wild public networks also theorize an unruly space wherein rational arguments blend with affective appeals in messy and moving configurations and dynamic connections between disparate users are constantly being drawn and redrawn. Understanding activism on wild public networks not only draws attention to the importance of media, especially in an authoritarian country, but also helps to trace multiple and diverse outcomes as they impact everyone from farmers to government officials to company executives.

As trust in social media sources grows alongside the number of users, the platforms become increasingly more wild, unpredictable, and threatening to a country that values harmony and stability. People's waning trust in officials that arose in the late 2000s occurred on a national scale and was intimately tied to the rise of social media, as is evident in a variety of cases wherein officials were exposed on platforms such as Weibo for taking bribes, having mistresses, and failing to follow official protocol and procedures (Link and Qiang 2013). Once exposed across wild public networks, the government had to take action. In one case, netizens found numerous mass media images of Yang Dacai, head of the work safety bureau in Shaanxi province, wearing expensive watches, including a Rolex, Bulgari, and Omega; they then posted these images to Weibo because the watches, users noted, were too expensive for someone in his position to afford. They, therefore, must have been purchased with bribe money. Though Chinese censors attempted to tamp down related conversations, they were often blinkered by creative netizens who invented new lexicons for critique. For example, some netizens used a picture of a "river crab" (which, in Chinese, is a homonym for "harmony," a word used to represent censorship) and wrapped several gold Rolex watches around it. This image came to represent both the scandal and the censorship of it. Yang was ultimately brought to court on charges of accepting bribes and sentenced to fourteen years in prison (Blanchard and Duncan 2013). In another case, officials were exposed for more egregious errors, such as attempts to literally cover up the aftermath of a massive high-speed train crash in Wenzhou in 2011 with bulldozers before the public became aware of the tragedy; had netizens not shared images of the accident on Weibo, officials likely would have been successful in suppressing the story (Bondes and Schucher 2014; Custer 2011).

The overall trend of decreased trust in officials is important to note, especially in the case of environmental issues because, as Li (2016) explains, many Chinese "often see environmental problems as the responsibility of the government" (11). Thus, people expect the government to address issues of pollution. Officials, however, have been less than transparent regarding pollution, as is evidenced by a series of high-profile food scandals in the late 2000s and early 2010s. For example, in 2008, infant formula tainted with high levels of melamine killed six infants and left 300,000 more ill. The government found out about the contaminated product but delayed a recall, which many believe would have saved children from illness and death (Ramzy and Yang 2008). Government officials in Xiamen in 2007 also failed to follow protocol that required them to create and release Environmental Impact Statements to the public and, instead, approved plans for a para-Xylene plant quietly (Ansfield 2013). When the public found out via online networks and a widely forwarded text message, they were outraged and organized street protests that resulted in plans for the plant being cancelled (Ansfield 2013). The same failure of officials to follow protocol incited anti-PX protests in Dalian, Maoming, Kunming, Chengdu, and numerous other larger cities (Brunner and Li 2018). Wild public networks, which connect hundreds of millions of people, can be used to organize outrage and expose corrupt officials.

The lack of trust in officials is closely tied to the turn to social media. According to Mou and Lin, "In contemporary China, social media are now the primary alternative platform for the Chinese Internet users to seek, learn and share risk information" (2017, 567). Authorities who remain quiet on a given health or environmental crisis topic only exacerbate the situation because "the less information the responsible authorities share with the public during a public health crisis, the more actively the public will utilize their social networks to appraise the risk and the greater their risk perception may become" (Mou and Lin 2017, 576). In the cases of food scandals in China, the government's lack of communication exacerbated citizen concerns.

Social media platforms, thus, became a place to which many turn for what they perceive to be more legitimate and unbiased information, making it an important alternative to state-controlled media. In the case of the Huangpu pigs, social media became an important place for people to exchange images, (mis)information, outrage, and disdain for official responses to the incident. It was not, however, a tool used by all people involved, as many lacked access to the technological devices needed to engage on these platforms at that time. For those with access, social media serve as a place wherein social trust is fostered among citizens in an authoritarian country. The resulting communities—some temporary and other lasting years—have been capable of bringing about social and water justice via protests, citizen-led initiatives, non-profit work, and by exposing corruption.

FOLLOWING THE PIGS AND POLLUTION

To achieve water justice across networks in this case, one needs to trace the complex pathways that led to it. By the time pig carcasses began floating down the river in 2013, "up to 40 percent of China's rivers were seriously polluted" and "20 percent were so polluted their water quality was rated too toxic even to come into contact with" (Yang 2012). The thousands of factories built along China's rivers played an important role in these alarming figures, but so too did agriculture—an industry that was growing exponentially in an effort to feed a population of more than one billion people. In 2010, agriculture surpassed industry as a leading contributor to water pollution (Watts 2010). The bloated pigs made the ongoing water pollution issue in Shanghai visible in a grotesque way that simply could not be ignored as was made easily recognizable across wild public networks. The sight of thousands of dead pigs floating down Shanghai's busiest river and a popular tourist destination day after day gave urban residents fodder to fight for cleaner water in a way they could not previously have done.

The river of dead pigs was not the result of a handful of irresponsible farmers, but rather numerous issues I outline here as I trace complex hydrosocial networks. The first important factor in play was the growing demand for pork. In 2013, China was the world's largest producer and consumer of pork. Pork has been credited with helping to alleviate hunger across the country, and the Chinese government played an important role in the booming market by subsidizing, investing in, and creating policies aimed at increasing pork production (Schneider and Sharma 2014). The rise in pork farms, however, also had numerous other consequences. It ruined the fishing industry in many rural fishing villages as pig farmers disposed of both manure and carcasses in the rivers. According to one villager in the "pig triangle" north of Shanghai, "No one has caught a living fish in the river system for more than a decade. And since 2003, villagers from nine fishery communities in Jiaxing . . . have petitioned the government over water pollution" to little or no avail (Di 2013). These citizens, who lacked access to the wild public networks that linked social media users with disparate populations, were forced to go through traditional channels that failed to produce the desired outcome. Government networks that invested heavily in the pork industry made fighting against it in the name of clean rivers a Sisyphean task.

As pork demand grew, it had other diverse consequences beyond alleviating hunger. Many farmers adopted problematic and dangerous farming practices. In 2011, 286 people from Hunan province sought medical help after ingesting pork that contained ractopamine, a drug designed to grow muscle and eliminate fat (Barclay 2011). Ractopamine is similar to clenbuterol, a drug pig farmers began using in the 1980s to increase yield, but which was banned in 2002 due to its adverse side effects. Some farmers also regularly use arsenic in their feed to improve the meat's appearance so that it sells faster and at a higher price at market (C. Li, Wu, and Wu 2017). A 2005 study found that of the twenty-nine large-scale pig farms tested in Beijing, 100 percent of the feed and manure samples contained arsenic and one-third were above the government limit (Lam et al. 2013). In the case of the Huangpu pigs, a number of livestock reportedly died due to arsenic poisoning. After a period of four to five months, organic arsenic becomes inorganic, killing the pigs, and if they are not sold within this short window, they must be disposed of (C. Li, Wu, and Wu 2017). A crash in the swine market left pig farmers with an overwhelming number of carcasses.

The sudden lack of demand for pork was attributed to the second major factor—President Xi Jinping's anti-graft campaign. After his election in 2012, Xi announced an "eight-point code" designed to reconnect officials with the masses and eliminate graft by banning the use of luxury cars, eliminating lavish gifts, and reducing "pomp, banquets, ceremony, bureaucratic visits, and meetings" (Yuen 2014, 41). Xi revived a Mao-era slogan, "four dishes and a soup" (*sicai yitang* 四菜一汤), to promote frugality (Yuen 2014, 41). This edict had rather immediate impacts and, in 2013 alone, approximately 180,000 officials were punished and more than 24,000 cases in violation of the "eight-point code" were filed (Yuen 2014, 41). Demand for certain products, including pork, decreased as lavish banquets were cancelled (C. Li, Wu, and Wu 2017). Though farmers are required by law to dispose of dead pigs at official disposal sites, some pig farmers clearly broke this law. The overwhelming number of carcasses contributed to the issue. One Jiaxing villager was quoted as saying, "According to our records, 10,078 pigs died in January, another 8,325 died in February. More than 300 pigs die every day in our village, and we barely have any space left to dispose of the dead pigs" (Hunt and Zhang 2013).

Another major factor at play was the recent arrest of those who bought dead pigs from farmers, butchered them, and sold the meat to restaurants. Several farmers in Jiaxing told reporters that a black market for dead pigs did, indeed, exist but was facing crackdowns (Yan and Zuo 2013; Zhou 2013). In October 2012, seventeen people were charged with illegally butchering carcasses and selling them to the public, with one of the suspects allegedly bringing in approximately eight million RMB from the transactions (Yan and Zuo 2013). As the Huangpu pig scandal was unfolding, *China Daily* published a story detailing how "46 people involved in processing and selling meat from diseased pigs" were sentenced to between six months and six and a half years in prison, ostensibly to make visible the government's effort to protect food safety (Xinhua 2013). Thus, some of the pigs in the river were there precisely because they could not be sold on the black market.

A fourth factor at play was the changing river. The river of 2013 was not the river older generations remember. In the course of just a few decades, the Huangpu river in Shanghai as well as many of its tributaries changed from a source of water people used to rinse their vegetables and send their kids to play in to one that sometimes turns an "inky black, covered in a slick lime green algae," according to a villager who lives on the Jiapingtang river, a tributary of the Huangpu (Davison 2013). The town of Jiaxing, the city to which many of the dead pigs were traced, sitting approximately one hundred kilometers upstream from Shanghai, was hit hard. One fisherman from the area told reporters that "things got worse in the early 2000s when more pig farms turned up. Dead pig carcasses and manure polluted the river. The water turned inky black, and in hot weather, you could smell the stink from far away" (Di 2013). The growing pork industry radically altered many fishing villages and those who remained often found their livelihood not fishing the rivers, but cleaning them (Davison 2013). Once famous for its fish and rice, industry now makes up close to half of the local GDP in Jiaxing. Thus, when farmers dumped the dead carcasses in the river, they were not dumping them into what they viewed as a pristine water source but rather a river that had, in many ways, died.

Finally, these issues, which threatened China's food security, did not directly impact officials, who, as a consequence, likely did not see it as a pressing matter. In 2011, a Chinese reporter for *Southern Weekly* revealed the existence of special organic farms dedicated to growing produce and raising safe livestock for Communist Party officials' consumption (Lu 2011). The farms are off limits to the common people, and some are protected by guards who stand outside the entrances (Demick 2011). Theses farms' existences were being exposed at the same time a report was published finding that 90 percent of the food in markets—where citizens buy their food—contained harmful additives (Fang 2011). The separation of food systems creates a distinct division between the people and elected officials, who are protected from the dangers to which the common people are exposed. Author Gao Zhiyong put it simply: "The officials don't really care what the common people eat because they and their family are getting a special supply of food" (Demick 2011).

Tracing the pigs upstream reveals complex hydrosocial networks composed of overburdened farmers engaging in dangerous farming practices, President Xi's anti-corruption campaign, a black market for dead pigs, degrading water, the separation of food systems, and the appearance of the Huangpu pigs across wild public networks of social and mass media. To achieve water justice among these hydrosocial networks requires significant changes to overlapping systems. While some changes are easy to identify—for example, prohibiting separate food systems for authorities—others are much more difficult to navigate. How, for example, can China go about providing an affordable source of protein for its 1.4 billion people without wreaking havoc on the land and water? While the upper echelons of society could ignore this crisis because they had stable access to safe food, urban populations had to consistently and arduously fight for cleaner resources using all available tools and resources at hand, including social media. Rural populations, meanwhile, were subject to the whims of the market and the broken systems in place that provided no relief from pollution.

MAKING HEADLINES

Media systems were integral to exposing the aforementioned issues to local, national, and international communities. First, if the pigs had not floated through Shanghai, they likely would not have been captured by smartphones and shared to social media, where they placed pressure on the local government to respond. People posting online were connected not only to other Shanghai residents, but friends, family, and acquaintances from across China. Business investor, Xue Manzi posted to his Weibo account, "Huangpu river is the source of drinking water for more than 20 million Shanghai residents. And this horrific incident was only made public when residents started posting pictures on Weibo" (Hunt and Zhang 2013). Without social media, the government could have easily prevented news of the story and/or more strictly controlled the framing.

Official media was slow to respond. The first *China Daily* story did not appear until three days after the pigs were first discovered. On March 12, the paper announced that officials were launching an investigation, but then fell largely silent. The city of Jiaxing held a press conference, during which officials reported that the carcasses found downstream were not all from Jiaxing and no evidence of an epidemic existed. Authorities also reiterated that the water quality remained normal. They attributed the pigs' appearance in Shanghai's Huangpu to high numbers of pig deaths, cold weather, and poor understanding of livestock laws that prohibited dumping carcasses in the river. This did not, however, satisfy the people.

A March 21, 2013, editorial in *China Daily* evidenced residents' growing frustrations with unanswered questions: "The longer it takes to discover the cause of the deaths, the more reasons people will have to believe that governments are covering up something crucial that's related to public health." Mou and Lin's work makes a similar argument ("Solve Crisis with Honesty and Transparency" 2013). The author of the editorial elaborated further:

> The governments in Jiaxing of Zhejiang province, where the dead pigs were thrown into the river, and Shanghai, where the dead pigs have been discovered, are obligated to find out the truth. Otherwise, a water-pollution crisis will evolve into a food-safety crisis, and finally a government credibility crisis. The best means to manage this crisis is to face it squarely. ("Solve Crisis with Honesty and Transparency" 2013)

When officials did speak, they lacked the kind of transparency for which citizens were calling. Social trust was decayed, and officials were distanced from the crisis because they had the resources to safe food and water, making water justice more difficult to achieve and driving more people to engage on wild public networks. User @LiMingsheng joined in the critique of state media on Weibo, writing "More than 2,000 dead pigs entering Huangpu

River is not news. More than 20 million Shanghai people drinking dead pig soup for half a month is not news" (Mudie 2013). The attention to the issue on social media and the specific calling out of officials forced a response as tens of thousands of users watched and waited. Critiques of officials accumulated across wild public networks as did anxiety, concern, and fear over food and water—two basic daily needs of citizens. A lack of communication from officials only increases the wildness of the networks as they speculate on water quality, food safety, and officials' trustworthiness.

After over a week, the chief veterinarian of the Chinese Ministry of Agriculture, Yu Kangzhen, announced that the pigs died due to cold weather and disease. Residents, invigorated and active over wild public networks, mocked the "official" cause of death, given that temperatures had not dropped below zero. Releasing daily water and food inspection reports on social media also failed to satisfy locals, who refused to believe that a river of rotting pig carcasses could provide safe drinking water. Yet, government reports repeatedly confirmed the water's safety. Shanghai's deputy environmental chief, Yu Hongwei, "said water samples collected from eight spots in rivers bordering Shanghai were of normal quality, and 'some indexes were better than the same period last year'" (Zuo 2013, para. 13). This was a claim many found difficult to believe, especially in the wake of conversations that had already unfolded across Weibo.

One Weibo user, a popular columnist and military expert by the name of Zhao Chu, wrote: "When you said more than 10,000 pigs froze to death, in my opinion, you are lying. [You] must be covering something up" (Luo 2013). A twenty-seven-year-old Shanghai resident reacted with the following questions: "What is our government doing every day? Are they too stupid to find out the reasons or do they want to cover up something from the public?" (Yan 2013). Another resident came up with a proposal: "I suggest the leaders fill a bowl with that water, and finish it" (Lasseter 2013). A similar request was posted by @litmustest, "Does that mean that the pigs are safe to eat? Why don't we serve the leaders some of the Huangpu river pork?" (Mudie 2013). All of these comments make apparent the lack of trust in local officials, who users alleged must be deceitful, stupid, and unconcerned with local citizens' access to safe food and water as well as the wildness of public networks, where concern accumulated. Though officials are expected to address environmental issues, many Shanghai people did not trust them to bring about water justice at home and turned to wild public networks to apply the extreme pressure needed to induce action. By posting on social media and drawing international attention, the people of Shanghai were exposing the incompetence of local officials to higher-level national officials and the world.

Shanghai officials addressed the Huangpu pig problem largely by initiating mass efforts to fish the pigs from the waters and erect barriers upstream that would prevent the carcasses from making it all the way to Shanghai. They announced on Weibo that the pig carcasses retrieved from the Huangpu "had either been incinerated or buried in seven-metre deep holes with quick-lime" (Zuo 2013a). People could feel safer knowing that the pigs were no longer in the river and had been properly disposed of so as to prevent their reappearance in restaurants. The Communist Party chief in Jiaxing stepped down from his post shortly after the city was identified as the main source of the pigs (though local officials insisted his resignation was not directly related). The pollution, according to these discourses, was not a product of myriad factors that needed to be addressed slowly and thoroughly but rather irresponsible farmers.

The government's anemic response to the issue created a void filled by wild public networks linking people across the country and world. As Mou and Lin (2017) demonstrate in their research:

> In a scenario where a series of food safety crises has occurred and continued, it is likely that the public may not possess sufficient knowledge to dealing with all of these crises. Under this circumstance, social trust will mostly likely erode, which will further lead to a greater perception of risk. (567)

The people of Shanghai were clearly concerned with the current water crisis, and their vocality linked the issue to other food and water crises. As one Shanghai citizen wrote, "There is no problem in the news and the water quality is good, but we certainly don't believe it. Many things are not true in the news. We don't dare to eat pork now. Living in Shanghai is helpless, Shanghai Cancer. The ratio is the first in the country, and it is terrible. There is no way. The people are very pitiful" ("3323头死猪惊现上海黄浦江" 2013). Another wrote, "A sluggish response, a lack of disclosure of official data and muddled information has done nothing to quell our doubts" (EJOLT 2019, para. 10). Rather, it drives people to engage and speculate on wild public networks.

Some Shanghai citizens, with their smartphones and social networks, helped to connect this crisis to larger issues and demand government action. Others focused on only the immediate issue in Shanghai. One Shanghai resident stated, "People in Shanghai think it's better not to eat pork, because they worry diseased pigs would end up on their dinner tables. . . . People say the water we use to brush our teeth is pork broth" (Langfitt 2013). Another Weibo user wrote, "I finally figured out why drinking boiled water makes me gain weight—because it is, in fact, pork soup" (EJOLT 2019). These comments as well as the likes and shares they received drew officials' attention while also creating concern among neighbors who were then drawn into the larger conversation about food safety and water justice. They also helped to foster a stronger online community that, together, made important demands for action public and much more difficult to ignore. The online discourse that brought about government critique, however, was largely limited to Shanghai residents because those in more rural areas lacked both the equipment and networks to participate.

HYDROSOCIAL NETWORKS AND WATER JUSTICE

The local governments' efforts to clean waterways and dispose of the pig carcasses solved only the highly visible portion of the problem that impacted Shanghai residents, for the people of Shanghai were the most vocal about the issue. It did not fix the larger persistent issue of poor water quality in Shanghai or the towns upstream of the bustling metropolis. Rather than becoming a focus of concern, Jiaxing became a scapegoat for the incident which led to a "slow violence" of accumulated injustices (Davies 2018; Mah and Wang 2019). The people of Jiaxing faced some of the harshest punishments while the city of Shanghai continued to, for example, release approximately half of its sewage untreated into the Huangpu. Those in Jiaxing who previously made their living fishing were still relegated to on-river trash collection after the pig cleanup (Di 2013). Despite citizen complaints in the town, the government did little to address the larger pollution problem. Instead, they focused their attention on pig farmers, which left them with a new set of problems.

After the news of the Huangpu pig incident drew massive global attention, authorities ordered pig producers in Jiaxing to scale down or close their farms. The local newspaper, *Jiaxing Daily,* announced that all farms within two hundred meters of the town's main river must shut down (Di 2013). Farmers in business for over two decades were forced to find a new way to earn a living. The government stated it would offer pig farmers assistance in transitioning to mushroom, flower, fruit, and vegetable farming, but this did not quell local concerns about their economic futures (Di 2013). Pig farmers could not translate twenty years of expertise with livestock to mushroom or vegetable cultivation. Further, those farms being forced to close did not necessarily play a role in the Huangpu pig crisis. They did not dump pig carcasses into the river, but their proximity to it forced them to close.

The farmers of Jiaxing, thus, bore the brunt of the responsibility for the pigs and the local authorities were charged with creating changes that they could report to those above them. Unfortunately, these initiatives were often delivered with a heavy hand that had severe negative repercussions for certain disadvantaged groups, including small pig farmers and did not take into consideration the complex factors that contributed to this one visible manifestation of pollution. Other equally as contaminating sources of pollution were largely ignored. Thus, while the people of Shanghai did help to draw attention to growing concern and civil unrest over food safety and water pollution, the political structures in place that both demand and privilege swift and dramatic action thwarted a more nuanced attempt to create water justice. In order to enact water justice, activists have to capitalize upon these online movements and devote themselves to mapping the complex array of rules, laws, regulations, corruption, economic disadvantages, government initiatives, and waterways that lead to crisis. This sort of careful tracing will likely not be conducted by those masses participating in the type of wild public protests that erupt over social media networks, for these are intended to achieve more immediate goals quickly. Rather, it is the delicate work of organizations with the time and resources to devote to mapping projects. The people can draw attention to an issue and help expose parts of the networks that make large-scale pollution possible, but they are not likely to engage in the slower work of tracing.

Authoritarian governments do have the ability to create swift and sweeping environmental change in a way that many democracies cannot, and these outbursts and uprisings across wild public networks make evident what is important to people. Authoritarian governments can, for example, shut down factories, limit the number of cars on the road in big cities, and take coal-burning stoves out of people's houses to prevent air pollution in a way democracies cannot. Authoritarian governments can act quickly, while democracies tend to function much more slowly. Such fast actions do have intense consequences that upset constituents who are left, for example, jobless and without heat. Rural residents who are most often more severely impacted by such measures oftentimes continually endure a sort of slow violence that can threaten their physical well-being, force them to relocate, and leave them jobless.

Cleaning up the Huangpu River was a step toward water justice and Shanghai residents armed with smartphones were integral to this step. They used the networks to which they were connected to demand immediate action and change. Shanghai officials ultimately did take the initiative to establish joint prevention and control measures to avert further incidents from reoccurring (EJOLT 2019). Water quality in Jiaxing, however, did not improve as a result of these efforts. Instead, the city was punished and more residents relegated to hardships. Water justice initiatives, therefore, need to consider the complex hydrosocial networks that link urban and rural areas, and urbanites must use their resources to help link their crises with larger crises. Recognizing, assisting, and helping rural populations to have a voice must become a priority in future water justice efforts. So, too, must dialogue and conversations between these seemingly disparate communities consequentially linked by waterways and pollution.

REFERENCES

"3323头死猪惊现上海黄浦江." 2013. Kaidi. March 11, 2013. http://m.kdnet.net/share-9043042.html.

Ansfield, Jonathan. 2013. "Alchemy of a Protest: The Case of Xiamen PX." In *China and the Environment: The Green Revolution*, edited by Sam Geall and Isabel Hilton, 136–202. London: Zed Books.

Barclay, Eliza. 2011. "Tainted Pork Is Latest Food Safety Scandal in China." NPR.org, April 29, 2011. https://www.npr.org/sections/health-shots/2011/04/29/135839397/tainted-pork-is-latest-food-safety-scandal-in-china.

Blanchard, Ben, and Maxim Duncan. 2013. "China's Grinning and Corrupt 'Watch Brother' Gets Time." *Reuters*, September 5, 2013. https://www.reuters.com/article/us-china-corruption-idUSBRE9840AC20130905.

Bondes, Maria, and Günter Schucher. 2014. "Derailed Emotions: The Transformation of Claims and Targets during the Wenzhou Online Incident." *Information, Communication & Society* 17, no. 1: 45–65.

Brunner, Elizabeth. 2017. "Wild Public Networks and Affective Movements in China: Environmental Activism, Social Media, and Protest in Maoming." *Journal of Communication* 67, no. 5: 665–77.

Brunner, Elizabeth. 2019. *Environmental Activism, Social Media, and Protest in China: Becoming Activists over Wild Public Networks*. Lanham, MD: Lexington Books.

Brunner, Elizabeth, and Hui Li. 2018. "Fragmented Arguments and Forces Majeure: The 2007 Protests in Xiamen, China." *Argumentation and Advocacy* 54, no. 4: 304–21.

Bullard, Robert D. (2000). *Dumping in Dixie: Race, Class, and Environmental Quality*. Boulder, CO: Westview Press.

China Internet Watch. 2014. "Sina Weibo User Demographics Analysis in 2013." China Internet Watch, January 9, 2014. https://www.chinainternetwatch.com/5568/what-weibo-can-tell-you-about-chinese-netizens-part-1/.

Custer, C. 2011. "The Wenzhou Crash and the Future of Weibo." *Tech in Asia*, August 1, 2011. https://www.techinasia.com/the-wenzhou-crash-and-the-future-of-weibo.

Cutter, Susan L. 1995. "Race, Class and Environmental Justice." *Progress in Human Geography* 19, no. 1: 111–22.

Davies, Thom. 2018. "Toxic Space and Time: Slow Violence, Necropolitics, and Petrochemical Pollution." *Annals of the American Association of Geographers* 108, no. 6: 1537–53.

Davison, Nicola. 2013. "Rivers of Blood: The Dead Pigs Rotting in China's Water Supply." *The Guardian*, March 29, 2013. https://www.theguardian.com/world/2013/mar/29/dead-pigs-china-water-supply.

DeLuca, Kevin M. 1999. *Image Politics: The New Rhetoric of Environmental Activism*. New York, NY: The Guilford Press.

Demick, Barbara. 2011. "In China, What You Eat Tells Who You Are." *Los Angeles Times*, September 16, 2011.

Di, Fan. 2013. "China's Dead Pig Scandal Ushers in Hard Times for Fishermen, Farmers." *South China Morning Post*. May 30, 2013. https://www.scmp.com/news/china/article/1249509/chinas-dead-pig-scandal-ushers-hard-times-fishermen-farmers.

EJOLT. 2019. "Environmental Justice Atlas." June 26, 2019. https://ejatlas.org.

Fang, Sophia. 2011. "The Party Has Its Own, Safe Food Supply in China." *Epoch Times*, May 15, 2011.

He, Huifeng. 2017. "In China's Richest Province, a Yawning Gap Exists between the Haves and the Have-Nots." *South China Morning Post*, September 5, 2017. https://www.scmp.com/news/china/economy/article/2109050/chinas-richest-province-yawning-gap-between-haves-and-have-nots.

Hunt, Katie, and Dayu Zhang. 2013. "Hundreds of Dead Pigs Fished from Shanghai River." CNN. March 11, 2013. https://www.cnn.com/2013/03/11/world/asia/china-pigs-river/index.html.

Lam, Yukyan, Jillian P. Fry, Emily Hu, Brent F. Kim, and Keeve Nachman. 2013. "Industrial Food Animal Production in Low- and Middle-Income Countries: A Landscape Assessment." Johns Hopkins Center for a Livable Future.

Langfitt, Frank. 2013. "Shanghai's Dead Pigs: Search for Answers Turns Up Denials." NPR.org. March 14, 2013. https://www.npr.org/sections/thesalt/2013/03/14/174302750/shanghais-dead-pigs-search-for-answers-turns-up-denials.

Lasseter, Tom. 2013. "For China's Pig Farmers, Carcasses in River a Sign of Government's Dysfunction." Mcclatchydc. April 4, 2013. https://www.mcclatchydc.com/news/nation-world/world/article24747730.html.

Lee, Melanie. 2012. "China's Twitter-like Weibo Poses Danger, Opportunity for New Leaders." *Reuters*, October 31, 2012. http://www.reuters.com/article/2012/10/31/us-china-congress-weibo-idUSBRE89U1N320121031.

Li, Chenxi, Kening Wu, and Jingyao Wu. 2017. "Environmental Hazard in Shanghai, China: Lessons Learned from Huangpu River Dead Pigs Incident." *Natural Hazards* 88, no. 2: 1269–72.

Li, Xinghua. 2016. *Environmental Advertising in China and the USA: The Desire to Go Green*. New York: Routledge.

Link, Perry, and Xiao Qiang. 2013. "From 'Fart People' to Citizens." *Journal of Democracy* 24, no. 1: 79–85.

Lu, Zhongshu. 2011. "Low-Key Farming." *Southern Weekend*, May 5, 2011.

Luo, Chris. 2013. "The Sacrificial Pigs and Other Weibo Jokes on Shanghai's Floating Hogs." *South China Morning Post*, March 12, 2013. https://www.scmp.com/news/china/article/1189207/sacrificial-pigs-and-other-weibo-jokes-shanghais-floating-hogs.

Mah, Alice, and Xinhong Wang. 2017. "Research on Environmental Justice in China: Limitations and Possibilities." *Chinese Journal of Environmental Law* 1, no. 2: 263–72.

Mah, Alice, and Xinhong Wang. 2019. "Accumulated Injuries of Environmental Injustice: Living and Working with Petrochemical Pollution in Nanjing, China." *Annals of the American Association of Geographers*. https://doi.org/10.1080/24694452.2019.1574551.

Mou, Yi, and Carolyn A. Lin. 2017. "The Impact of Online Social Capital on Social Trust and Risk Perception." *Asian Journal of Communication* 27, no. 6: 563–81.

Mudie, Luisetta. 2013. "Officials Call for Calm Over Dead Pigs in River." *Radio Free Asia*, March 12, 2013. https://www.rfa.org/english/news/china/calm-03122013145025.html.

Murphy, Mollie K. 2017. "What's in the World Is in the Womb: Converging Environmental and Reproductive Justice through Synecdoche." *Women's Studies in Communication* 40, no. 2: 155–71.

Olesen, Alexa. 2012. "Sina Weibo, China's Twitter, Leads Microblog Craze." *Huffington Post*, August 2, 2012. http://www.huffingtonpost.com/2012/08/02/sina-weibo-chinas-twitter_n_1732300.html.

Ramzy, Austin, and Lin Yang. 2008. "Tainted-Baby-Milk Scandal in China." *Time*, September 16, 2008. http://content.time.com/time/world/article/0,8599,1841535,00.html.

Schneider, Mindi, and Shefali Sharma. 2014. "China's Pork Miracle? Agribusiness and Development in China's Pork Industry." Institute for Agricultre and Trade Policy.

"Solve Crisis with Honesty and Transparency." 2013. *China Daily*, March 21, 2013. http://www.chinadaily.com.cn/opinion/2013-03/21/content_16332006.htm.

Watts, Jonathan. 2010. "Chinese Farms 'Cause More Pollution Than Factories.'" *The Guardian*, February 9, 2010. https://www.theguardian.com/environment/2010/feb/09/china-farms-pollution.

Xinhua. 2013. "46 Sentenced for Selling Diseased Pork." *China Daily*, March 13, 2013. http://www.chinadaily.com.cn/china/2013-03/13/content_16306430.htm.

Yan, Alice. 2013. "Two Weeks on and Still No Answers on Source of Dead Pigs Found in Shanghai River." *South China Morning Post*, March 19, 2013. https://www.scmp.com/news/china/article/1194046/two-weeks-and-still-no-answers-source-dead-pigs-found-shanghai-river.

Yan, Alice, and Mandy Zuo. 2013. "Pig Carcasses in River Result of Weather and Illnesses, Not Epidemic: Officials." *South China Morning Post*, March 14, 2013.

Yang, Jian. 2012. "China's River Pollution 'a Threat to People's Lives.'" *People's Daily,* February 17, 2012. http://en.people.cn/90882/7732438.html.

Yeh, Chieh-Ting. 2012. "Translation: A Blogger's Sober Thoughts on the Qidong Protests." *Tea Leaf Nation,* July 31, 2012. http://www.tealeafnation.com/2012/07/translation-a-bloggers-sober-thoughts-on-the-qidong-protests/.

Yuen, Samson. 2014. "Disciplining the Party: Xi Jinping's Anti-Corruption Campaign and Its Limits." *China Perspectives* 2014, no. 3: 41–47.

Zhou, Mandy. 2013. "Dead Pigs Pulled from Rivers in and near Shanghai Top 13,000." *South China Morning Post,* March 18, 2013. https://www.scmp.com/news/china/article/1193193/dead-pigs-pulled-rivers-and-near-shanghai-top-13000.

Zuo, Mandy. 2013. "Pig Toll from Shanghai, Jiaxing Rivers Climbs above 12,000." *South China Morning Post,* March 17, 2013. https://www.scmp.com/news/china/article/1192519/pig-toll-shanghai-jiaxing-rivers-climbs-above-12000.

Zwarteveen, Margreet Z., and Rutgerd Boelens. 2014. "Defining, Researching and Struggling for Water Justice: Some Conceptual Building Blocks for Research and Action." *Water International* 39, no. 2: 143–58.

Chapter Twelve

Sun, Sand, and Satire

A Rhetorical Analysis of the Great Barrier Reef's Obituary

Catherine J. Bruns

On October 11, 2016, author Rowan Jacobsen posted an online article titled "Obituary: Great Barrier Reef (25 Million BC–. . .)" in *Outside* magazine to announce the loss of "Australia's crown jewel," the Great Barrier Reef. In it, Jacobsen (2016) stated that the reef's twenty-five-million-year life had come to an end after it succumbed to "the most catastrophic bleaching event in its history." The obituary spread quickly on social media, with millions of readers expressing sadness and disappointment at the passing of one of the world's most treasured icons (D'Angelo, 2016).

The only problem? The reef was not dead.

Scientists and environmentalists struck back at the "wildly irresponsible" commentary, arguing that Jacobsen failed to communicate that his article was satirical and misled readers about the Great Barrier Reef's (GBR) live-but-threatened status (D'Angelo 2016). Russell Brainard, chief of the Coral Reef Ecosystem Program at NOAA's Pacific Islands Fisheries Science Center, likened the obituary to describing an entire forest as destroyed after chopping down only half the trees (D'Angelo, 2016). Similarly, coral scientists contended that Jacobsen neglected to consider the damage misinterpretation of the article could cause, such as dampening public interest in conservation efforts (D'Angelo, 2016). Overwhelmingly, critics maintained that if Jacobsen's article was intended satirically, "you wouldn't know it at first glance" because nothing in the text branded it as such (D'Angelo, 2016).

As of 2019, large portions of the GBR continue to endure damaging bouts of coral bleaching, yet the reef—and a growing list of other environmental issues—is still in need of protection (D'Angelo, 2016). Pezzullo (2016) explains that the burden of climate advocacy has contributed to an increase in public denial and an overload of strategies to "energize the world, both literally and figuratively" (803). Additionally, framing environmental issues as emergencies has led to environmental communication being described as a "crisis discipline" where researchers must attempt new strategies to create a sense of urgency for the environment (Cox 2007). Past scholars have explored humor as one such environmental advocacy strategy (Bore and Reid 2014; Branagan 2007; DeLaure 2011; Nisbet and Scheufele 2009), but attention to more specific forms of humor in environmental communication, such as satire, has been minimal (Stewart and Clark 2011). Because Jacobsen's obituary was widely misinterpreted as a literal death notice and accrued significant criticism for its satirical presentation, an exploration of the article's failed satirical strategy is essential.

This chapter examines Jacobsen's article as a failed environmental justice intervention for the GBR. Specifically, it asserts that Jacobsen's close parody of the obituary genre negatively impacted the article's satirical appearance and confused readers as to the obituary's gravity and intended message. To support this claim, the chapter examines Jacobsen's article to assess the clarity of its satirical parody. To begin, I make the case for reimagining the GBR's survival as an environmental justice issue. Next, I consider communication scholarship overviewing the GBR, satire, and the obituary genre. I then propose a methodological framework and conduct a rhetorical analysis. The chapter concludes with a discussion of disciplinary impacts and future environmental justice considerations.

THE GREAT BARRIER REEF AND ENVIRONMENTAL JUSTICE

The relationship between the environmentalist and environmental justice movements is fraught with tension. The environmentalist movement has been accused of furthering a classist, racist, and limited agenda (Pezzullo and Sandler 2007); in contrast, the environmental justice movement has faced criticism over its seeming disregard for the continued marginalization of animal and wilderness rights (Taylor 2000). These disagreements have sparked vitriol between parties whose goals too often operate in opposition rather than collaboration. In speaking at the First National People of Color Environmental Leadership Summit in October 1991, Dana Alton, senior program officer of the Panos Institute in Washington, DC, articulated the need for a "just partnership" between the environmentalist and environmental justice movements, saying "It is not just ancient forests; it is not just saving the whales or saving other endangered species. These are all very important. We understand the life cycle and the inter-connectedness of life. But our communities and our people are endangered species, too" (as cited in Pezzullo and Sandler 2007, 7).

It is within this complex backstory that this chapter's subject exists. On one hand, the longevity and health of the GBR is an environmentalist issue, a wilderness in need of protection for the purposes of ecological sustainability. Coral reefs like the GBR shelter coastlines from the impacts of waves and tropical storms, house and host marine animals, and produce nitrogen and nutrients that sustain marine food chains (Queensland Museum, n.d.). The GBR is a core component of oceanic life, yet it is a constant victim of global human practices enacted with an anthropocentric focus. This anthropocentrism also positions the survival of the GBR as an environmental justice issue, whereby the reef functions as a commodity and part of indigenous culture. The GBR is vital to commercial fishing industries that depend on its support of popular market species of fish, crab, and prawn (Queensland Museum, n.d.). The reef also provides more than sixty-four thousand jobs, generates $6.4 billion for the Australian economy, and "is valued at $56 billion as an Australian economic, social and iconic asset" (Deloitte Access Economics 2017, 7). Furthermore, the GBR is "embedded in Indigenous culture, spirituality and wisdom," with Aboriginal and Torres Strait Islander Traditional Owner clans dating their connections to the reef to more than sixty thousand years ago (Deloitte Access Economics 2017, 45).

Both of these approaches to the GBR have value. This chapter does not attempt to prioritize one over the other, nor does it seek to choose one perspective at all; rather, this chapter proposes combining these approaches into a new conceptualization of the meaning of "environmental justice."

In his essay "A Wilderness Environmentalism Manifesto: Contesting the Infinite Self-Absorption of Humans," Kevin DeLuca (2007) highlights two oft-referenced definitions of environmental justice. In one, Gibbs (1993) declares, "The Environmental Justice Movement is about people and the places they live, work and play" (2). In another, Bullard furthers that the movement "basically says that the environment is everything: where we live, work, play, go to school, as well as the physical and natural world" (as cited in Schweizer 1999). By defining environmental justice in this way, DeLuca (2007) argues, "environmental justice activists have successfully shifted the meaning of environmentalism from a wilderness focus to a human and human habitat focus" while still chastising the environmentalist movement for its hesitancy to prioritize human problems (30).

These definitions have consequences. Specifically, this interpretation of environmental justice crafts a master frame linking environmentalism with anthropocentric concerns such as racism and human injustice (Taylor 2000). If, as DeLuca (2007) warns, the environmentalist movement is eventually pushed into adopting the same "human-centered perspective of the environmental justice movement, they will be unable to make the hard decisions that increase human suffering, that require putting other beings and ecosystems, not humans, first" (34) and will allow the rights of animals and the natural world to be placed "even further on the back burner" (Pezzullo and Sandler 2007, 11).

The current chapter addresses this dilemma by strengthening the environmental element of environmental justice and recognizing certain environmentalist issues as nonanthropocentric issues of environmental justice. This choice is contentious, yet rooted in a deep ecology perspective in which nature is valued independently of its utility (Sessions 1995) and humans and nature are seen as existing on an equal playing field (Hendry 2010). This interpretation does not diminish the need for human social justice as it relates to the environment, but, rather, advances that the environmental portion of environmental justice should be highlighted just as strongly as the term's justice component. In this reimagining, environmental justice gains an environmentalist layer: deep ecologists' belief that nature should receive the same ethical considerations given to humans (Hendry 2010). Combining this environmentalist attitude with environmental justice in the traditional sense thus enables the term "environmental justice" to reflect a goal of attaining equal justice for *both* humans and the environment.

My decision to perceive specific environmentalist issues as environmental justice issues is supported by multiple public actions to grant legal status to natural areas—to give the environment the right to seek justice, as would any human. As of today, legal rights have been granted to rivers in India and Ecuador (Tanasescu 2017), and the Australian state of Victoria has introduced the Yarra River Protection (Wilip-gin Birrarung murron) Act to legally establish the Yarra River as "one living, integrated natural entity for protection and improvement" (Fawcett, n.d.). In 2017, New Zealand decreed that Mt. Taranaki was privy to the same protections as a New Zealand citizen (Cheng 2017), and in 2019, a Lake Erie bill of rights was approved with 61 percent public support from residents in Toledo, Ohio (Williams 2019). Yet perhaps most important to this chapter is an ongoing campaign launched by the Environmental Defenders Office of Northern Queensland (EDO NQ) to grant legal standing to the Great Barrier Reef (Fawcett, n.d.).

These progressive efforts demonstrate that global publics are coming to view the environment as worthy not just of protection but also of true justice in its own right. The following analysis relies on this rising public sentiment and a reimagined definition of environmental justice in its exploration of the vitality of the GBR as not merely an environmentalist issue, but also as an issue of environmental justice.

CONCEPTUAL FRAMES

To understand the GBR as an environmental justice issue, I will first lay out a few conceptual frames: previous research on the reef, satirical humor, and the obituary genre.

The Great Barrier Reef

Much of the existing research on the GBR analyzes the reef from a scientific perspective, including studying the reef's water quality (Devlin et al. 2015), growth (Dechnik, Webster, Webb, Nothdurft, and Zhao 2017), and marine organisms (Luter, Duckworth, Wolff, Evans-Illidge, and Whalan 2016). In contrast, communication scholarship primarily explores human interest in conserving the reef.

Past research has addressed stakeholder engagement in GBR protective efforts (Caillaud, Damiens, Salvat, and Wilkinson 2012; Davey and Gillespie 2014; Wynveen and Sutton 2017). As a Marine Protected Area, it is vital the GBR interact with stakeholders key to its conservation (Ballantyne, Packer, and Hughes 2009; Ferse, Costa, Máñez, Adhuri, and Glaser 2010; Trenouth et al. 2012), such as publics near the Gladstone Harbour (Davey and Gillespie 2014) and residents along the GBR Marine Park coastline (Wynveen and Sutton 2017). Semi-structured interviews with fishing, conservationist, and researcher stakeholders have elucidated fears of environmental degradation, with many indicating that conservation management of the GBR is not ideal (Davey and Gillespie 2014). However, studies of the environmental behaviors of fishers and recreational visitors to the GBR have found that providing stakeholders informational materials on climate change can increase stakeholder knowledge of climate change's impact on coral and marine life (Wynveen and Sutton 2017).

Another way to increase human interest in reef conservation is through ecotourism. Scholarship identifies the GBR as a popular last chance tourism (LCT)—or "see it before it's gone" tourism (Lemelin et al. 2010)—destination (Piggott-McKellar and McNamara 2017). Unfortunately, LCT is a catch-22: it increases public interest in protecting environmental destinations while simultaneously intensifying the impact of population pressures and climate change on vulnerable areas (Huebner 2011). While studying LCT and the GBR, Piggott-McKellar and McNamara (2017) analyzed tourist motivations for visiting the GBR and their concern for the reef's health. Unsurprisingly, they found that most tourists were aware of the reef's declining appearance and health but were still motivated "to see the reef before it's gone," leading them to perceive their visit as a "last chance to experience" rather than an opportunity to aid reef protection and conservation efforts (Piggott-McKellar and McNamara 2017, 410).

Most recently, research has studied strategies of communicating GBR policy and conservation efforts (Foxwell-Norton and Konkes 2019; Foxwell-Norton and Lester 2017; Lester 2016). An analysis of GBR media content produced by industry representatives, environmental campaigners, politicians, and journalists has enabled Lester (2016) to examine discursive construction of the reef as an emergent transnational spectacle. Additionally, media examinations produced by Foxwell-Norton and Lester (2017) and Foxwell-Norton and Konkes (2019) have examined "Save the Reef" and "Fight for the Reef" campaign efforts and news media communications, respectively. In doing so, Foxwell-Norton and Konkes (2019) found that Australian news media "struggle to articulate the merits of protecting the Reef as a national or international goal," thus opening the door for alternative communication approaches that may see more success (230).

Humor, Satire, and Parody

Although environmental communication scholars have broadly explored humor in environmental advocacy (Branagan 2007; DeLaure 2011; Stewart and Clark 2011), Nisbet and Scheufele (2009) further argue that satirical humor specifically can help make science more accessible to nonelite audiences. Bore and Reid (2014) used this supposition as justification for examining the relationship between satire and climate change communication, thereby providing this chapter the grounds to conduct research in other environmental communication areas.

"Satire" as a term is evolved out of the Latin *satura*, meaning "a mixed bag or mixed ingredients," and involves humor strategies such as irony, sarcasm, and parody (Peifer 2013, 158). Satire exists to critique social and political beliefs, actions, organizations, or individuals in a way that is entertaining for audiences (Bore and Reid 2014), making it "playful-yet-aggressive" (Peifer 2013, 158). However, satire's varied appearance makes it not only difficult to identify but also risky to enact depending on the audience's sense of humor and feelings toward the target of critique. In exploring the role of satire in the climate change theater production *U: The Comedy of Global Warming*, Bore and Reid (2014) stress that for satire to succeed, it must demonstrate awareness of the audience's humor preferences, which differ based on criteria such as "cultural knowledge," "style," and "taste" (Kuipers 2006, 11). They also recommend that satirical works target niche audiences so that the humor is more likely to match the audience and the social critique component less likely to fail (Bore and Reid 2014).

Another oft-explored humor strategy is parody (Bakhtin 1981; Dentith 2000; Morson 1989). Dentith (2000) defines parody as a "range of related cultural practices, all of which are imitative of other cultural forms, with varying degrees of mockery or humour" and envisions it as a battle between two texts, one original and one imitative (193). Bakhtin (1981) further constitutes parodic conflict as "double-voiced discourse":

> It [double-voiced discourse] serves two speakers at the same time and expresses simultaneously two different intentions: the direct intention of the character who is speaking, and the refracted intention of the author. In such discourse there are two voices, two meanings and two expressions. (324)

Morson (1989) adds that to be considered parody, three criteria must be fulfilled: an imitative text must evoke an original target text, readers must perceive the imitative text as conflicting with the target text, and readers must understand that they are to agree with the imitative text rather than the target text.

It is important to note that parody sometimes imitates style for purposes of comedic effect rather than social criticism (Kaye and Johnson 2016), meaning that parody does not always function satirically (Gray, Jones, and Thompson 2009). However, "satirists [can] choose to use parodies of the most familiar of texts as the vehicle for their satire in order to add to the initial impact and to reinforce the ironic contrast" (Hutcheon 1985, 58). In these instances, satirists identify texts representative of aspects of the social world that are most in need of criticism, such as popular genres (Druick 2009).

Genre parodies of Western film (Turner 2003), pornography (Booth 2014), and sitcoms (Gray 2005, 2006) have been previously explored, yet Peifer (2013) notes that most scholars focus on news parody, or satirical humor that analyzes "the naturalistic illusion that news is (or could be) an unmediated window on the world" (Baym and Jones 2012, 5). The most frequently explored news parodies are the television programs, *The Daily Show* and *The Colbert Report*, as well as the "Weekend Update" portion of *Saturday Night Live* (Baym 2005; Baym and Jones 2012; Day and Thompson 2012; Druick 2009; Kaye and Johnson 2016). Of these studies, arguments have been made that news parody can lead to strong self-efficacy (Kaye and Johnson 2016), reveal media excess and deceit (Baym and Jones 2012), and propose alternative solutions for the problems the satirical hosts highlight on-air (Day and Thompson 2012).

In environmental communication, Stewart and Clark (2011) have examined the use of satirical parody in environmentally themed episodes of the cartoon *South Park*. They analyze the use of parody in three *South Park* episodes to demonstrate "the condescension and intolerance of environmental advocates and their equally strident opponents" (333). They concluded that parodies within popular mediums can teach viewers how negative stereotypes of environmental advocates damage the environmental movements they promote and teach environmentalists how to better communicate their green agendas to the public.

The Obituary Genre

Obituary—also referred to as obits (Al-Ali 2005) or death announcements (Afful 2012)—is a text genre common in print journalism (Stein, Paterno, and Burnett 2006). Today, obituaries are recognized as public announcements of death published in newspapers and accompanied by a short biography of the deceased (Ondimu 2014). These advertisements are often written by family or close friends of the departed (Afful 2012) and typically emphasize the best qualities of the person's life and death (Ondimu 2014).

As a genre, obituary falls within a category Miller (1984) refers to as "homely" discourses, or "easily recognizable socially constructed text genres of everyday life," such as wedding invitations and birth announcements (Al-Ali 2005, 6). Obituaries reveal how a society perceives death (Karimnia and Barbareh 2017) and can provide insight into a culture (Moses and Marelli 2004). Due to obituary's large influence in society, obituary semantics, pragmatics, schematic structure, linguistic structure, and gender encoding have been thoroughly examined by researchers (Ondimu 2014).

Crespo-Fernández (2007) states that there are two obituary types: informative and opinative. Informative obituaries use impersonal language to communicate facts about the deceased (Crespo-Fernández 2007) for the purpose of conveying that an individual has died and sometimes to share funeral details (Nwoye 1992). In contrast, opinative obituaries are subjective and use emotional, intimate, or figurative language to express the writer's feelings (Crespo-Fernández 2007). The goal of an opinative obituary is to highlight "the social relevance, exemplary conduct or religious fervour of the deceased" (Ondimu 2014, 5). In his examination of Arabic death announcements, Al-Ali (2005) proposes that obits can follow slightly different typology, such as in Jordanian newspapers. His identification of normal death notice and martyr's wedding types demonstrates that the purpose of an obituary is strongly linked to the culture in which it is shared (Al-Ali 2005).

The content of an obituary can vary, but with some key elements. A discourse analysis of death announcements published in *The New York Times* during 2002 and 2003 identified a predictable obit "template" with an obligatory order:

> The first paragraph of the text begins with a simple statement of who died, why he or she was an important person, and when and where he or she died. . . . There is often a cause of death given, especially in the case of younger persons. The cause is often medically detailed and is attributed to a source. . . . Life and career history are recounted in reverse order, that is, from the present back. The life history section is expandable and it is here that the

> length of an obituary is extended, with more detailed life stories and longer lists of accomplishments. . . . Finally, there is a list of family members usually starting with the spouse or partner. (Moses and Marelli 2004, 125)

The inclusion of a photograph is also customary, whereas elements such as donation requests for those who would like to pay their respects, the deceased's date of birth, and use of titles are perceived as optional (Matiki 2001; Moses and Marelli 2004).

In summary, environmental communication research on the GBR has addressed stakeholder engagement (Caillaud et al. 2012; Davey and Gillespie 2014; Wynveen and Sutton 2017), the reef's role as an ecotourist destination (Piggott-McKellar and McNamara 2017), and the GBR in the media (Foxwell-Norton and Konkes 2019; Foxwell-Norton and Lester 2017; Lester 2016). Research conducted by Nisbet and Scheufele (2009) and Bore and Reid (2014) suggests that satirical humor may be useful for environmental engagement, but examinations of satirical parody specifically have been minimal (Stewart and Clark 2011). Finally, although genre analyses of obituary are frequent (Afful 2012; Al-Ali 2005; Elekaei, Faramarzi, and Tabrizi 2014; Jalilifar et al. 2014; Ondimu 2014), scholars have not yet examined an obituary created as satirical parody. To fill this gap, the following section analyzes Jacobsen's GBR obit as a failed satirical attempt at environmental justice.

METHODOLOGY

Sillars and Gronbeck (2001) define a communication text as "an interpretive weaving made by someone who is 'reading a work'" (23). They further that it is the responsibility of the interpreting scholar to identify features within a text that are worthy of analysis and judgment and will provide readers a critical understanding of a phenomenon (Sillars and Gronbeck 2001). Although there are multiple rhetorical elements worthy of analytical consideration (e.g., theme, attribution, style), this chapter analyses the genre of the GBR obituary to determine if the obituary elements imitated were done so appropriately given the article's satirical goal (Sillars and Gronbeck 2001).

I lean in to Sillars and Gronbeck's claim that criticism is interpretative, meaning that multiple readings of a text are necessary to consider all possible perspectives. Thus, Jacobsen's article was first read for coherency and understanding. Next, the article was read in accordance to the obituary genre, with attention directed toward its obituary type and use of obituary elements such as name of deceased, cause of death, life and career history, and surviving family. Finally, I read the article through a satirical lens to examine Jacobsen's establishment of the obituary as a satirical parody text.

ANALYSIS OF "OBITUARY: GREAT BARRIER REEF (25 MILLION BC–. . .)"

Jacobsen applies obituary genre elements in three segments: introduction, elaboration, and conclusion. I elaborate on each in the following pages.

Introduction: Establishing the GBR

Both Matiki (2001) and Moses and Marelli (2004) establish particular introductory elements as common to the obituary genre. These elements ground readers in the details of the deceased and include details such as name of deceased, time and place of death, cause of death, and general importance. As revealed below, Jacobsen uses many of these elements in the first two introductory paragraphs to craft an obituary that appears authentic.

The obituary begins by naming the deceased: "The Great Barrier Reef of Australia passed away in . . . after a long illness." Two details emerge in this short segment. First, Jacobsen's inclusion of "of Australia" constructs the reef not as its own entity, but as part of Australia. This naming decision is critical in constituting the GBR as a national rather than international environmental area and attributes Australian-specific loss in the reef's death. Second, Jacobsen breaks away from obituary templates by inserting an ellipsis ("passed away in . . .") where a date of death is normally used, which matches the use of an ellipsis in the article's title. Intentionally replacing the death date with an ellipsis opposes a key element of obituaries and signals to readers that while the surface of the article may look like an obituary, something underneath is different. The paragraph continues with references to the GBR's cause of death and age at time of death. Moses and Marelli (2004) explain that it is typical to reveal a cause of death for younger persons but optional for those who survive past normal life expectancy. On the surface, dying at twenty-five million years far surpasses a normal lifespan, but the accompanying descriptor "after a long illness" complicates this understanding: did the GBR succumb to illness after a respectable twenty-five million years, or did the hard-fought illness cut short a life that was just beginning? With no human age for reference, the tone of the introduction remains ambiguous.

The introduction culminates in a second, longer paragraph with facts about the reef. Some of the details are well-known, such as the GBR's visibility from space; others, such as the reef's length, composition, and area, are less familiar. This segment also lists the exact number of marine species the reef shelters, including fish, mollusks, whales, and birds. This emphasis on facts rather than emotions sets up the forthcoming obituary as informative rather than opinative and suggests that the author is communicating objectively (Crespo-Fernández 2007). Furthermore, the impressive mentions of biodiversity and size are framed by Jacobsen as life "achievements." Moses and Marelli (2004) assert that overviewing life accomplishments is common in the intro, further supporting the article's close imitation of a genuine obituary text.

Despite its close replication, the parody does not utilize all introductory obituary elements. Although the article mentions that the GBR is "of Australia," it neglects to specify the reef's exact place of death. This is important because the reef's substantial length (1,400 miles) means the death could have occurred outside of Australian limits, thus enlarging the scope of international responsibility. Furthermore, with no exact place of death, readers are left to assume that the entire reef has died, when in reality, only portions of the GBR's 2,900 individual reefs are deceased.

Elaboration: The Life of the Reef

In accordance with the obituary genre, Jacobsen's article also summarizes the life of the deceased reef. Moses and Marelli (2004) note that the length of this section can vary, a flexibility Jacobsen's obit needs given the GBR's extensive lifespan. To accomplish this, the article highlights three stages of the reef's life: birth, discovery, and sickness.

Jacobsen first compares the birth of the GBR to that of a baby, writing "The reef was born on the eastern coast of the continent of Australia during the Miocene epoch." He then attributes a childlike innocence to the reef, writing, "Its first 24.99 million years were seemingly happy ones" before describing its fast growth "into a vast labyrinth of shallow-water reefs." The article's language also shifts from detached to expressive via mentions of the reef's "thrilling marine adventures" and exhibition of "otherworldly colors and patterns." Although expressive language is common in opinative obituaries, the article's earlier introduction suggested an informative obituary format. This failure to maintain obit typology, while minor, opposes genre norms and further suggests to readers that Jacobsen's obituary is different.

Following the reef's birth, the article continues the human comparison when explaining the GBR's role in the community. Phrases like "active member" with a "generous spirit" advance the reef's selfless and charitable nature, but the arrival of actual humans juxtapose these anthropomorphic qualities against the reef's true environmental body. The article quotes a source close to the reef—much like an obituary would quote a friend or family member of the deceased—named Charlie Veron, a chief scientist at the Australian Institute of Marine Science, who remembers the reef as a "Stone Age Utopia" where Aboriginal clans would hunt and fish "right up to its demise." For the first time, Jacobsen's obituary clearly establishes itself as imitation by clarifying that the deceased subject being remembered is not *actually* human, but a space "comprised of limestone cliffs and innumerable caves." In doing so, the parody text reassigns audiences' growing feelings of attachment—feelings usually reserved for humans—to the environmental icon.

The life history next segues into the GBR's discovery by Captain James Cook. Jacobsen elucidates that although the reef's "deadly maze" nearly sunk one of Captain Cook's ships, it was nevertheless soon flush with explorers, scientists, artists, and tourists. The article suggests that this attention made the GBR's life complicated, as "Australia's crown jewel" was "rarely out of the spotlight" in the years that followed. These celebrity descriptors are effective at capturing the reef's popularity, but similar to the earlier child comparisons, Jacobsen's text neglects to clarify if these humanistic qualities are intended ironically or seriously.

The obituary's satirical tone peaks during an overview of the reef's use by the Queensland government. Jacobsen explains that outrage over a 1960s decision to lease a portion of the GBR to oil and mining companies initiated a major Australian conservation movement and "Save the Reef" campaign that successfully limited fishing and development in the GBR and led publics to believe that the reef would survive "for all time." If the GBR had lived on for eternity, however, it would not need an obituary, an inconsistency that begins Jacobsen's divergence from the usual obituary genre and asks readers to consider when this life-altering damage occurred.

The reef's life history thus transitions into its final movement: sickness. Despite being designated a Marine Park and selected as a UNESCO World Heritage Site, the article explains that rising water temperatures in 1981 bleached two-thirds of the coral in the GBR's inner ring. Two paragraphs then detail the cause and extent of this mass bleaching, likening it to the spread of cancer, and remind readers that the reef was alive when increasingly acidic oceans started "to dissolve the living reef itself." The article continues by adding that while many were "concerned for the reef's health" and "attempted interventions," a surge in atmospheric carbon dioxide concentrations ultimately led to its "demise" when the ocean reached toxic levels in 2025—nine years after the obituary's publication.

The subtle inclusion of this death year carefully balances uncertainty with unavoidability. The year 2025 is nearly a decade away from the obituary's 2016 publication year, and real obituaries are rarely written (or published!) before death. Additionally, omitting a date of death in an obit's introduction but including it during the life history breaks significant obituary norms. Although these deviations from the obituary genre suggest to readers that there is still time to intervene and save the GBR, assigning the reef a death date at all—be it a year, a month, or a specific day—labels the reef's expiration an inevitability. "No one knows if a serious effort could have saved the reef, but it is clear that no such effort was made" Jacobsen laments, crafting an argument that embraces optimistic change even while admitting defeat.

Readers in 2016 are thus left to decide their future course, which the obituary addresses in a hypothetical scenario:

> [In] 2016, shortly after approving the largest coal mine in its history, [the government of Australia] successfully pressured the United Nations to remove a chapter about the reef from a report on the impact of climate change on World Heritage sites. Australia's Department of the Environment explained the move by saying . . . if you tell people the reef is dying, they might stop coming. By then, the reef was in the midst of the most catastrophic bleaching event in its history, from which it would never recover.

This biting critique of eco-tourism is a clear example of satire's "playful-yet-aggressive" nature (Peifer 2013, 158). Jacobsen's premonition of the mine's impacts asks readers to envision a GBR that "looks like a war zone," as Veron explains a few sentences later, and situates 2016 as a turning point. Yet despite painting a passionate image of the injustice the GBR faces, the text fails to clarify which elements of the satirical scenario are mocking and which are still alterable—what options do publics have the ability to change?

Conclusion: Saving the Survivors

The conclusion of an obituary is used to connect a deceased individual to their remaining family (Moses and Marelli 2004) and communicate funeral or charitable arrangements (Matiki 2001). Jacobsen's obituary follows suit in its final paragraph, which provides coral relations and donation info.

The article identifies two types of relations: predeceased and surviving. First, the obituary reports that "the Great Barrier Reef was predeceased by the South Pacific's Coral Triangle, the Florida Reef off the Florida Keys, and most other coral reefs on earth." Connecting the GBR to past relatives continues Jacobsen's framing of the reef as human, but including "most other coral reefs on earth" in this list is an important satirical exaggeration that stresses the GBR is not the first reef to expire and will likely not be the last. However, tacking on this dark satire at the end of the sentence leaves readers, particularly those unfamiliar with the status of coral reefs, unsure if the sarcastic hyperbole extends to the two deceased reefs mentioned earlier. If readers incorrectly assume that the Coral Triangle and Florida Reef are already dead rather than endangered, they unknowingly attribute a death sentence to reefs that could otherwise be saved.

This potential for misinterpretation makes the article's inclusion of a second relational category, survivors, pivotal. The obit emphasizes how significant the death of the GBR is to its small group of surviving family, which it lists as only "remnants of the Belize Barrier Reef and some deepwater corals." Reef "remnants" and coral in the deep ocean lack individual names and are less recognizable than the supposedly deceased Coral Triangle or Florida Reef, which makes their protection even more imperative.

The obituary then ends succinctly: "In lieu of flowers, donations can be made to Ocean Ark Alliance." Including information on flower or monetary donations is an optional obituary element Jacobsen's article employs to end the obit on a genre-authentic note. At the same time, the final sentence serves as a call to action for audiences, which is critical to the success of Jacobsen's environmental justice intervention.

Obit readers could reasonably assume that Ocean Ark Alliance (OAA) is some sort of marine protection organization that supports reef conservation efforts—and they would be right. In 2016, clicking on the phrase "Ocean Ark Alliance" directed audiences to OAA's homepage, which was crowded with photos and information clarifying the reef's live status. As of summer 2019, however, the site is just one page with a few lines of text that starts, "Ocean Ark Alliance are ceasing operation on 30 June this year after 10 years of providing marine footage and resources to the community" (Hannan 2019). The page directs "lovers and defenders of our ocean environment against climate change and pollution" to the website of the Australian Marine Conservation Society, but it does not address readers of Jacobsen's obituary or specifically signal that the GBR is alive (Hannan 2019). Jacobsen's obituary is still available online and still ends with the OAA weblink, but, with the organization now discontinued, it is possible that today's readers may never receive direct confirmation that the GBR is indeed alive.

DISCUSSION

Analysis of Jacobsen's obituary offers key insight as to where the article went wrong. Jacobsen seems to have believed that a satirical obituary would inspire audiences into action, but the article's commitment to imitating the obituary genre detracted from its satire and overshadowed its environmental justice message. Specifically, while the obit did make an effort to distinguish itself as parody, its underutilization of humor and sarcasm failed to communicate its parody as satirical, which confused readers who recognized the imitation but missed the underlying intent. Usually a satirical misfire would be a disappointment but not a deal breaker; unfortunately, Jacobsen's misfire caused real damage. The negative impact of Jacobsen's article on environmental justice issues such as the GBR makes a final reflection of the article's shortcomings imperative.

First, the likeness of Jacobsen's imitation to an original obituary text created mixed messages about its parody. Dentith (2000) claims that to qualify as parody, texts must imitate other cultural forms and exhibit various "degrees of mockery or humour" (193). Jacobsen's use of satirical exaggeration, humanistic metaphors, and critical tone, though muddied and infrequent, did fulfill these requirements. The existence of a double-voiced discourse, as Bakhtin (1981) advances, was also noticeable in the article's subtle rebukes of the obituary genre, such as replacing the death date with an ellipsis and publishing the obit in advance of the reef's supposed expiration. According to Morson's (1989) parody criteria, however, the obituary had a significant flaw. Per Morson's criteria, effective parodies must evoke the memory of an original text—in Jacobsen's case, a text from the obituary genre. Readers should then recognize the imitative parody as conflicting with the original text and agree with the imitation. Regretfully, Jacobsen's obituary evoked the original text so well that some readers failed to understand the work as imitation, let alone that they should take action in agreement. Jacobsen's intervention failed the most in this regard, as a misconstrued parody risks that readers may take a text at face value—in Jacobsen's case, as a death announcement. This risky choice of genre was especially damning for the already vulnerable GBR, which desperately needed the parody to incite action, not dampen it.

The risks of parody lead to a second shortcoming in the obituary: ultimate trust in audience comprehension of satire. Time and again, the parody relied on underlying satire to further its environmental justice message, but time and again, audiences failed to see past the surface of the text, which Jacobsen had largely trusted would not be a problem. Although the article did include a "big reveal" of its truth in the final sentence, veiling the reveal as a common obituary ending dissuaded readers from perceiving the statement as more than common obituary practice, and Jacobsen failed to consider the impacts if audiences did not finish the article or neglected to click through the OAA link. This is even more challenging today, as the culminating fail-safe of OAA is now entirely gone.

In the end, most readers misidentified the obituary first as a serious text, then a parody text, and finally a satirical text, which buried the most important aspect of the article—its intended message—too far beneath the surface. As a result, the attention the article intended to use on advocating for the GBR was squandered on misplaced mourning and audience correction, neither of which benefited the reef. Although satire can be useful in generating audience interest, the misinterpretation of Jacobsen's article is a cautionary tale to those considering satire in environmental justice: should a satirical intent be missed, audiences may become not only disinterested in participating, but misinformed, which can be a far worse problem to combat.

Finally, trusting that audiences would comprehend the obituary's satire ignored the possibility of disconnect in author-audience humor preferences. Bore and Reid (2014) recommended that satirical works target narrow audiences with a similar humor style to decrease the likelihood of satirical misunderstanding, but Jacobsen's article was available to anyone with access to the internet—including those with differing perceptions of satire.

Though narrowing the audience scope may have controlled this spiral of misinformation, it would be unfair to entirely blame Jacobsen for reader error considering that satire itself lacks a clear definition. With no universal satirical criteria, it is the author who decides that their work is satirical and the audience who decides its reception. However, some satirical failures have greater repercussions than others—after all, Jacobsen's obituary still exists, and the damage left in its wake still remains. This begs the question: are authors absolved of all responsibility if their satire is misunderstood by audiences? Furthermore, if a satirical failure addresses a matter of substantial importance, is a formal clarification of intent ever warranted?

This chapter explored Jacobsen's article principally as a satirical imitation of the obituary genre. However, this approach overlooked the presence of another communication strategy: apocalyptic rhetoric. The employment of apocalyptic rhetoric in environmental advocacy has seen mixed results, yet it remains a prominent tool in environmental intervention. In a world filled with pressing ecological concerns, the challenge remains of how to inspire change and attain justice without scaring or depressing audiences into inaction. Future scholarship could analyze Jacobsen's obituary from an apocalyptic perspective to examine if application of a comic rather than tragic apocalyptic frame would have been a more effective intervention strategy, and moreover, if a change in apocalyptic tone was possible given the constraints of the obituary genre.

Although Jacobsen's article received considerable criticism, its failure has created a learning opportunity for satirists, communication scholars, and environmental justice advocates. Successful satire can capture audience interest, which makes it an attractive tool for environmental justice issues that need publicity to bring attention to their struggles. When satirical efforts backfire, however, vulnerable environmental issues such as the GBR need time to regroup and shift tactics—time that they do not always have. These potential drawbacks do not mean that satire should not be used in issues of environmental justice, but that communicators must acknowledge the risks they are undertaking when using satire as an advocacy tool.

REFERENCES

Afful, Joseph B. A. 2012. "A Genre Analysis of Death Announcement in Ghanaian Newspapers." *Language Discourse & Society* 1, no. 2: 118–37.

Al-Ali, Mohammed N. 2005. "Communicating Messages of Solidarity, Promotion and Pride in Death Announcements Genre in Jordanian Newspapers." *Discourse & Society* 16, no. 1: 5-31.

Bakhtin, Mikhail. 1981. *The Dialogic Imagination: Four Essays.* Translated by Caryl Emerson and Michael Holquist. Austin: University of Texas Press.

Ballantyne, Roy, Jan Packer, and Karen Hughes. 2009. "Tourists' Support for Conservation Messages and Sustainable Management Practices in Wildlife Tourism Experiences." *Tourism Management* 30, no. 5: 658–64.

Baym, Geoffrey. 2005. "The Daily Show: Discursive Integration and the Reinvention of Political Journalism." *Political Communication* 22, no. 3: 259–76.

Baym, Geoffrey, and Jeffrey P. Jones. 2012. "News Parody in Global Perspective: Politics, Power, and Resistance." *Popular Communication* 10, nos. 1–2: 2–13.

Booth, Paul. 2014. "Slash and Porn: Media Subversion, Hyper-Articulation, and Parody." *Continuum: Journal of Media and Cultural Studies* 28, no. 3: 396–409.

Bore, Inger-Lise K., and Grace Reid. 2014. "Laughing in the Face of Climate Change? Satire as a Device for Engaging Audiences in Public Debate." *Science Communication* 36, no. 4: 454–78.

Branagan, Marty. 2007. "Activism and the Power of Humour." *Australian Journal of Communication* 34, no. 1: 41–54.

Caillaud, Anne, Florence Damiens, Bernard Salvat, and Clive Wilkinson. 2012. "Preventing Coral Grief: A Comparison of Australian and French Coral Reef Protection Strategies in a Changing Climate." *Sustainable Development Law & Policy* 12, no. 2: 26–31.

Cheng, Derek. 2017. "Mt. Taranaki Will Be Granted Special Legal Status Similar to Te Urewera and the Whanganui River." *New Zealand Herald,* December 21, 2017. https://www.nzherald.co.nz/nz/news/article.cfm?c_id=1&objectid=11963982.

Cox, J. Robert. 2007. "Nature's 'Crisis Disciplines': Does Environmental Communication Have an Ethical Duty?" *Environmental Communication* 1, no. 1: 5–20.

Crespo-Fernández, Elicier. 2007. "Linguistic Devices Coping with Death in Victorian Obituaries." *Revista Alicantina de Estudios Ingleses* 20: 7–21.

D'Angelo, Chris. 2016. "Great Barrier Reef Obituary Goes Viral, to the Horror of Scientists." *The Huffington Post*, October 25, 2016. http://www.huffingtonpost.com/entry/scientists-take-on-great-barrier-reef-obituary_us_57fff8f1e4b0162c043b068f.

Davey, Madeline, and Josephine Gillespie. 2014. "The Great Barrier Reef World Heritage Marine Protected Area: Valuing Local Perspectives in Environmental Protection." *Australian Geographer* 45, no. 2: 131–45.

Day, Amber, and Ethan Thompson. 2012. "Live from New York, It's the Fake News! *Saturday Night Live* and the (Non)Politics of Parody." *Popular Communication* 10, nos. 1/2: 170–82.

Dechnik, Belinda, Jody M. Webster, Gregory E. Webb, Luke Nothdurft, and Jian-Xin Zhao. 2017. "Successive Phases of Holocene Reef Flat Development: Evidence from the Mid- to Outer Great Barrier Reef." *Palaeogeography, Paleoclimatology, Palaeoecology*, 466: 221–30.

Deloitte Access Economics. 2017. "At What Price? The Economic, Social and Icon Value of the Great Barrier Reef." https://www2.deloitte.com/content/dam/Deloitte/au/Documents/Economics/deloitte-au-economics-great-barrier-reef-230617.pdf.

DeLaure, Marilyn. 2011. "Environmental Comedy: *No Impact Man* and the Performance of Green Identity." *Environmental Communication* 5, no. 4: 447–66.

DeLuca, Kevin. 2007. "A Wilderness Environmentalism Manifesto: Contesting the Infinite Self-Absorption of Humans." In *Environmental Justice and Environmentalism: The Social Justice Challenge to the Environmental Movement,* edited by Ronald Sandler and Phaedra C. Pezzullo, 27–55. Cambridge, MA: The MIT Press.

Dentith, Simon. 2000. *Parody: New Critical Idiom*. New York: Routledge.

Devlin, Michelle J., Caroline Petus, Eduardo Da Silva, Dieter Tracey, Nicholas H. Wolff, Jane Waterhouse, and Jon Brodie. 2015. "Water Quality and River Plume Monitoring in the Great Barrier Reef: An Overview of Methods Based on Ocean Colour Satellite Data." *Remote Sensing* 7, no. 10: 12909–41.

Druick, Zoë. 2009. "Dialogic Absurdity: TV News Parody as a Critique of Genre." *Television and New Media* 10, no. 3: 294–308.

Elekaei, Atefeh, Sajad Faramarzi, and Hossein Heidari Tabrizi. 2014. "Genre-Based Discourse Analysis of Death Announcements and Obituaries in Iranian Newspapers." *The Iranian EFL Journal* 11, no. 1: 37–52.

Fawcett, Grant. n.d. *Rights of Nature: A Vision to Protect Our Great Barrier Reef.* Environmental Defenders Office of Northern Queensland. Retrieved from https://edonq.org.au/wp-content/uploads/2019/06/EOFY-appeal-letter-for-web.pdf.

Ferse, Sebastian C. A., María Máñez Costa, Kathleen Schwerdtner Máñez, Dedi S. Adhuri, and Marion Glaser. 2010. "Allies, Not Aliens: Increasing the Role of Local Communities in Marine Protected Area Implementation." *Environmental Conservation* 37, no. 1: 23–34.

Foxwell-Norton, Kerrie, and Claire Konkes. 2019. "The Great Barrier Reef: News Media, Policy and the Politics of Protection." *The International Communication Gazette* 81, no. 3: 211–34.

Foxwell-Norton, Kerrie, and Libby Lester. 2017. "Saving the Great Barrier Reef from Disaster, Media Then and Now." *Media, Culture & Society* 39, no. 4: 568–81.

Gibbs, Lois Marie. 1993. "Celebrating Ten Years of Triumph." *Everyone's Backyard* 11, no. 2: 2.

Gray, Jonathan. 2005. "Television Teaching: Parody, *The Simpsons*, and Media Literacy Education." *Critical Studies in Media Communication* 22, no. 3: 223–38.

Gray, Jonathan. 2006. *Watching with The Simpsons: Television, Parody, and Intertextuality.* New York: Routledge.

Gray, Jonathan, Jeffrey P. Jones, and Ethan Thompson. 2009. "The State of Satire, the Satire of State." In *Satire TV: Politics and Comedy in the Post-Network Era,* edited by Jonathan Gray, Jeffrey P. Jones, and Ethan Thompson, 3–36. New York: New York University Press.

Hannan, D. 2019. Ocean Ark Alliance. Retrieved from https://www.oceanarkalliance.org.au.

Hendry, Judith. 2010. *Communication and the Natural World.* State College, PA: Strata Publishing, Inc.

Huebner, A. 2011. "Tourism and the (Un)expected: A Research Note." *Pacific News* 36: 25–28.

Hutcheon, Linda. 1985. *A Theory of Parody: The Teachings of Twentieth-Century Art Forms.* New York: Methuen.

Jacobsen, Rowan. 2016. "Obituary: Great Barrier Reef (25 million BC–. . .)." *Outside*, October 11, 2016. https://www.outsideonline.com/2112086/obituary-great-barrier-reef-25-million-bc-2016.

Jalilifar, Ali Reza, Masoud Varnaseri, Soheil Saidian, Saeed Khazaie. 2014. "An Investigation into the Genre Features of Funeral Announcements: People's View in Social Communications in Focus." *Procedia—Social and Behavioral Sciences* 98, no. 6: 747–56.

Karimnia, Amin, and Leila Barbareh. 2017. "Moves, Poetic Types, Poetic Schemes, and Cultural Themes in Iranian Obituaries." *Journal of Multicultural Discourses* 12, no. 3: 290–305.

Kaye, Barbara K., and Thomas J. Johnson. 2016. "Restoring Sanity through Comic Relief: Parody Television Viewers and Political Outlook." *Atlantic Journal of Communication* 24, no. 3: 131–43.

Kuipers, Giselinde. 2006. *Good Humor, Bad Taste: A Sociology of the Joke*. Berlin: Mouton de Gruyter.

Lemelin, Harvey, Jackie Dawson, Emma J. Stewart, Pat Maher, and Michael Lueck. 2010. "Last-Chance Tourism: The Boom, Doom and Gloom of Visiting Vanishing Destinations." *Current Issues in Tourism* 13, no. 5: 477–93.

Lester, Libby. 2016. "Containing Spectacle in the Transnational Public Sphere." *Environmental Communication* 10, no. 6: 791–802.

Luter, Heidi M., Alan R. Duckworth, Carsten W. Wolff, Elizabeth Evans-Illidge, and Steve Whalan. 2016. "Recruitment Variability of Coral Reef Sessile Communities of the Far North Great Barrier Reef." *PLoS ONE* 11, no. 4: 1–16.

Matiki, Alfred. 2001. "The Pragmatics of Obituary Notices in Malawian Newspapers." *Language Matters: Studies in the Languages of Southern Africa* 32, no. 1: 29–52.

Miller, Carolyn R. 1984. "Genre as Social Action." *Quarterly Journal of Speech* 70, no. 2: 151–67.

Morson, Gary Saul. 1989. "Parody, History, and Metaparody." In *Rethinking Bakhtin: Extensions and Challenges,* edited by Gary Saul Morson and Caryl Emerson, 63–86. Evanston, IL: Northwestern University Press.

Moses, Rae A., and Giana D. Marelli. 2004. "Obituaries and the Discursive Construction of Dying and Living." *Texas Linguistic Forum* 47: 123–30.

Nisbet, Matthew C., and Dietram A. Scheufele. 2009. "What's Next for Science Communication? Promising Directions and Lingering Distractions." *American Journal of Botany* 96, no. 10: 1767–78.

Nwoye, Onuigbo G. 1992. "Obituary Announcements as Communicative Events in Nigerian English." *World Englishes* 11, no. 1: 15–27.

Ondimu, Jacquiline. 2014. "A Socio-Cultural Understanding of Death: A Genre Analysis of Obituaries in a Kenyan Newspaper." *Language Matters* 45, no. 1: 3–22.

Peifer, Jason T. 2013. "Palin, *Saturday Night Live*, and Framing: Examining the Dynamics of Political Parody." *The Communication Review* 16, no. 3: 155–77.

Pezzullo, Phaedra C. 2016. "Hello from the Other Side: Popular Culture, Crisis, and Climate Activism." *Environmental Communication* 10, no. 6: 803–6.

Pezzullo, Phaedra C., and Ronald Sandler. 2007. "Introduction: Revisiting the Environmental Justice Challenge to Environmentalism." In *Environmental Justice and Environmentalism: The Social Justice Challenge to the Environmental Movement,* edited by Ronald Sandler and Phaedra C. Pezzullo, 1–24. Cambridge, MA: The MIT Press.

Piggott-McKellar, Annah E., and Karen E. McNamara. 2017. "Last Chance Tourism and the Great Barrier Reef." *Journal of Sustainable Tourism* 25, no. 3: 397–415.

Queensland Museum. n.d. "Human Impact on the Reef." https://www.qm.qld.gov.au/microsites/biodiscovery/05human-impact/importance-of-coral-reefs.html.

Schweizer, Errol. 1999. "Environmental Justice: An Interview with Robert Bullard." *Earth First! Journal*. http://www.ejnet.org/ej/bullard.html.

Sessions, George. 1995. *Deep Ecology for the Twenty-First Century*. Boston, MA: Shambhala Publications.

Sillars, Malcolm O., and Bruce E. Gronbeck. 2001. *Communication Criticism: Rhetoric, Social Codes, Cultural Studies*. Prospect Heights, IL: Waveland Press, Inc.

Stein, M. L., Susan F. Paterno, and R. Christopher Burnett. 2006. *Newswriter's Handbook: An Introduction to Journalism*, 2nd ed. Oxford: Blackwell.

Stewart, Julie, and Thomas Clark. 2011. "Lessons from *South Park*: A Comic Corrective to Environmental Puritanism." *Environmental Communication* 5, no. 3: 320–36.

Tanasescu, Mihnea. 2017. "Rivers Get Human Rights: They Can Sue to Protect Themselves." *Scientific American*, June 19, 2017. https://www.scientificamerican.com/article/rivers-get-human-rights-they-can-sue-to-protect-themselves/.

Taylor, Dorceta E. 2000. "The Rise of the Environmental Justice Paradigm: Injustice Framing and the Social Construction of Environmental Discourses." *American Behavioral Scientist* 43, no. 4: 508–80.

Trenouth, Amy L., Cher Harte, Chloe Paterson de Heer, Kanwaljeet Dewan, Anna Grage, Carmen Primo, and Marnie L. Campbell. 2012. "Public Perception of Marine and Coastal Protected Areas in Tasmania, Australia: Importance, Management and Hazards." *Ocean and Coastal Management* 67: 19–29.

Turner, Matthew R. 2003. "Cowboys and Comedy: The Simultaneous Deconstruction and Reinforcement of Generic Conventions in the Western Parody." *Film & History* 33, no. 2: 48–54.

Williams, Timothy. 2019. "Legal Rights for Lake Erie? Voters in Ohio City Will Decide." *The New York Times*, February 27, 2019. https://www.nytimes.com/2019/02/17/us/lake-erie-legal-rights.html.

Wynveen, Christopher J., and Stephen G. Sutton. 2017. "Engaging Great Barrier Reef Stakeholders: Mediation Analyses of Barrier among the Antecedents of Pro-Environmental Behavior." *Human Dimensions of Wildlife* 22, no. 2: 126–41.

Section IV

Posthuman and More-Than-Human Justice

In 2017, New Zealand passed the Te Awa Tupua (Whanganui River Claims Settlement) Bill, which recognized the Te Awa Tupua, the Whanganui River, as an indivisible and whole being with rights unto itself (Davison 2019). A decade earlier, Ecuador set a model by writing fundamental rights of nature into its national constitution (Humphreys 2017). Increasingly, in the Anthropocene, human actors and organizations are recognizing that human concepts like "justice" may and ought to extend beyond the human, granting distributive justice, procedural justice, and justice of social recognition to plants, animals, physical landscapes, environments, and waterways themselves. An environment that includes plants, animals, and water, they suggest, cannot be just unless those aspects of the environment are afforded justice equal to that enjoyed by and protected for human communities (Low and Gleeson 1998).

Recent years have demonstrated how plants and animals and bodies of water are, indeed, frequently the victims of water injustice. Aquifers and bodies of water are exhausted for human use. Human actions affect climates and environments, rerouting rivers and evaporating water bodies, like the Great Salt Lake (Penrod 2019) and Aral Sea (Howard 2014). In recent years, red tide blooms of toxic algae have become annual occurrences on the Florida coast, infecting and killing sea life (Allen 2019).

The material and ecological turns in rhetorical scholarship are defining turns for environmental communication in the early twenty-first century (McGreavy et al. 2018) and they are prompting researchers, critics, and activists to rethink our foundational investigative definitions. In this collection, already, themes of posthuman and more-than-human justice have appeared (see chapters 4, 8, and 12, especially), suggesting questions of posthuman and more-than-human justice (and injustice) are central to all other rhetoric, water, and justice conversations in the current moment.

This final section of the collection extends the conversation of water justice into such novel areas by both examining rhetorical invention in campaigns for more-than-human justice and meditating on the theoretical and practical implications of applying a more-than-human justice in water justice discourses. Each chapter highlights a rhetorical concept useful in conceptualizing this new realm of discussion.

In chapter 13, Joshua Trey Barnett engages more-than-human ecological rhetorics by theorizing "*ecological grievability*" conveyed in photographic images of a water catastrophe: the 2015 three-million-gallon spill of acid mine drainage into Colorado's Animas River. Following the spill, the water of the river took on a spectacular yellowish-orange color. Contrary to the usual invisibility of toxins, the plume made toxicity strikingly visible and, in the wake of the disaster, many people reported that they were in "mourning" for the watershed. The chapter, extending themes introduced by Bruns earlier in the collection, uses theories of mourning to guide its rhetorical analysis of images from the Animas River spill, with particular attention to a photo syndicated by the Associated Press and circulated widely online.

In chapter 14, Mark Pedelty uses personal activism experience to consider the tangible and intangible legal boundaries at play in the Salish Sea, connecting Canada and the United States. The Salish Sea and other water bodies, he argues, are "*boundary objects*," interpretable bridges between social worlds that "facilitate communication across groups" (White et al. 2010, 221). The chapter examines the Sea's function as boundary object encountered as Pedelty produced and directed *LOUD*, a music video concerning noise pollution in the Salish Sea region. Pedelty reflects on challenges and failures in the production and reception of the video, looking specifically at intersections and boundaries in symbols of nationality, gender, species, and genre. The chapter concludes that *LOUD* had retreated from these borders when creatively transgressing them might have been more efficacious and artful, and encourages future activist campaigns to take note.

In chapter 15, Christopher Ingraham reflects on a 2017 court ruling in the Indian state of Uttarakhand, ruling that the Ganges and Yamuna rivers could be granted the legal status of "living human entities" and afforded the full set of rights thereby implied. Articulating this case of a river's human rights within the recent history of a broader legal movement advocating the rights of nature at large, the chapter discusses how such movements can be understood as endeavors to *codify the posthuman*, engaging a series of questions begged by the new legislation. It argues that in a time when posthuman thinking is sweeping the academy, the legal assignation of human rights to a river can help human communities think more clearly about what is at stake in moving beyond anthropocentrism, particularly when it comes to environmental justice and those rhetorical measures undertaken to achieve it. Building initially from the case of the Ganges and Yamuna, Ingraham expands to reflect more broadly on how the attempt to give nature rights can be understood to operate rhetorically in service of environmental justice beyond its standard meaning as a project of improving uneven distribution of natural resources and human access to clean, thriving environments for all people and incorporating a commitment to the immanent justice due to the environment itself, irrespective of what people might look to gain or take from a relationship with it.

In chapter 16, Jacob A. Miller considers the Great Pacific Garbage Patch (GPGP), a nearly one-million-square-mile floating patch of trash in the Pacific Ocean that contains seventy-nine thousand tons of plastic. The chapter identifies a need for analysis on the *inert discourses* imbued in the GPGP. In attuning to an *Ecospheric* rhetoric, which addresses humanity's impact on Earth systems, Miller argues that the GPGP represents the material accumulation of the effects of the spatiotemporal rift between the ecosphere, its local manifestations, and popular discourses. In doing so, it presents a theoretical frame that challenges the distinction between signifier and signified in posthuman and more-than-human material rhetorics.

These final chapters of the collection are followed by a summary reflection and conclusion, inviting future research criticism and meditation on the confluence of water, rhetoric, and social justice.

REFERENCES

Allen, Greg. 2019. "After 16 Months of Dead Fish, Manatees and Dolphins, Florida's Red Tide Ebbs." National Public Radio. February 14, 2019. https://www.npr.org/2019/02/14/694479180/after-16-months-of-dead-fish-manatees-and-dolphins-floridas-red-tide-ebbs.

Davison, Isaac. 2019. "Whanganui River Given Legal Status of a Person under Unique Treaty of Waitangi Settlement." *New Zealand Herald*, March 15, 2017. https://www.nzherald.co.nz/nz/news/article.cfm?c_id=1andobjectid=11818858.

Howard, Brian Clark. 2014. "Aral Sea's Eastern Basin Is Dry for First Time in 600 Years." *National Geographic*, October 2, 2014. https://www.nationalgeographic.com/news/2014/10/141001-aral-sea-shrinking-drought-water-environment/.

Humphreys, David. 2017. "Rights of Pachamama: The Emergence of an Earth Jurisprudence in the Americas." *Journal of International Relations and Development* 20: 459–84.

Low, Nicholas, and Brendan Gleeson. 1998. *Justice, Society and Nature: An Exploration of Political Ecology*. London: Routledge.

McGreavy, Bridie, Justine Wells, George F. McHendry, Jr., and Samantha Senda-Cook. 2018. *Tracing Rhetoric and Material Life: Ecological Approaches*. Cham, Switzerland: Palgrave Macmillan.

Penrod, Emma. 2019. "Scientists Say the Great Salt Lake Is Disappearing, but Could Utah Residents Save It?" *Salt Lake Tribune. Updated May 1, 2019*. https://www.sltrib.com/news/environment/2017/12/16/scientists-say-the-great-salt-lake-is-disappearing-but-could-utah-residents-save-it/.

White, Dave D, Amber Wutich, Kelli L Larson, Patricia Gober, Timothy Lant, and Clea Senneville. 2010. "Credibility, Salience, and Legitimacy of Boundary Objects: Water Managers' Assessment of a Simulation Model in an Immersive Decision Theater." *Science and Public Policy* 37, no. 3: 219–32.

Chapter Thirteen

Grievable Water

Mourning the Animas River

Joshua Trey Barnett

> I read the stories, and look at the pictures, and I mourn the river. I mourn for the animals that may not survive. I mourn for the communities whose lives and livelihoods are built on its shores. I know that the river will outlive this disaster, and outlive many of us, but I mourn for it all the same.
> —Josh Stearns, "Memories of the Animas River Before It Ran Yellow" (2015)

> Without grievability, there is no life, or, rather, there is something living that is other than life.
> —Judith Butler, *Frames of War: When Is Life Grievable?* (2009, 15)

Perched some 11,200 feet (more than two miles) above sea level, Animas Forks is one of the highest mining camps in the United States. Although prospectors abandoned Animas Forks in the early twentieth century, the camp endures as one of countless Ghost Towns scattered about the American West. Near the deserted mining camp, the confluence of the West and North Forks of the Animas River—the site's namesake—forms the Animas River, which descends some 126 miles from southwestern Colorado's alpine forests to northwestern New Mexico's shrubby steppe, where it spills into the San Juan River near Farmington. After catching water from the Animas, the San Juan meanders westward through New Mexico, Colorado, and Utah before dumping into Lake Powell, a reservoir formed during the 1960s when the U.S. Bureau of Reclamation dammed the Colorado River near present-day Page, Arizona.

The splendor of the Animas River belies a toxic history. Along the banks of the Animas near Durango, Colorado, for example, the U.S. Vanadium Corporation began milling uranium for the Manhattan Project during the

1940s, but continued its operations well into the 1960s. That single mill in Durango, Jonathan Thompson (2018b) reports in *High Country News*, "kicked out some 997,000 pounds of tailings per day" for about two decades. In addition to the tailing ponds near the mill, there was a "separate liquid waste stream that flowed directly into the [Animas River] at about 340 gallons per minute, carrying with it at least 15 tons per day of spent ore solids, leftover acid leaching chemicals, and an iron-aluminum sludge." Everywhere the waters of the Animas River went—downstream to become irrigation and drinking water in settlements and indigenous reservations alike—so, too, did the radioactive material dredged up out of the earth in wartime frenzy. It was not until 1986 that the U.S. Department of Energy initiated efforts to clean up and decontaminate the abandoned Vanadium mill. Thompson (2018b) writes, "The site of the once-sprawling industrial complex is now a dog park, and is slated to become the city's official homeless camp. The place where the radioactive water once poured into the river is the biggest rapid in town, and a favorite of rafters and kayakers alike."

Yet, these efforts to clean up and, in some cases, to cover over the Animas River's toxic history could not conceal the devastatingly spectacular consequences of the Gold King Mine spill in 2015. A report authored by researchers at the University of Arizona's Superfund Research Program notes that "the Gold King Mine was discovered in 1887 and operated until 1922," making it "only one of many abandoned gold and silver mines [. . .] within the Upper Animas watershed," where, they continue, "the water is often acidic and contaminated with heavy metals" (Chief et al. 2016). On August 5, 2015, during an investigative visit to "assess the ongoing water releases from the mine, treat mine water, and assess the feasibility of further mine remediation" ("Emergency Response" 2015), U.S. Environmental Protection Agency staffers accidentally released three million gallons of backed-up acid mine drainage into the north fork of Cement Creek, which flows into the Animas in Silverton, Colorado. Thompson relays the events that led up to the spill: "Workers from the [EPA] were hoping to install a pipe to drain the water so that they could eventually plug the mine," but "they ended up accidentally breaching the dam, releasing the water" (Thompson 2015). The acid mine drainage released into Cement Creek "contained a number of metals and salts totaling about 190 tons of solids," including some toxic metals like lead, arsenic, mercury, and cadmium (Chief et al. 2016). On its own, acid mine drainage is rusty red or dark orange; when it mixes with freshwater, the metallic liquid becomes less acidic and turns the water a spectacular yellowish-orange.

While the acidic effluent rushed from Cement Creek in the Animas River, a bright orange plume declared the catastrophe to downstreamers—to the residents of the cities of Silverton and Durango, Colorado, and to the towns of Cedar Hill, Aztec, and Farmington, New Mexico; to the indigenous peoples living on ancestral lands in the Southern Ute Reservation; and, in due course, to the Diné people to the west dwelling on Navajo lands straddling the San Juan River. Contrary to the usual invisibility of toxins, the plume made toxicity strikingly visible—at least for a few days until the acid mine drainage became so diluted that it once again went incognito, settling into the sediment on the riverbed. As Thompson, describing a throng of onlookers who "gathered on a bridge over the deep-green waters of the Animas River" the day after the spill, put it:

> They were not here to cheer anyone on, but to mourn, gathered to watch a catastrophe unfold in slow motion. Soon, the waters below would become milky green, then a Gatorade yellow, before finally settling into a thick and cloudy orangish hue—some compared it to mustard, others Tang. (Thompson 2015)

The discolored water was unsettling, to say the least. In the first few pages of his book on the spill, Thompson (2018a) characterizes the plume as "the orange menace," "the terrible color," "electric-orange water," "psychedelic smoke," and "a bright orange incision" (3–4). So disconcerting was the Animas River's hue that many indigenous farmers downstream immediately stopped drawing on the river to irrigate their crops; some, weary of official reports of the river's return to health, refused to draw on the river for the remainder of the growing season. Members of the Navajo Nation eventually sued the U.S. EPA for damages (see Smith 2018).

Although the orange plume dissipated within a few days, the feelings it exposed gesture toward a deep-rooted sense of interconnection with the waterways most directly impacted by the spill: both the Animas and San Juan rivers are intimately interwoven not only into the landscape but into the ways of life of those who dwell—who live, work, play, and pray, to adopt a slogan from the environmental justice movement—in the Four Corners region. In the immediate wake of the disaster, many people reported that they were in mourning for the watershed. In the epigraph to this chapter, for instance, journalist Josh Stearns (2015) registered his own emotional response: "I mourn for the river." In an *ABC News* report, past president of the Navajo Nation Russell Begaye shared similar feelings:

> Our soul is hurting. [. . .] I see people daily who weep when they see me, asking me, 'How do I know the water will be safe?' The Animas River and the San Juan rivers are our lifelines. Water is sacred to us. The spirit of our people is being impacted. (Tan 2015)

As the plume made its way through the Animas River watershed, others expressed a sense of loss and distress. One downstream farmer, Kim Howe, told *The Washington Post*, "I had nothing else to do but cry" when the plume reached Shiprock, New Mexico, her hometown (Kaplan 2015). Inspired by the Animas River spill and others like it, Winona LaDuke (2016) poignantly asked, "How do we grieve the death of a river?"

How *do* we grieve the death of a river? LaDuke's query resonates with my own interest in "ecological grief," which Ashlee Cunsolo and Neville Ellis (2018) describe as "the grief felt in relation to experienced or anticipated ecological losses, including the loss of species, ecosystems, and meaningful landscapes due to acute or chronic environmental change" (275). Contributors to *Mourning Nature*, the first scholarly volume to directly engage the question of ecological grief, demonstrate both the pervasiveness and complexity of peoples' experiences of mourning for the loss and transformation of the natural world (Cunsolo and Landman 2017). As a rhetorical scholar, my own approach to ecological grief has been to ask under what discursive conditions we become capable of mourning for other beings, species, places, and ecosystems. Elsewhere, for instance, I consider how the discourse of "naming [often] foreshadows mourning" for our more-than-human kin (Barnett 2019, 291). "To know and to utter the other's name is," I argue, "to render the life or way of life it refers to grievable" (298). Important as it may be as a precondition for experiencing grief, though, naming is not the only condition that sets mourning into motion. To begin to answer LaDuke's question, we must also consider the visual rhetoric of ecological trauma and loss.

At the time of the Gold King Mine spill, I was dwelling in Salt Lake City, Utah, several hundred miles northwest of where the San Juan River diffuses into Lake Powell within the Glen Canyon National Recreation Area. Photographs of the orange plume captured and sustained my attention for several days as I worried about the toxins' effects on the larger Colorado River watershed: I could not look away as new photographs of the drifting cloud emerged to reveal the extent of the catastrophe. Although I knew very little about the Animas, I felt (and still feel) a sense of loyalty to southern Utah's red rock country, and I worried about what the plume might mean for the waterways in my own adopted state. I was apparently not alone in being struck by the images. As Thompson (2018a) writes, "In the days to come, that distinctive shade of orange [would] strike a lot of other people—millions of them, in fact" (4). Writing in *The New York Times*, Gwen Lachelt (2015)—a commissioner for La Plata County, Colorado, through which the Animas River flows—called the photos "jaw-dropping." For some of us, they were heart-wrenching.

Durango, Colorado–based photographer Jerry McBride's image of three kayakers floating the intoxicated river, in particular, left me feeling bereft, as though I was witnessing a watershed on the verge of destruction and, possibly, irreparable loss. Vibrant orange water floods the frame as the river slips through verdant forest and drifts around ancient rocks. Although

McBride's photo is typical of post-spill images in that it focuses on the discolored river, it is somewhat unusual in that it includes human subjects within the frame: three kayakers, adrift with paddles in hand, pose in the foreground, dwarfed by the waterway that both buoys them along and, now, exposes them to a range of toxins. McBride's photograph was syndicated by the Associated Press and circulated extensively in print and online, appearing first in the *Durango Herald*, and then in *The New York Times*, *CBS News*, *The Wall Street Journal*, *ABC News*, *The Salt Lake Tribune*, *The Seattle Times*, *Fox News*, and *The Washington Times*, to name just some of the venues in which it was published. Almost four years after the spill occurred, as I was revising this chapter, McBride's photograph was still the first result in a Google Image search (see durangoherald.com/articles/1532 for image).

A later section of this chapter is devoted to a reading of McBride's photograph, which, I contend, forcefully exposes the precarity and grievability of the Animas River watershed. Exploiting the drama inherent in the discoloration, McBride's photo makes visible what is otherwise likely to remain unseen and unsensed by those of us who are not, owing to distance, intimately involved with this particular watershed. The photo asks viewers to understand the river as precarious, vulnerable, fragile, and, thus, in need of care and support. Such care, when and where it is extended, is animated by a sense of grievability, an awareness that one would grieve for the river if it were lost or destroyed. To render these claims intelligible, a brief theoretical detour is in order. Thus, in the following section I briefly theorize the relationship between ecological grievability and the visual rhetoric that sometimes sets it into motion.

ECOLOGICAL GRIEVABILITY

In "Prairie Birthday," an entry in his *Sand County Almanac*, the American ecologist Aldo Leopold (1949) laments the loss of a single *Silphium laciniatum* plant, "a pin-point remnant of the native prairie," tucked away in a forgotten corner of a country cemetery (45). Knowing that this particular plant survived in that spot, Leopold routinely visited the graveyard in the summer to observe the compass plant until, one August day, he found it had been cut. Reflecting on his own sense of bereavement in the wake of this particular loss, Leopold concludes that "we grieve only for what we know" (48).

For many who dwell within the Animas River and San Juan River watersheds, who "know" the rivers and their tributaries well, whose ways of life depend—not only economically but also ecologically and spiritually—on these waterways, the sight of the toxic plume was grief-inducing. It represented one more in a long line of assaults on their home waters (Thompson

2018a). For those of us who dwell in other watersheds, however, such feelings are less likely or, at the very least, less directly registered: it is hard to grieve for something so distant from one's everyday life. In their introduction to *Mourning Nature*, Ashlee Cunsolo and Karen Landman (2017) write, "Although it is hard to imagine the loss of a human, it is harder still to think through the loss of non-humans, particularly if the non-humans we are mourning are large assemblages or systems, such as a body of water or an ecosystem or a forest" (16). That, of course, was Leopold's lesson: bereavement is strongest when born out of intimacy. It is also the subject of much writing that deals with the question of whether and how we might be moved to care for suffering others whose painful fates seem detached from our own lives (see, for instance, Sontag 2003). There is a limit, it seems, to one's capacity to grieve for those one does not know.

And yet, this limit is often breached by the loss of human and more-than-human others whom we do not know but whose (potential) loss nevertheless makes an ethical claim upon us. In such moments, what capacitates us to feel that someone or something important, meaningful, and valuable has been or could be lost, and that this loss deserves (or would deserve) to be mourned?

Judith Butler's work (2009) on precariousness, precarity, and grievability provides a useful starting point for responding to this question. For Butler, *precariousness* is an existential condition: there is no life that is not lived precariously. "Precariousness," she writes, "implies living socially, that is, the fact that one's life is always in some sense in the hands of the other" (14). We are precarious from the moment we are born and remain so until the moment we die because we live not independently but interdependently; we depend upon, and are in turn depended upon by, others, not all of whom are human. Whereas precariousness conditions all life, *precarity* "designates that politically induced condition in which certain populations suffer from failing social and economic networks of support and become differentially exposed to injury, violence, and death" (25). Butler's conception of precarity emphasizes the extent to which precariousness is differentially distributed by political and economic forces not entirely within one's control: although all lives are precarious "by definition" (25), not all lives are equally precarious.

While Butler's notion of precarity assumes a narrow conception of life as *human* life (Taylor 2008, 63), the condition of precarity need not be so anthropocentrically conceived. Elsewhere, in my own attempt to bridge Butler's work with ecological thought, I have offered the following extension:

> Not only are human bodies and lives at stake in the unfolding of politics; so, too, are the bodies and lives of our more-than-human cohabitants on the line. The ecosystems that provide shelter and nourishment, and that supply the basic conditions of a livable life for human and more-than-human creatures, are

> continually threatened and exposed to chaotic and disruptive actions. (Barnett 2018, 36)

The Gold King Mine spill is just one example of how "chaotic and disruptive actions" expose the more-than-human to forms of injury, violence, and, if not death, then certainly transformation and degradation. If the terms *precariousness* and *precarity* are to explain the forms that vulnerability takes today, they must themselves be able to account for the ways in which the more-than-human in its manifoldness is existentially and politically exposed to both forms of harm and modes of care. In other words, such concepts must heed the lesson of the Anthropocene—namely, that human and natural history are but two names for one story (Chakrabarty 2008).

Apprehending the precariousness of a life is the precondition of caring for and supporting that life. Because something is vulnerable or fragile, it requires care if it is to be sustained. "Precisely because a living being may die," Butler (2009) writes, "it is necessary to care for that being so that it may live." And yet, not all precarious lives receive the care they need. "Only under conditions in which the loss would matter," Butler argues, "does the value of the life appear" (14). Only if the potential loss registers, in advance of any actual loss, as a significant loss will we be moved to care for the other. Butler calls this presupposition of the potential loss's significance *grievability*. A grievable life is one that matters: if and when it is lost, we can assume that its loss will be mourned. Without this assumption, the life is unlikely to receive the kinds of support it requires to survive, much less to thrive. And so, grievability names an imagined future response to loss that animates care in the present. This imagined response is related to but different from the question of whether one actually grieves the loss of another. This means that "grievability is a condition of a life's emergence and sustenance" rather than an activity carried out in the wake of an actual loss (15). Grievability, then, names an anticipatory stance one takes up in the face of another.

Like precarity, grievability is differentially allocated: the presupposition, "this life will be grieved if and when it is lost," does not attend all lives equally. The domain of grievability, then, presupposes a shadow realm of *un*grievability. Butler (2015) writes:

> The reason that someone will not be grieved, or has already been established as one who will not be grieved, is that there is no present structure of support that will sustain that life, which implies that it is devalued, not worth supporting and protecting as a life by dominant schemes of value. (197–98)

Thus, we are able to ascertain whether someone or something is living a grievable life by considering whether there are supports in place to sustain that life. If there are supports in place, this implies that there are some who value this life and who wish to preserve it. If there is no support in place, however, we must accept that this life is not valued. Though Butler's discussion and examples focus on human lives, plants and animals and places are also subject to the differential allocation of grievability: whereas some will be mourned if they are lost (pets and charismatic megafauna, national parks, and sublime vistas), others are relegated to the realm of what Butler calls the "ungrievable" (mosquitoes and other "pests," damaged lands, and mundane sites). The question is whether and how we might be made to apprehend the differential allocation of precarity and grievability when it is not human but more-than-human entities on the line.

Whether we understand an entity as grievable and precarious, and therefore in need of care, depends upon how that entity appears to us. Some forms of representation invite us to apprehend precariousness and grievability better than others. At the end of *Precarious Life* (2004), Butler asks, "What media will let us know and feel that frailty, know and feel at the limits of representation as it is currently cultivated and maintained?" (151). In *Frames of War* (2009), Butler turns to the leaked photos of the tortured Abu Ghraib prisoners as an example of images which, after breaking with their original context, helped Americans to apprehend the detainees' lives as ungrievable, as exposed to unjustifiable injury and unethical state violence. The photographs demonstrated that, within the reigning scheme of values in the United States, those Iraqi lives did not matter. And yet, in exposing the vulnerability of these lives, Butler argues, the images also enabled Americans to question and to resist the very norms governing whose lives become grievable and whose do not (63–100). So, at one and the same time, the photographs reveal the ungrievability of some lives *and*, by exposing their frailty, invite us to envision those lives as grievable. Butler's question about media animates my analysis of McBride's photograph of the Animas River, for it seems to me that we desperately need new ways of seeing that enable us to apprehend the (un)grievability and precarity of more-than-human beings, species, places, ecosystems, and, as we collectively live into the Anthropocene, the planet itself. While certainly not the only medium capable of inciting new modes of viewing the more-than-human world, photography is sometimes capable of framing the more-than-human in ways that help us to "know and to feel [its] frailty."

Toxins pose certain challenges for visual representation. Indeed, a key element of toxins' nefariousness is that they frequently *do not appear* at all and, when they do, they often do not look dangerous. In *Toxic Tourism*, Phaedra Pezzullo (2007) observes that "when touring toxins, the limits of the visual become all the more apparent" since one is able to draw on senses other than sight to perceive toxicity (29). Although Pezzullo challenges ocularcentrism throughout her book, she does devote part of the penultimate chapter to a reading of a documentary film of a toxic tour produced by the Sierra Club. While she admits that the documentary "is politically useful" since, "as a durable, mobile, and relatively affordable mode of communication," it "provide[s] a means by which to witness places, people, and behaviors that might not be otherwise witnessed," Pezzullo quickly reminds readers that "pollution makes a difference to our bodies in ways that images of pollution cannot—and vice versa" (168). Pezzullo privileges *in situ* tours over images, but acknowledges that since most people will not join a toxic tour, images and videos are also significant sites where the meanings of contamination are constructed. Following Pezzullo, several environmental communication scholars, myself included, have considered what sorts of differences photographs of pollution might make.

"Toxic images," as Jennifer Peeples (2013) defines them, are "*visual representations that are found in print or digital media of people, places, or toxins, which are used to make claims of human-produced contamination causing the degradation of the natural (the body or environment)*" (193). Through an analysis of 184 photos accompanying news stories about Agent Orange—a defoliant containing the toxic chemical dioxin, which American forces used during the Vietnam War to destroy tree cover and crops—Peeples identified several characteristics of toxic images. Toxic images, Peeples claims, are usually situated within larger cultural contexts and alongside other images that clarify their significance; often enable victims of contamination to testify to the consequences of toxins; sometimes reduce the complexity of contamination events; and frequently lessen the uncertainty that clouds accusations of contamination by "clarify[ing] risks that have been purposefully obfuscated" by powerful stakeholders (207). One kind of toxic image is what I have called the "toxic portrait," close-up, *in situ* photographs of people in toxically assaulted places that articulate the precariousness of life to specific forms of pollution by suggesting the links between the human subject's health and their surrounding environment (Barnett 2015, 411). In theorizing toxic portraits, my claim was that these photographs "suggest that toxic exposure might lead directly to the deaths and dying of individuals" (406), and thus invite viewers to respond (or, at least, to imagine responding) in ways that might safeguard these precarious lives against further harms from contamination.

Yet, it is not just humans whose lives are rendered vulnerable because of toxins. As McBride's photograph of the Animas River forcefully suggests, watersheds, too, are subject to the unequal distribution of precarity and grievability. In her otherwise excellent essay theorizing the "Toxic Sublime," Peeples (2011) approaches but ultimately does not pursue a line of thought that takes seriously the precariousness of land communities themselves. Reflecting on the general absence of human subjects in Edward Burtynsky's photographs of contaminated places, including polluted and discolored rivers, Peeples sadly concludes that "there are no victims" depicted in his photos. "The purposeful omission of a victim," she claims, "leaves open the space for viewers to question whether they themselves might be the casualties of the toxins, since no other victims are present" (385). The reduction of "victims" to humans prevents Peeples from recognizing the incredible vulnerability of the landscapes themselves. It is not, then, that "there are no victims" in Burtynsky's work. I would argue, rather, that many of his photographs actually demand that we reimagine what we mean when we use the word "victim." Emphasizing the huge toll of human acts on the planet, Burtynsky's photographs invite us to apprehend the precarity of the earth itself and, thus, to see the earth as a potential victim of our violent deeds.

My interest in noticing and nurturing ways of seeing that appreciate the precarity and grievability of the earth emerges out of a conviction that, as Butler (2009) puts it, our ethical "obligations are precisely to the conditions that make life possible" (23), especially the earthly conditions upon which all lives depend. If we are responsible to and for the conditions of life, in addition to specific living beings or ways of life, then we must come to see those conditions, including hydrological conditions, as themselves grievable. In some cases, this will entail exploiting opportunities to make visible what usually remains invisible. McBride's photograph of the intoxicated Animas River does just this; working in tandem with the spectacular display produced by the chemical reaction of acid mine drainage and freshwater, the photograph constructs a scene in which the precarity and (un)grievability of a watershed come into view. Although, as Thompson (2018b) reports, "The EPA and state and county health departments assured everyone along the Animas and San Juan rivers that all was back to normal" just "a few weeks after the spill," the shocking spectacle provided viewers an opportunity to witness the long history of abuse faced by the Animas and to imagine its potential loss. We often "attend to environmental problems not because they are the most dire, pressing, or dangerous," Peeples (2011) notes, "but because they are the most evocatively articulated" (374). In the next section, I demonstrate how McBride's photograph evocatively articulated the impacts of the Gold King Mine spill in ways that enabled viewers to see the Animas River as grievable.

ENVISIONING GRIEVABILITY

Slick orange water floods the frame, spilling out to the left and right and slipping into the distance. It is unclear where the contamination begins or ends; as far as we can see, the orange water continues downstream as well as up. In the immediate foreground, grey, brown, and red rocks—some smooth, others jagged—plunge into the technicolor river upon which three kayakers linger. With helmets on their heads and paddles in their hands, the boaters pose for the photographer perched overhead, their brightly painted vessels all that separates them from the intoxicated water which spreads out in all directions around a central rocky island. Look closely and you may notice that the boaters are perfunctorily performing for the camera, their mouths contorted into a smile amidst the chaos that surrounds them. On the left and the right, an impenetrable verdant curtain of plants and trees, deciduous and evergreen alike, lines the river's stony shores. In the upper third of the photograph, low lying mountains in the distance reach lushly up into a slice of the firmament, barely blue and streaked with wispy white clouds.

Taken on August 6, 2015, one day after the EPA crew dislodged the plug behind which acid mine drainage had been building up for nearly one hundred years since the Gold King Mine closed in 1922, McBride's photograph was first published in the *Durango Herald* in tandem with an article entitled "Catastrophe on the Animas." The authors report that not only was the river "closed to tubers, rafters, and kayakers," but that Durango had ceased drawing drinking water from the Animas River, "an important secondary source of water for the city during the summer" (Olivarius-Mcallister, Shinn, and Benjamin 2015). Two days after the photograph went viral via social media and the Associated Press, *Durango Herald* editor Shane Benjamin (2015) estimated the impact of McBride's shot: "It is now the iconic photo of a mine cleanup gone wrong," he wrote, "a photo that has been seen by countless people and one that defines the enormity of an environmental disaster." More significantly, it is a photograph which reveals to viewers the historical *un*-grievability of the Animas River and, thus, invites us to envision the watershed anew by enabling us to imagine its potential loss. Since, as Cunsolo and Landman (following Butler) contend, "who and what we do not mourn tells us as much about ourselves as who and what we do mourn" (2017, 11), it is worth considering how the photo reconstructs a sense of the river's historical ungrievability and moves viewers toward an understanding of the watershed as grievable.

Rendered in full color, the orange water dominating the frame summons viewers' attention. The river's hue is jarring, unexpected, unsettling: in a word, uncanny. Although rivers often run murky, and sometimes appear brownish-orange, the saturation of the yellowish-orange water is a bit too vibrant, too bright, too flamboyant to be the product of a heavy rain or upstream rapids, either of which could easily churn up sediment and discolor the water. Some described the color of the river as akin to the color of Tang (Thompson 2015), the artificially flavored and colored powdered drink mix that astronaut John Glenn drank on his NASA-sponsored earth orbit in 1962. The analogy is apt, for not only do the colors of the two liquids resemble one another, but the comparison accentuates the apparent artificiality of the hue. The water is closer in color to the kayaker's ostentatiously painted plastic vessels than it is to any expectation one might harbor for the color of water—a palette that usually includes blues and greens and browns, but never orange. "Whatever you called [the color of the water]," Thompson (2015) wrote, "it was clearly not right."

"Not right," it seems, means "not natural." Indeed, the orange hue of the Animas as pictured in McBride's photo aligns the river not with nature but with culture and, more specifically, with the boom-and-bust economies of mineral mining. As Jeff Snowbarger (2015) lamented in *Orion Magazine*, "In mining country this color is common, from retention ponds to the waste rock piles that pock the Rocky Mountains." But, Snowbarger cautions, familiarity is not necessarily soothing: "the raw color we saw last month wasn't simply 'mustard-colored.'" Rather, he wrote, the "color reflected a nation's worth of short-sighted missteps. The hue was so riveting and sulfuric you could almost picture the Devil himself, dropping his towel and giddily flapping his arms, before landing one wicked bellyflop after another" in the Animas.

Snowbarger's florid description captures the dis-ease engendered by the orange river, for this particular shade viscerally evokes a history of ecological abuse that encompasses yet far exceeds any specific spill. Try as we might to forget or cover over the (ongoing) legacy of extraction in the American west, this surge of Tang-colored liquid reveals that the repressed really does return. The intoxicated river reflects back to us our forebears' choices and, thus, uncovers the ecological consequences with which we must now grapple. It makes plain, in other words, that our inheritance is anything but pure.

The unsettling quality of the river's orange hue is heightened by its contrast with the adjacent rocks, plants, trees, and sky, all of which synecdochally represent "nature" in the photo. While the orange of the river gestures toward artificiality and contamination, the grey of the rocks and the green of the trees symbolizes naturalness and environmental continuity. We associate some colors, but especially green, with nature—as I scan my shelf of environmental histories and philosophies, for instance, I notice that many of the books feature green covers—and others with culture. Given such chromatic connotations, Dylan Wolfe (2008) notes, color is often an essential element of ecological rhetoric. Color "sets tone, defines time, demonstrates difference." "In short," Wolfe writes, "color functions with rhetorical force" (4). This may be especially true when it comes to visualizing contamination. As Peeples (2011) notes, contrasting color is a means of "visually constituting the presence and extent of [. . .] contaminants" since toxins can be "represented through unexpected, sometimes startling color and/or streaks," which pronounce the presence of "something contrasting, unnatural, or unwanted" (383). In McBride's photograph, the orange water disrupts a "natural" landscape. And so, the color of the river makes conspicuous that which might otherwise go unnoticed: the risks of mining to the broader ecosystem through which it slices. The color interrupts the order of things.

The predominance of the orange water in McBride's photograph simultaneously constructs and troubles the borders between nature and culture. On the one hand, as I have already begun to suggest, the color itself serves to demarcate the contamination as a product of extractive cultural practices: the strong orange contrasts with the greens and greys and blues and browns of the other entities in the photograph, thus distinguishing those elements which are affected by human activities and those which seemingly are not. On the other hand, the orangeness of the river indicates the imbrication of the natural and the cultural. It is, after all, a river (a symbol of the natural) that has become contaminated and, albeit temporarily, changed colors. What we assumed was natural (the river) turns out to be thoroughly cultural. It is not a matter of choosing one or the other interpretation. Rather, the discoloration of the Animas invites us to hold both of these perspectives at once. The event that caused the spill is cultural; had humans not mined minerals in the San Juan Mountains, no spill would have occurred. And yet, it makes no sense to draw sharp lines between the natural and cultural because, as the photograph illustrates, what we do "culturally" cannot be confined. Because we are earthly beings, whenever we act we act into nature. The orange river is a startling reminder of this truth.

It is also, and for the same reason, an unsettling reminder of the historical *un*grievability of the Animas River watershed. The overwhelming presence of the orange water makes clear that the river has been and remains exposed to avoidable risks. Accumulations of acid mine drainage are not inevitable; neither are spills. Such risks are manufactured, albeit unintentionally. In the Anthropocene, a time in which humans are impacting not just ecosystems but the entire planet, it is difficult to say whether any watershed is unexposed to such risks. Indeed, any serious consideration would mitigate against making such claims. Yet, it is nevertheless possible and necessary to distinguish among and to evaluate kinds and degrees of exposure. Butler teaches us that although precariousness is existential, precarity is politically produced. So, we have to ask why some entities are exposed to risks in ways that others are not. The spectacular display of the Gold King Mine spill suggests, without confirming, that the Animas is especially precarious, that it has been exposed to injury and violence in ways that other rivers—those that have not turned orange—have not been. Thus, the discoloration dramatically demonstrates that the forms of support and care necessary to safeguard this watershed from abuses have not been in place.

If the discoloration marks the Animas River as ungrievable, as lacking the support necessary to preserve the watershed, the role of place in the photograph complicates such an interpretation. Indeed, besides the orangeness of the water, all other signs in the image signify not just "nature" but, more specifically, "wilderness." We do not see the mine from which the toxic sludge emerged. The riverbanks are lined with rocks and foliage rather than sidewalks and gardens. Mountains, not rooftops or skyscrapers, frame the horizon. No smokestacks rise above the tree line; no drainage pipes spill into the river. If there is industry nearby, it is beyond the photograph's borders. Although the photograph was taken near a road, neither the road nor the many cars that traverse it are visible. There is no inkling of the narrow-gauge railroad that follows the river and carries tourists from Silverton to Durango, Colorado. Even the kayakers—whose presence in the frame breaks with the tradition of wilderness photography, which has historically kept humans out of images (DeLuca and Demo 2000)—function as a symbol of wilderness, for their gear marks them as sophisticated visitors to, but not inhabitants of, this watery, seemingly remote environment. The kayakers' presence on the river contrasts with and even heightens the sense that wherever they are is *out there*, a place where humans might venture, yes, but not remain—which, of course, is the very definition of wilderness as outlined in the 1964 Wilderness Act.

McBride's photographic construction of place—his decision to downplay signs of human incursion into the surrounding landscape, but not in the intoxicated river—plays into popular ideas not only about what wilderness looks like (and, indeed, about what wilderness *is*) but, specifically, about what the American west looks like. For many, visions of the west have long been conjured by landscape photography—the landscapes of the west feature heavily in oversized coffee table books, posters, hanging calendars, and television shows, many of which promote tourism. Kevin DeLuca and Anne Demo (2000), in their study of Carleton Watkin's early landscape photographs from Yosemite, argue that Watkins helped construct an image of nature that is "devoid of human markings, a pristine wilderness where one could glimpse the sublime face of God"—in short, "a nature apart from culture" (254). This idea, this image, of nature has been a key element in wilderness preservation: many demands for protection are predicated on a (deceptive yet persuasive) claim of pristine, untouched, unpeopled nature—a place where people visit but do not dwell (DeLuca 2007). The stories and captions surrounding the photo further identify its location in the American west, specifying "Durango, Colorado" as the nearest town. Yet, though Denver or Boulder register as urban, the less populous areas of Colorado are likely unfamiliar to outsiders except through the myths and narratives they have heard and the images they have seen. McBride's photograph presents the Animas River as both *out there* and *within reach*.

In many ways, the vibrant orange plume put Durango and the Animas on the map for national audiences. By juxtaposing signs of the mythic western wilderness (e.g., an isolated landscape where heroic individuals, here represented by the kayakers, go to conquer nature) with those of toxic contamination (the startlingly unnatural color of the river), McBride's photograph offered national audiences a new vision of a landscape they may have thought they understood. Things like this happen in industrial sacrifice zones, to be sure, but in the wilds of Colorado? This, after all, is supposed to be the land of national parks and untrammeled wildness—not as wild as neighboring Wyoming and Montana, perhaps, but close. Unlike Burtynsky's "manufactured landscapes," upon which Peeples bases her conceptualization of the toxic sublime, the landscape presented in McBride's photograph seems given, not created. The rocks were not placed there by hand, nor were the trees planted by people: everything is too haphazard, too random, too organic for humans to have been involved. Providing the backdrop and surroundings for the orange river, this naturalized landscape invites viewers to see the spill as an incursion into otherwise untouched land, an adulteration of an otherwise Edenic paradise. Viewers understand not only that something terrible is happening in the frame, but that this event is unfolding not in some place that has already been forfeited to industry but, rather, in a place which—despite the historical evidence to the contrary—is still often presumed to be relatively pristine, untroubled. McBride's photograph thus challenges our conceptions of nature and wilderness in the American west. *This*, the photo asserts, *is what the wild west now looks like*. There is, then, a sense of dissonance between taken-for-granted ideas of what nature/wilderness look like and what McBride's photograph shows us.

I wish to make one final point about McBride's photograph: It is not only the orangeness of the water but the presence of people in the photo that constructs viewers' understandings of the Animas River as grievable. The kayakers' presence within the photo further establishes the Animas as part of a particular kind of environment—one that is neither agricultural nor industrial but, rather, recreational. It is not immaterial that it is kayakers—themselves a symbol, like campers and fly fishermen, of a kind of urbane tourism—we see in the photograph, as opposed to farmers or miners or ordinary citizens or children whose daily lives bring them into contact with the Animas in more mundane ways. Clad in technical gear—plastic helmets, waterproof clothing, slim life jackets—the boaters are both a part of and yet apart from the river. They are tourists, traversers: they glide across the water's surface but do not immerse themselves. They are exposed to the water beneath, certainly, but ostensibly don't depend upon the Animas for drinking or irrigation water. They have chosen to float the river, and they can also, presumably, choose to leave. Regardless of how else these kayakers may in fact relate to the Animas (perhaps they live nearby or are also farmers), within McBride's photo they symbolize the river's recreational and aesthetic values. Their presence implies that the river is a destination, a place where adventurers from elsewhere choose to visit. At the same time, the color of the water suggests that those qualities of the river that attract tourists are under threat.

Thus, by picturing the Animas River as a recreational destination, a sort of wild retreat in peril, McBride's photograph perhaps resonates with the values and anxieties of viewers witnessing the catastrophe from afar, viewers for whom the scene reveals the remarkable risks of pollution in a place where they might themselves consider taking a vacation. Although the text surrounding McBride's photograph clarifies that the stakes of the spill were much more than aesthetic, and indeed that the discoloration was only a fleeting indicator of a potentially much more persistent problem, the kayakers' placement in the center of the photograph implies that aesthetic risks are themselves central. While viewers may wonder whether the kayakers were in harm's reach as they floated down the intoxicated river, whether splashing water might have put them at risk, the degradation of the water quality registers first and foremost as a dimunition of the beauty of the river. For those who dwell near the Animas River and who, perhaps, benefit from the tourism industry along its shores, the aesthetic qualities matter financially. The beauty of the river is a free gift, something which humans have appropriated for their own ends. A threat to the beauty of the watershed presumably makes a difference, then, not only to those who might see this place as a destination but to those whose livelihoods depend upon such a perspective.

Curiously, the kayakers' presence highlights potential losses that exceed aesthetic dimunition. If the boaters represent the river's value to recreationists, they also represent the not entirely instrumental values of the Animas. While much of the commentary about the spill focused on the potential risks of drinking contaminated water and of irrigating crops after the spill, this image of the kayakers amidst the intoxicated river opens up other interpretations of the river's value. If some us desire to run the Animas River, or to cast a line into its riffles and pools, or to spot birds along its shores, or to walk amidst the trees that line its banks, this is so because we implicitly perceive this river as grievable, as a waterway that matters because of the many forms of life (and ways of life) it supports. Indeed, the watershed is one of the conditions of life in the Four Corners region. The river gathers together fish and foul, insects and mammals, trees and people—farmers and fishers and miners and swimmers and sightseers, too. Although the only people we see in the frame are kayakers, their presence suggests to viewers that this place matters—that despite the alarming color of the water, and the historical *un*-grievability it implies, there are many for whom the destruction of the Animas River would register as a significant loss.

CONCLUSION

Throughout this chapter, I have attempted a partial answer to LaDuke's query, "How do we grieve the death of a river?" My contention has been that we must cultivate a sense of grievability for watersheds, but that such efforts depend to a certain extent on the ways in which those watersheds appear to us. As Peeples (2011) observes, the question of appearances is fraught when it comes to toxins: "With the invisibility, complexity, and uncertainty of many toxins and an image-fixated western audience, constructing a compelling and potentially transformative visual frame for toxins is an ongoing challenge" (388). Given that the risks and perils of toxins are all too often invisible, I suggested that it is worthwhile to consider those moments when toxicity surges forth into unconcealment. Such moments, exemplified here by the Gold King Mine spill and the subsequent discoloration of the Animas, provide important opportunities to witness the precarity and (un)grievability of hydrological systems, which too many of us too typically take for granted as unexamined conditions of our existence. McBride's photograph provides an illustration of how photos of contamination sometimes reveal the historical ungrievability of the more-than-human world by exposing its exposure to avoidable, anthropogenic harms, and, at the same time, evoke a sense of grievability, engendered by care and concern, for the more-than-human by articulating its value beyond its instrumentality.

My response to LaDuke's question, then, has been to ask under what conditions those of us who are not intimately entwined with (who, as Leopold might put it, do not "know") a particular waterway may nevertheless become capable of apprehending it not only as something losable but as something for which we would mourn if it were lost. It matters whether we collectively understand watersheds, or any ecosystem, as grievable, for it is from out of our anticipatory grief that we harness the affective energy needed to enact the forms of support and care necessary to safeguard those conditions upon which all forms of life depend. Trained as we are to trace influence, rhetorical scholars should locate and analyze those texts that make it more possible for more of us to experience the more-than-human world as grievable, as well as the ethical and political implications of being affectively opened up to the precarity and grievability of what exceeds the human. "In this [anticipatory] grief," Cunsolo and Landman (2017) write, "we can come together through the shared experience of understanding and feeling loss, and allow for a shared transformation" (14). I am (perhaps irrationally) hopeful that such transformations in vision might lead us to take better care of this earth, our only home, which we have inherited and for which we are responsible.

REFERENCES

Barnett, Joshua Trey. 2015. "Toxic Portraits: Resisting Multiple Invisibilities in the Environmental Justice Movement." *Quarterly Journal of Speech* 101, no. 2: 405–25.

Barnett, Joshua Trey. 2018. "Thinking Ecologically with Judith Butler." *Culture, Theory & Critique* 59, no. 1: 20–39.

Barnett, Joshua Trey. 2019. "Naming, Mourning, and the Work of Earthly Coexistence." *Environmental Communication* 13, no. 3: 287–99.

Benjamin, Shane. 2015. "Durango Kayaker Reacts to Animas River Photo That Went Viral." *Durango Herald*, August 8, 2015. https://durangoherald.com/articles/1532.

Butler, Judith. 2004. *Precarious Life: The Powers of Mourning and Violence.* London: Verso.

Butler, Judith. 2009. *Frames of War: When Is Life Grievable?* London: Verso.

Butler, Judith. 2015. *Notes toward a Performative Theory of Assembly.* Cambridge, MA: Harvard University Press.

Chakrabarty, Dipesh. 2008. "The Climate of History: Four Theses." *Critical Inquiry* 35: 201–7.

Chief, Karletta, Janick F. Artiola, Sarah T. Wilkinson, Paloma Beamer, and Raina M. Maier. 2016. "Understanding the Gold King Mine Spill." April 2016. https://superfund.arizona.edu/sites/superfund.arizona.edu/files/u70/understanding_the_gold_king_mine_spill_v13_preamble_final.pdf.

Cunsolo, Ashlee, and Neville R. Ellis. 2018. "Ecological Grief as a Mental Health Response to Climate Change-Related Loss." *Nature Climate Change* 8: 275–81.

Cunsolo, Ashlee, and Karen Landman, eds. 2017. *Mourning Nature: Hope at the Heart of Ecological Loss and Grief.* Montreal: McGill-Queen's University Press.

DeLuca, Kevin Michael. 2007. "A Wilderness Environmentalism Manifesto: Contesting the Infinite Self-Absorption of Humans." In *Environmental Justice and Environmentalism: The Social Justice Challenge to the Environmental Movement*, edited by Ronald Sandler and Phaedra C. Pezzullo, 27–55. Cambridge, MA: The MIT Press.

DeLuca, Kevin Michael and Anne Teresa Demo. 2000. "Imaging Nature: Watkins, Yosemite, and the Birth of Environmentalism." *Critical Studies in Media Communication* 17, no. 3: 241–60.

"Emergency Response to August 2015 Release from Gold King Mine." *U.S. Environmental Protection Agency*, October 1, 2015. http://www2.epa.gov/goldkingmine.

Kaplan, Sarah. 2015. "'I Had Nothing Else to Do but Cry': Farmers Mourn Toxic River Plume While EPA Copes with Fallout." *The Washington Post*, August 12, 2015. http://www.washingtonpost.com/news/morning-mix/wp/2015/08/12/i-had-nothing-else-to-do-but-cry-farmers-mourn-toxic-river-plume-while-epa-copes-with-fallout/.

Lachelt, Gwen. 2015. "When a River Runs Orange." *The New York Times*, August 20, 2015. http://www.nytimes.com/2015/08/20/opinion/when-a-river-runs-orange.html.

LaDuke, Winona. 2016. "How Do We Grieve the Death of a River?" *Indian Country News*, May 31, 2016. https://www.indiancountrynews.com/index.php/columnists/14275-how-do-we-grieve-the-death-of-a-river.

Leopold, Aldo. 1949. *A Sand County Almanac and Sketches Here and There.* London: Oxford University Press.

Olivarius-Mcallister, Chase, Mary Shinn, and Shane Benjamin. 2015. "Catastrophe on the Animas." *The Durango Herald*, August 6, 2015. http://www.durangoherald.com/article/20150806/NEWS01/150809765.

Peeples, Jennifer. 2011. "Toxic Sublime: Imaging Contaminated Landscapes." *Environmental Communication* 5, no. 4: 373–92.

Peeples, Jennifer. 2013. "Imaging Toxins." *Environmental Communication* 7, no. 2: 191–210.

Pezzullo, Phaedra C. 2007. *Toxic Tourism: Rhetorics of Pollution, Travel, and Environmental Justice.* Tuscaloosa: University of Alabama Press.

Smith, Noel Lyn. 2018. "Navajo Farmers, Ranchers Sue EPA and Contractors for Gold King Mine Spill." *Farmington Daily Times*, August 10, 2018. https://www.daily-times.com/story/news/local/navajo-nation/2018/08/10/navajo-farmers-ranchers-sue-epa-and-contractors-gold-king-mine-spill/961536002/.

Snowbarger, Jeff. 2015. "On the Dirty, Deep-Seated Origins of the Animas River Spill." *Orion Magazine*, August 26, 2015. https://orionmagazine.org/2015/08/letter-from-colorado-on-the-dirty-deep-seated-origins-of-the-animas-river-spill/.

Sontag, Susan. 2003. *Regarding the Pain of Others.* New York: Picador.

Stearns, Josh. 2015. "Memories of the Animas River before It Ran Yellow." *Josh Stearns*, August 9, 2015. https://stearns.wordpress.com/2015/08/09/memories-of-the-animas-river-before-it-ran-yellow/.

Tan, Avianne. 2015. "Navajo Nation Mourning, Pleading for Help after Toxic Mine Spill Contaminates Rivers." *ABC News*, August 11, 2015. http://abcnews.go.com/US/navajo-nation-mourning-pleading-toxic-mine-spill-contaminates/story?id=33011914.

Taylor, Chloë. 2008. "The Precarious Lives of Animals: Butler, Coetzee, and Animal Ethics." *Philosophy Today* 52, no. 1: 60–72.

Thompson, Jonathan. 2015. "When Our River Turned Orange." *High Country News*, August 9, 2015. https://www.hcn.org/articles/when-our-river-turned-orange-animas-river-spill.

Thompson, Jonathan. 2018a. *River of Lost Souls: The Science, Politics, and Greed behind the Gold King Mine Disaster.* Salt Lake City: Torrey House Press.

Thompson, Jonathan. 2018b. "The Dark Secrets of the Animas River." *High Country News*, May 28, 2018. https://www.hcn.org/issues/50.9/water-the-dark-secrets-of-the-Animas-River-Gold-King-Mine-spill.

Wolfe, Dylan. 2008. "The Ecological Jeremiad, the American Myth, and the Vivid Force of Color in Dr. Seuss's *The Lorax*." *Environmental Communication* 2, no. 1: 3–24.

Chapter Fourteen

Singing across the Sea

The Challenge of Communicating Marine Noise Pollution

Mark Pedelty

Waters present a paradox. From moats to oceans, polities have historically used water to limit communication. However, water has also been used to facilitate community exchange and contact. For millennia, rivers and seas have been major conduits for the exchange of goods, people, and information. Well into the digital era, water remains a medium of communication.

As a case in point, the Salish Sea connects Canada and the United States to each other and the world while, at the same time, it is used to regulate and limit the movement of people, ships, and material. Comprised of the Puget Sound, Strait of Juan de Fuca, and Georgia Strait, the Salish Sea is the aquatic medium, border, and metaphor around, over, and through which environmental communication takes place in western Washington and British Columbia. The region is home to 52 First Nations. Governments, industries, movements, nonprofit organizations, and animals (including humans), communicate about, over, and through the Salish Sea.

Given their dual functions, water bodies like the Salish Sea are often "boundary objects." Drawing on Star and Greisemer's original conception of boundary objects (1989), White et al. (2010) define these "as material or abstract objects that simultaneously inhabit independent but intersecting social worlds, are flexible to the needs of multiple communities, yet durable enough to maintain an identity" (221). Boundary objects are "interpretable, bridge social worlds, and facilitate communication across groups" (221). To be considered a boundary object, a concept and/or conception of space must be polysemic enough to elicit multiple interpretations, yet coherent enough to retain shared meanings across its borders.

The force of the Salish Sea as a boundary object became apparent to me, personally, during the production of LOUD. LOUD is an experimental music video concerning noise pollution, a part of my ongoing work with Ecosong.net, a collective of musicians and educators using song and music videos to advance environmental stewardship. As with past Ecosong.net productions, LOUD involved a number of media makers, actors, musicians, scientists, and volunteers performing specialized roles. I was the producer, director, and, as a member of the band, Hypoxic Punks, composer, which deeply colors the perspective I bring to bear here. Along with colleagues in a grant project entitled *Field to Media: Applied Ecomusicology for a Changing Climate*, I approach media production as an extension of my previous ethnographic fieldwork (Pedelty 2016) and as a form of participant observation in and of itself, heavy on the participation. LOUD was also the most problematic Ecosong.net production to date, evidenced among other things by its less successful entries into film festivals when compared to past Ecosong.net productions. Nevertheless, LOUD was a particularly revelatory field experience despite the numerous challenges we faced during production, or perhaps because of them. LOUD illuminated the border questions explored in this chapter.

This chapter's analysis of the Salish Sea as a boundary object begins and ends with the most obvious boundary, the international border separating Canada and the United States. The imaginary line drawn across the Salish Sea exerts a very real material and symbolic force for the people living on both sides, including media productions, as we discovered during the making of LOUD. Although sea life crosses the border with relative ease, the frontier inhibits movement of humans and radically affects communication between institutions on each side of the border. Thus, the border has profound ramifications for stewardship of the regional ecosystem. For example, nonprofit organizations, policy makers, environmental advocates, and artists collaborate far less across the border than is necessary to solve shared and mounting problems, from the decline of Southern Resident Killer Whales and Chinook Salmon to runaway residential and industrial development.

Of course, a boundary object is not just a thing. It is also a sign. As a sign it is therefore articulated with other signs and boundary objects, in this case nationality, gender, species, and genre. In producing LOUD, we encountered each of these border questions while working in a literal, liminal, and metaphoric space wherein they intersect. In fact, nationality, gender, species, and genre are so deeply imbricated in this story that disentangling them for the sake of analysis was difficult. However, it was also useful for the sake of explication, so this chapter will step through each boundary question separately before bringing it all together under the theme of the Salish Sea as an aquatic boundary object.

Although LOUD was designed to challenge dysfunctional communication dynamics—namely, anthropogenic noise—and promote alternative policies, as the producer and director I mostly failed to successfully negotiate the boundary objects explored here when producing and directing LOUD. For example, difficulties in financing an international team led to unanticipated impacts on gender representation in the video. Furthermore, the musical recording of the song raised questions in regard to genre, especially in terms of volume dynamics, timbre, and texture. The metaphoric ecology of media production paralleled and was affected by the political ecology of conservation in the region. However, this chapter is not a confessional. There are many excellent elements in LOUD, including outstanding performances by the talented singers, body artist, and video editor. Most importantly, as media studies fieldwork, the project was revelatory. Taking part in the project allowed me to develop a clearer understanding of how each of these complex boundary identities functions in practice, in production, and in relation to one another. Therefore, in the conclusion, I will argue against letting the challenge of boundary objects keep us from taking creative risks.

Within the context of this collection, this chapter serves as a firsthand reflection on advocating for justice to plants, animals, and water itself, concluding in a set of recommendations for future activists, educators, and advocates. As multimedia formats like music videos, dramatization, and viral documentaries become increasingly notable sources of social and environmental justice discourses, it's important that media producers take the time and thought to get the message and the media right the first time. In cases of water justice and media with limited funding, there isn't always a second chance.

THE MUSIC VIDEO

LOUD (2018) is a community-based music video project about noise pollution in the Salish Sea region of Washington State and British Columbia and an Ecosong.net production. Ecosong.Net is an applied environmental communication project involving media production and media research, bringing together musicians, media makers, scientists, and community organizations in collaborations that combine outreach, education, art, and entertainment, all aimed at advancing local stewardship. As part of an outreach campaign, LOUD brings together music and imagery to argue for stricter noise policies and regulations (it would be helpful for readers to watch the video in order to more fully understand the following analysis; see https://www.ecosong.band/loud-1).

Finally, LOUD is also part of an academic collaboration supported by the Mellon-funded *Humanities Without Walls* grant program as part of their "Music in a Changing Climate" initiative. That initiative is based at the University of Illinois. Our multidisciplinary grant-project is entitled *Field to Media: Applied Ecomusicology for a Changing Climate*. I have joined four colleagues to form the research collaborative: two ethnomusicologists, Rebecca Dickersen and Tara Hatfield; a composer-choreographer, Yan Pang; and Elja Roy, an environmental communication scholar. Dickersen is working in Haiti, Elja Roy in Bangladesh, Yan Pang in China, and Tara Hatfield in China. They have had essential input into the formation and especially the postproduction analysis of LOUD.

LOUD was a revelatory ethnographic experiment. The production and circulation of LOUD provided insights into the political economy of environmental communication at the local level. It was particularly revealing in terms of communication across boundaries, especially frontiers between genres, genders, species, and nations. Each will be explored in turn before bringing them all together in focusing on the unifying theme of the essay: the Salish Sea as boundary object.

Boundary One: International

Shooting a film in two nations is very difficult. In the planning stages I underestimated just how difficult it would be to shoot talent in both countries, having traveled across the border often in the past. Unanticipated border issues required me, as producer and director, to completely redesign the storyboard and script in the days leading up to the one-week shoot. University of Minnesota rules and procedures in regard to getting the Canadian talent paid as well as the difficult logistics of getting community-based singers in British Columbia into the United States to perform stymied the previous plan. The original plan was to gather all talent and our body artist, Catriona Armour, on Orcas Island in Washington State. The plan was to have all the singers occupy a single liminal space—literally the "littoral" zone between water and land—which would also represent the liminal territory between the human and more-than-human. Here was the original plan as I wrote about it in my preproduction notes in preparation for a journal manuscript:

> Through makeup, settings, and lyrics, singers will evoke the orca in performance. As has been common in the Salish Sea area for some time, orca whales will serve as "condensation symbols" (Cox and Pezzullo 2017, 51–66) in the video, semiotic stand-ins for all living creatures in the local biome much like polar bears have been employed to communicate climate change in the arctic.

Note the use of the term "evoke." Rather than being directly representational, human singers would visually inflect the orca. If successful, having people serve as liminal stand-ins for whales would avoid a clichéd wildlife film motif. If unsuccessful, it would present a comical tableau of people dressed up like whales. In that regard I am glad we decided, following Catriona's lead, to abandon an orca-inflecting motif for more abstractly marine influenced body art. However, I greatly regret that we were not able to execute the single-site shoot with all singers adorned with similar body art and coherent costuming. We never got a chance to succeed or fail with that original plan. Only one character could take on that role due to international logistics and resulting financial stresses. The result was more pastiche than purposeful presentation. While the body art Catriona created for Felicia Harding is remarkable, in isolation it does not fulfill its full potential. A more holistic commitment to that "between worlds" theme might have worked better artistically than the singular figure atop a cliff.

Live performances of the song are now reinforcing that belief. Quite literally, when sung at events by the shore, the main message of the song is being delivered more effectively to audiences that seem to be more readily understanding, and feeling, the point of the piece.

Boundary Two: Species

The production's international challenges, as outlined above, mirror similar difficulties experienced by conservation groups around the Salish Sea as they attempt to negotiate the international border between the United States and Canada. Ultimately this inhibits their ability to make changes that might reduce the impact of development and human actions on more-than-human communities. There is no more fundamental form of interspecies communication than humans' constant and growing threats to other species. The message is being received loud and clear, as evidenced by declining biodiversity and extinction.

As the producer, director, and composer of LOUD, my goal was to actualize the decentering ethic Nuria Almiron develops in her article "Beyond Anthropocentrism" (2016). In arguing for the Political Economy of Communication (PEC) approach, Almiron argues that "decentering humanity to embrace a truly egalitarian view is the next natural step in a field driven by moral values and concerns about the inequality triggered by power relations" (57).

As Paolo Freire (2000) noted, it is necessary to "codify" a group's aspirational values so that an organization can be mobilized to create institutional change (114–21). That is the goal of LOUD, to aesthetically codify the decentering ethic. In the video we feature human and more-than-human voices and images in an effort to communicate a basic message: excessive anthropogenic noise is disruptive to the lives and survival of many living creatures. Latour (2018) presented the most recent reminder that the more-than-human and abiotic worlds exert agency. However, often forgotten in such discussions is that for us, as humans, to understand such messages they must be received and interpreted by human communicators. It is inaccurate to claim that the orca, for example, simply speak for themselves. Therefore, in choosing to emphasize charismatic megafauna, as we have done somewhat in LOUD, we have consciously chosen to take on both the negative and positive communicative possibilities in emphasizing the orca as symbols for a much larger living world and problem. Hausmann et al. (2017) note that the emphasis on charismatic megafauna can lead to an "underappreciation of other biodiversity" (91). In the case of sound, however, far more is known about the effects of noise on whales than other marine life, so it is perhaps a reasonable choice. At this point, whales are a required cipher for policy-relevant communication about sound in the Salish Sea ecosystem.

Given that, as producer and director I am comfortable with the balance that we struck between representing whales and other animals, including the human singer-interlocutors. Nevertheless, my decision not to include text explaining the issue was, in retrospect, a poor one. Past Ecosong.Net video messaging has been assisted by the presentation of facts and framing information at strategic moments in the visual and aural narrative. After watching LOUD, the audience might understand the basic message that living creatures in the Salish Sea would benefit from a quieter environment. However, without a few well-placed facts and quotes culled from the bioacoustics literature, that pitch might seem a bit vague and less convincing. For example, Sorrel North, a leader of the Southern Resident Protection group, noted that she "loved" the video, but noted that it was a bit too "avant-garde" to more directly assist their advocacy efforts. Such a criticism might not phase colleagues in acoustic ecology and the soundscape composition movement. In fact, such colleagues often complain when art more directly engages environmental politics and popular culture. However, I take audiences' criticisms about messaging and meaning to heart. And I wholeheartedly agree with North. Without reference to bioacoustics research demonstrating the deleterious effects of noise on the orca (Houghton et al. 2015), fish (Radford, Kerridge, and Simpson 2014), birds (Buxton et al. 2017), humans (Ma et al. 2018), and other animals (Shannon et al. 2016), there is no way for the average audience member to understand just how severe the problem is or how to convert that knowledge into meaningful action. Without that, LOUD comes off as a more generalized plea for quiet.

In the case of LOUD and work like it, the best way to instantiate Almiron's "Beyond Anthropocentrism" ethic is by following Freire's advice and to more effectively codify it. Sound and image are part of that symbolic effort, but in this case selective use of text would have completed the act.

Again, sonic purists might find such recourse to text and visual representation as lacking, and I fully understand their genre-based preference for soundscape composition, but I believe that there is also ample room in art for an aesthetic that holistically integrates sound, image, text, narrative, politics, policy, and popular culture. With that kind of holistic esthetic as our goal, LOUD fell short. In fact, "falling short" is an apt description for communicative failure across a boundary object. By not introducing the textual method from previous Ecosong.Net videos, LOUD shrinks back from the communicative borderline where humans encounter each other and engage other species.

That is unfortunate, because the whale-watching industry has shown no such restraint. The image of the orca, the Salish Sea's charismatic megafauna, is used for just about everything in the region, especially to stoke consumption and development. For every actual orca, there are thousands of pictorial, animated, and abstracted images of killer whales throughout the Salish Sea used to sell objects, ideas, and the place itself as tourist, business, and development destination. Naomi Klein's description of "tree museums" in BC (2009, 156) comes to mind. Klein notes that special forest preserves are used to convince visitors to BC that the province's forests are doing well and carefully stewarded. Similarly, the whale-watching industry and their paid advocates use ritualistic media imagery and live viewing sessions to lead the casual observer to believe that they are leading the way in protecting the orca, when in fact whale watching is a threat to cetacean survival in the Salish Sea. The same people exploiting, corralling, and harassing the orca—the \$144-million-per-year motorized whale-watching industry (Kivitso 2018)—are those most loudly claiming to protect them. In such a place and time, the stakes of representation could not be higher. Although I know that one small music video project cannot do much, I do wish that the counterimagery and rhetoric we crafted in LOUD could have been more successful as a communicative tool. Fortunately, again there are reasons to believe that the current stage of the project, live performance, is more successfully achieving that aim.

Boundary Three: Gender

My concerns over gender representation started brewing during preproduction, rose to the surface on site, and appeared as a full-blown problem by the time LOUD was completed. The original plan was to have a transgressive, transgendered theme. Four or five singers of varying shapes and sizes would be costumed in androgynous fashion, including a unified body art motif. The final result was very different. In the final video, an attractive, scantily clad woman could be viewed, decorated in brilliant body art while the rest of the musicians remain fully clothed. Rather than a challenge to what Morgan

Kuecker and I referred to as "scopocentric sexism" in a study we completed on advertisements (2014), LOUD could be seen as an example of the male gaze. In addition to the matter of Harding's costuming, the video features attractive young women, along with one middle-aged man, Dana Lyons, a famous figure in the environmental movement, and a child, Hazel Macpherson. The intent here is not to stigmatize attractive young women—from pop stars on through to the incredibly talented singers in LOUD, the power and talent of such women is underappreciated—but rather to note that as a class the images of attractive women are conspicuously emphasized in media meant to sell consumer goods, downloads, events, and so on.

In addition to my own sense that the imagery in LOUD could be read in that way, I was also led to that interpretation by a participant in the singing contest who complained after not being selected. She complained that only attractive young women seemed to make the cut. There was some truth to that. An independent group of judges, evenly split between men and women, made that decision when asked to evaluate the contestant's voices. To what extent the judges were swayed by look as well as voice, I do not know, but I do know that all of singers that they chose have exceptional voices. Further complicating matters was the fact that the majority of the contestants were young women. There was a high level of self-selection involved. It might be because of the environmental subject matter. Or, perhaps attractive young women see other attractive young women emphasized in popular music and music video (symphony promo posters, films, fashion, and so on) and thus view these women as their "possible selves" (Fisherkeller 1997). Other performers, although perhaps equally talented, might not think of themselves as potential talent for a music video. While we were thrilled with the exceptional talent involved in LOUD, in the future we will need to find or return to methods that can further diversify the talent, drawing from a deeper and wider community pool.

However, featuring talented young women is not the aspect that struck me as sexist *per se*. Ecosong.Net works under the premise of free association, and in this case those who chose to take part were mostly young women. Furthermore, a middle-aged indigenous singer and drummer was among those slated to appear, but the flu removed her at the very last minute, further skewing the demographic of the on-screen talent. Instead, what concerns me is the presentation of one individual, scantily clad woman on the cliffs. As it turned out, her look is ultimately incongruent with her surroundings, in part because she is isolated but also because we did not anticipate it being such a cold and foggy day. I get cold just watching it. Again, what was supposed to be an ensemble of singer-actors dressed in Catriona's fantastical body art instead became just one model type in isolation, more of a common trope than a transgressive message. In that way LOUD wanders into the centuries-old tradition of artists' work representing nude women as objects of the "male gaze" (Mulvey 1989).

Again, this matter of gender representation was affected by border issues and the financial stress it caused. In retrospect it would have been better to abandon that element rather than have just one figure attempt to fulfill that entire concept. Harding was amazing in that role, and her fellow singers were also exceptional, but the contrast between them was and is problematic to my eye. If I could redo one aspect of the video, it would be that one. Either all or none would be in body art. Better yet, I would have preferred to recruit a wider range of singers so that more members of the human community could be represented. By placing individuals of varying sizes and shapes together in a liminal, littoral zone between sea and land, water and sky the focus could be more on their sound, and the visual concept, that they sing and speak for the place, might come through more successfully.

One more anecdote is worth relating in regard to fashion, makeup, and gender. With each video we have added a new dimension. For example, having experimented with movement in *You Can Build a Garden* (2016), we fully dove in to dance with *Watershed* (2017). Having done little by way of costuming and makeup in our projects, I thought that body art would be a good turn for LOUD. It seemed like a promising aesthetic for a music video about sound and the sea. In theory it was and would be. However, when thinking about body art, it did not occur to me that there might be a problem in terms of nudity. Body artists tend to prefer to work with the entire canvas, or as much of it as possible. Yet, community-based, activist art needs to circulate widely. In this case, there was a tension between a body art aesthetic and community activism. That problem did not occur to me until Catriona asked whether I wanted Harding partially clothed or naked. My first impulse was to provide Catriona the same artistic autonomy that all, or at least most, artists are accorded in Ecosong.Net productions. "You choose," was my first response, only to realize later, after thinking more about it, that having one bare breasted woman amidst a group of otherwise clothed singers could be problematic in terms of circulating the video on YouTube and to family audiences in the region.

Having reconsidered, I asked Catriona if she could paint over a wetsuit or swimsuit and she did so, purchasing a bra that could take the makeup. Again, that would not have been as much of a problem if a diverse cast was all costumed in such body art, with nudity more purposefully presented from an artistic and messaging perspective, rather than patterned by gender norms. What was to be a transgressive and carnivalesque became an obtuse reference to convention.

Few viewers have overtly criticized the video as sexist, perhaps watching the video with a greater sense of charity and good faith than I am extending it. For my part I feel as though I compromised a work of art, an exceptional artist (Catriona), and the production as a whole. I struggled in the boundary zone rather than more successfully negotiating it with the aid of my collaborators, who undoubtedly could have helped improve the gendered messaging in the video. In retrospect, it would have been good to have been able to delay filming and have time to collectively work out a new plan. Unfortunately, budget, time, and logistics would not allow for that sort of time, reflection, and a project reset. All we can do is apply these lessons to the next Ecosong.Net project, and hopefully reading this is instructive to fellow community-based media producers and directors.

Boundary Four: Genre

Economy, policy, and ideology are integrally connected to aesthetics. That is among the many reasons why the arts and popular culture need to be part of our work as environmental communicators. Music is essential to that effort. Mark Mattern made that clear in *Acting in Concert* (1998), demonstrating through case studies how small groups have successfully used music to come together to form communities, communicate shared values, and mobilize.

However, not all songs are created equal. A song that no one enjoys will do little to communicate an environmental message. That is not to say that it must be upbeat and "fun," but rather that pleasure is always integral to art, whether the pleasure derives from critical reflection, catharsis, or joy. In material terms, sound is little more than cycles of compression and rarefaction, vibrations then experienced by the ear and brain as meaningful. Of course, enculturation and experience heavily condition and overdetermine that sound experience, as do all of our senses. Few sounds are so overwhelming as to create a preternatural response. A predator's roar, for example, might cause one to jump, but to experience a song as pleasant or distasteful, or even as "music," is in the ear of the beholder. We are culturally conditioned to interpret certain complex patterns of sound—harmonics, timbre, textures, and rhythms—as music. As audiences we then accept only a certain range of violation to those patterns for a sound to be considered compelling, yet we expect a certain level of transgression for a work to be considered artful. And even the acceptable range and style of transgression is culturally conditioned.

Central to that cultural conditioning is genre. Some folks learn to like folk, for example, whereas others would rather hear fingernails scratching a chalkboard. At a finer-grained level, the individual's experiences, personality, and enculturation lead each listener to have a slightly different constellation of sonic preferences, within and across genres. Given that, I will certainly not make an argument for any single set of universal principles for what constitutes artful and effective environmentalist music, nor even attempt to do so within the relativistic confines of particular genre. Genre and style cultures such as folk, rock, hip-hop, classical, jazz, electronic dance music, Tibetan throat singing, or any other form each have their own rules. Such distinctions are always relative to an aural tradition, genre-bound aesthetics, the moment, geographic context, purpose, and so on. As Daniel Levitin (2006) explains throughout *This Is Your Brain on Music*, musical communication requires a recognition of pattern shared by composer, performer, and audience. When that precondition is met, purposeful variations from that pattern can be viewed as artful and interesting. "Composers imbue music with emotion by knowing what our emotions are," writes Levitin, "and then very deliberately controlling when those expectations will be met, and when they won't" (117). Ecosong.Net involves a group of musicians who normally perform in a punk-tinged folk-rock style. We fairly consciously switched to a more generic pop-folk sound for our first three videos, a sound that implicitly welcomes everyone to the dance. LOUD was, by contrast, written to be more sedate and critically contemplative; it evolved into a pop music video.

With that in mind, it is fair to say that for LOUD's probable audience, purpose, and time, the musical genre of pop was a poor choice. Yet again, this eventuality developed more by accident than through purposeful fulfilment of a directorial plan. If nothing else, this chapter should be a cautionary tale against critics that apply an auteur theory to media. No text, certainly not a popular song or music video, can be traced back to the authorial intent of a single individual. Media texts are always products of collective creativity and labor, an artifact of the moment, the productions of place, culture, and of course ideology. Yet, individual choices matter, and some more than others. In the case of LOUD, my original choice, as the composer, to write a sedate, reflective piece calling for quiet places, with a discordant and contrasting metal bridge near the end, was perhaps the best set of choices that I made. Live performances have reinforced that belief in the greater power of a more reflective mode for this song about silence. A call for sound policy change and deep listening is hard to make via pop. It is also the most remarked upon aspect of the video. Several viewers and listeners have commented upon or asked about the choice of featuring a pop style in the video. Among those commenters was Gregory Melchor-Barz, president of the Society for Ethnomusicology, who asked in an email correspondence why LOUD was presented in "a band style typically associated with 'loud' performance."

The answer is that soundtrack development was partly accidental and largely a question of inertia. The pop sound had worked fairly well for our prior videos. Upbeat and entertaining, they were aimed at mobilizing movement around pressing issues and to motivate very specific organizational efforts. However, the sound genre that gets a mostly middle-class American, environmentally-inclined audience to think, feel, and act a bit more around specific water problems is different than that which is needed to get people to reflect upon soundscape. Artful contrast might do the trick, as in the "noise music" approach. Or, a softer, slower, more reflective sound could model the aspirational soundscape called for in the lyrics. However, the bouncy pop sound is more often associated with consumption and mindless fun, and nothing in the video implies that it is ironically referencing pop. More overtly, semiotically signaling the call for quietude in the music would have been a better choice.

How did the contemplative song become a pop soundtrack? It all started out as a quiet singer-songwriter reflection, as it has become again since. Then my band, the Hypoxic Punks, started playing it, relatively sedately at first. As is our way, we started slipping into a faster-paced mode. In studio it took another move in the folk-tinged pop direction. My highly skilled musical partners, drummer Robert Poch and guitarist Tim Gustafson (the composer of *Watershed* and *You Can Build a Garden*) along with our Grammy-winning sound engineer, Karl Demer, looked to me as the director to bring continuity and the proper style to this complex project. I was not thinking enough about what came before (the song as written), or what was coming next (the video), when recording the song in studio. By the time I was out of studio, I was hoping that the pop-rock recording could seem more ironically discordant than it was, in performance. By the time the problem was apparent, the musical soundtrack had been set. Only Harding's a capella intro and outro reference the original intent for the song and video. Again, with better direction the entirety of the piece would have cohered to that consistent aural motif.

Finally, there have been useful comments from the audience in regard to length. Both the song and music video are too long for the genre and purpose. The song is nearly six minutes, far too long for a pop song that seeks more immediate impact and social circulation. And the music video is nearly ten minutes, far too long for the genre and format. Listeners expect much shorter works. Film festivals demand much shorter entries into the music video genre category.

All of that is to say that Ecosong.Net jumped the shark with LOUD. Up to this point, one strength of our community-based media making has been to match methods to means. Small productions, small budgets. That allows for greater attention to detail. Instead of limitations, limitations have inspired creativity. It is no coincidence that much of the most compelling popular culture comes from impoverished neighborhoods and nations. While we are by no means poor, our small scale and relative lack of resources have inspired us to use our resources creatively.

As Karl pointed out near the end of our weeklong shoot for LOUD, the scale of such an operation would normally involve an assistant director and additional crew members. Like Karl, I was having to wear many hats at the same time. I was directing while catering, dealing with location landlords and logistics, wrangling talent, making calls, and so on. In other words, with

LOUD we started moving from a nimble, community-based organization based on the principles of free association, to more of a full production company. Short of fully achieving or even desiring to become the latter, it would have been better for LOUD to better match the storyboard, script, plan, and production methods to our relatively small budget, schedule, and labor capacity. In other words, LOUD would have probably been a better recording and music video if less complex as a production.

Colloquialisms like "jumping the shark" and "falling short" nicely reflect the concept of getting caught between, of launching into a liminal space without the means to make it to the other side. And such language works well for a discussion of communication within and across boundary objects. We were getting stuck inside. In terms of genre this made us neither fish nor fowl, neither folk nor rock.

Nor was LOUD an example of genre bending. Instead of more faithfully complying with genre expectations, the recording could have merged and creatively transgressed genre expectations to artful effect. Instead, an all-too-safe pop texture and tempo developed. LOUD is an earworm, perhaps, but not necessarily the kind of sound that moves listeners toward sonic contemplation. That might be asking too much of an environmentally themed song or music video, but I continue to believe that art and popular culture hold a distinct power to communicate ideas about water and environment. For example, Sting's composition and recording of "Fragile" propels the soundtrack of an IMAX film, *The Living Sea,* a work that won the Academy Award for Documentary Short Subject (MacGillivray, Streep, and Sting 1995). As another case in point, Beyoncé's Grammy-nominated music video, *Formation* (2016), drew connections between the catastrophic flooding following Hurricane Katrina, racism, and environmental justice for millions of listeners. However, even in these exceptional cases there is no way to demonstrate communicative effect on the audience, much less in terms of policy and practice. There is no explicit evidence that audiences were moved to thought or action by watching and listening to these popular examples of environmental media. Therefore, we certainly cannot make that claim for a community-based production like LOUD.

Of course, we are in fine company in that regard. Some of the most definitive scientific findings in regard to climate change, for example, have failed to motivate effective social, political, especially ecological change. Art and popular culture are not alone in their ecopolitical indeterminacy. We craft environmental inquiries and interventions in the hope that they might matter. We have to accept small signs as to their effect and meaning—a few positive comments, selections and awards at film festivals, screenings at pollution control board meetings, and so on. In the case of LOUD those signs of social resonance have mostly been missing when compared to past Ecosong.Net projects. To wrap up discussion of this fourth and final boundary object, LOUD demonstrated the power of genre and the need to meet an audience halfway when communicating environmental messages. It is important to understand the power of genre even when, or perhaps especially when, attempting to transgress convention.

CURRENTS: CIRCULATING MEDIA WITHIN A CONTESTED SEASCAPE

The Salish Sea is an aquatic boundary object where people (genre, gender), places (nations), and living beings of all sorts (species) encounter each other. LOUD had trouble communicating in and across those boundaries of identify and place. However, the challenge of negotiating boundary objects might not fully explain LOUD's uneven impact. Some organizations have enthusiastically welcomed LOUD, sharing it with their membership and asking for closer partnering via live events. Most of these are place-based stewardship groups, such as Friends of Saltwater State Park, Friends of Lime Kiln State Park, and Friends of Moran State Park. The Whale Trail organization has also been supportive. LOUD reflects their interest in getting people to watch from shore rather than taking part in the whale-watching industry's sonic and physical harassment of wildlife.

The power of that industry was recently revealed when Pacific Whale Watch Association lobbyists watered down Washington State legislation that would have put a moratorium on whale watching, a ban that would have held until the seventy-five remaining Southern resident killer whales started to rebound. Before engaging in this field-to-media project, I would have had no idea just how powerful the whale-watching industry is financially, culturally, and politically. For example, industry representatives, advocacy groups, and even a few scientists that accept significant funding from the industry have effectively deflected from the noise issue for years. Based on bioacoustics research and what is known about orca behavior, physiology, and the properties of sound through the medium of water, it is clear that motorized craft are impacting whale communication, organization, hunting, reproduction, and movement (Holt et al. 2009). As for what remains to be fully researched, the precautionary principle should be applied when dealing with whale watching. Sadly, whale-watching industry funding and control over access to the whales has prefigured what research foci will be completed. Only now are some industry-funded groups and scientists being called to task for their dual roles as researchers and industry spokespeople (Pailthorp 2019).

Washington Governor Jay Inslee's task force recommended cessation of whale watching, but the legislature listened to the lobbyists and it will continue apace, moved back a mere 100 yards further from the whales, from the 200-yard limit to the 300-yard limit. The Fish and Wildlife officers in charge of regulating the industry are overstretched and often not present to enforce that distance. Whale-watching boats run from sunup to sundown throughout the long whale-watching season and there is no way to monitor them all the time. In fact, I first started writing LOUD after watching the whale-watching flotilla from a kayak as they harassed a group of transient orcas, interspersed amongst the pod and far closer than legally allowed. I was paddling near Point Lawrence on Orcas Island, near shore, when a transient pod (i.e., not the Southern Resident Killer Whales) came swimming by, surrounded by large whale-watching boats, recreational boaters, and their passengers who cheered each time an orca would breach. As a convergence of currents, the waters around Point Lawrence are a great place to feed. Even the amateur observer can see that a coordinated hunt is made more difficult by the boaters, businesses, and passengers trying to outdo each other to capture the closest possible glimpse and picture of the whales. I went ashore to get out of the way, a bit embarrassed to be there at all. This is a scene replicated day in and day out all over the Salish Sea. Media and music could play a role in helping audiences and policy makers to better understand what whale-watching and other motorized craft do to the whales.

Once LOUD was done and in circulation on YouTube, the hidden outlines of the local nonprofit advocacy system became clearer as well. That is a central goal of ethnographic research: to better understand tacit cultural rules and worlds. Such truths might remain occluded in more ritualistically rendered interviews, public texts, and surveys. LOUD was very successful in that regard. Via distribution the video revealed a tension between the manifest goal of many orca conservancy groups—to protect the orca and increase their numbers—and the latent truth that many of these groups hit the limits of permissible discourse when it comes to whale watching and noise. Quite simply, calling attention to the loudest proximate source of noise pollution, whale watching, would threaten a major source of their funding. The latest evidence of that is a strategic lawsuit against public participation (SLAPP) lawsuit brought against the Southern Resident Protection group. They gathered and submitted enough signatures to qualify a ballot measure that would enact new rules for the San Juan Islands, moving the whale-watching flotilla back 650 yards from the orca. The whale-watching industry wants to end their efforts before they reach the court of public opinion.

In such a loud and combative context, it has been difficult to communicate the ethic Jeff Todd Titon refers to as "The Sound Commons" or to help actualize Almiron's "more-than-human" concept. When setting out to make LOUD, I did not realize how strongly the political economy of the region overdetermined the political ecology of the region, including in regard to orca issues. However, engaging the matter via LOUD has led to a new understanding of regional ecopolitics. For example, it was apparent in distributing LOUD just how profoundly competitive the nonprofit economy is, often to the detriment of a more cooperative ecopolitics. Groups often promote their own programs, solutions, and fund-raising efforts to the exclusion of and not in cooperation with ostensibly allied organizations. On a local scale this parallels the dynamic that Clifford Bob (2005) discovered in his research concerning nongovernmental organizations (NGOs) on a global scale. In the case of the orca, inordinate funding received from industry sources—in this case from the whale-watching and related tourism consortia—blunt and redirect efforts to deal with anthropogenic noise and harassment, two of the most obvious and immediate threats to the dwindling southern resident killer whale population.

Nevertheless, knowledge resulting from applied efforts at creative environmental communication like LOUD could potentially assist in the effort to navigate complex political ecologies like the Salish Sea. It is a world that involves a great deal of regulatory capture, not only imbricating industry and governance, but also influencing the popular culture and conservationist organizations.

CONCLUSION

Debates in the humanities often seem to be about finding rhetorical purity. That is evident in the way that scholar-critics skewer media for failing to live up to normative standards, such as gender equity, racial diversity, and other questions of representation. The world needs such critics. However, if that is all that we do, then we miss opportunities to discover new truths, learn about media from unexpected angles, and most importantly, to create more equitable, interesting, engaging, ethical, artful, and *just* media. However, to make media is to risk making mistakes, perhaps even to fall prey to the same forces that bedevil professional directors, the sorts of people and media we routinely criticize. The scholar-producer often experiences an uncomfortable form of critical engagement. Yet, in so doing we can see media from new vantage points and intersubjective angles. In other words, there are far worse things than making mistakes if the objective is to learn, including learning how to affect policy, behavior, and political ecology. Therefore, rather than succumbing to a paralyzing fear of boundaries and shrinking back from borders, for me the LOUD experience has brought home the value of creative risk-taking, critical reflexivity, polysemy, and transgression.

Matthew Nisbet (2010) suggests that scientists need to advocate more actively in relation to their research-based knowledge. Perhaps those of us who study environmental communication should take that message to heart as well. We can engage in the channels of public communication we typically study rather than simply critiquing others' utterances and productions. Creatively interpolating the audience might be just as important as critically decoding others' media texts. Granted, making media, even for the sake of academic inquiry, puts us at risk of getting things just as wrong as those whom we critique. Yet, as in science, art advances mainly through making mistakes (Firestein 2012; Langer 2007, 74–98). Media production, as a form of fieldwork, is a good way to apply what we know and, by bringing practice into conversation with theory, learn via praxis. In the case of LOUD, I believe that I learned as much or more about how sexist imagery develops in media by navigating and tripping through dominant structures and conundrums as I have through studying media sexism from a purely analytical vantage point (Pedelty and Kuecker 2014).

Granted, producing media is "fraught" with danger in a period of "precarity," to use two currently popular terms in the academy. Engaging in meaningful media production can get us hung up on the rocks of misrepresentation. We can end up pandering to the popular, for example, rather than challenging it. Or, at the other extreme, the academic media maker might practice safe, academically sanctioned arts without making much of a public impact at all or facing the sort of critical scrutiny that follows. In music studies, at least, more scholar-artists choose the latter tact, thus failing to communicate with anyone but ourselves. LOUD comes closer to the former end of the spectrum, however, replicating popular expectations rather than challenging them, and in the end does too little to communicate a critical message. The trick, as with all boundary objects, is to meaningfully engage the popular while at the same time artfully stretching or transgressing the space between a mostly esoteric form of "resistance" on one side, and hegemonic incorporation, on the other.

Unfortunately, the communicative terrain of the Salish Sea, when it comes to environmental stewardship, is even more complex and dysfunctional. Limiting discourses and simplistic messaging (e.g., destroy the Snake River dams and the orca will be saved; see chapter 4 of this book), combined with sublimated competition between advocates, have created a communication soundscape in which proactive public engagement is challenging and less effective than might be required to protect the remaining seventy-five resident orcas. Again, I point my fingers back at myself as producer and director. In LOUD we somewhat shrank back when encountering border challenges and boundary objects. Largely due to my choices, we failed to represent a positive alternative.

With the help of my collaborators, I am trying to help instantiate each of the field-based lessons outlined above in the next Ecosong.net project. LOUD has also helped me to understand past projects better, including successful elements that I had begun to take for granted. Those include the selective use of titling; matching sound genre to message; avoiding gendered clichés; communicating emotion honestly; and focusing our efforts, time, and budget on building community in place. Those qualities have helped our projects to build organizational esprit-de-corps and mobilize community while also earning the respect of film festival juries. Perhaps as scholars we should not be seeking such external validation, nor as artists should we think we require exogenous measures of success. However, when it comes to the environmental arts and especially environmental communication, art and efficacy are interrelated.

As for transgression, that might require more of a dual consciousness. We can recognize, respect, and at the same time critically transgress boundaries. In an era of climate crisis with incalculable levels of pollution (e.g., noise, plastics, endocrine disruptors), decreasing biodiversity, and diminishing environmental justice, we need communication that fosters creativity, diversity, depth, and critical polysemy. That means that we cannot work solely within safe and simplistic binaries. Efforts to limit, norm, and shame others for the sake of "winning" rhetorical battles against an ostensible opposition can sometimes work against creating more inclusive, divergent, and effective ecopolitical cultures. Conversely, the objective of some environmental communication, and much of the criticism thereof, seems to be "impression management" (Goffman 1959), placing oneself and one's group safely and rhetorically on the good side, rather than getting entangled in the inevitable messy engagements, compromises, and creativity that might actually lead to the achievement of ecological objectives.

I came to those conclusions not only through production-based fieldwork and analysis of the resulting media and outreach, but also with the aid of my colleagues in the *Field to Media* project. LOUD was produced as part of that initiative, and my understanding of boundary objects has been advanced by receiving their feedback, watching and discussing their work, learning from their collective example, and exchanging ideas with each of these scholar-producers. Each of these colleagues is similarly grappling with complicated matters of production and representation concerning nationality, genre, gender, species, and tying it all together: water. My ultimate takeaway from that ongoing experience and this subproject, LOUD, is a reminder that boundary objects cannot be successfully traversed alone. The communicator, producer, and organizer only succeeds in making meaning across boundaries to the extent that they successfully communicate (i.e., "make common") with others. Among other things, that means becoming a more "skillful listener" (Von Glahn 2013). Looking back at the work to date of Ecosong.Net, all of which has focused on communicating water-related challenges, our videos have succeeded both artistically and organizationally to the extent that we have critically negotiated and creatively engaged difference. LOUD has thus far been more bogged down in boundary objects. There are signs in the live, postproduction afterlife (e.g., live performances) and intertextual rearticulations of LOUD (e.g., building in the missing elements through conversation about the video), that the project might yet be a successful act of environmental communication in reference to regional noise pollution.

Hopefully, everything I learned as a producer and everything we learned as a team while producing LOUD will enhance Ecosong.Net's future productions. Meanwhile, LOUD's afterlife is going fairly well. Audiences are responding well to live performances. For example, on June 30, 2019, Dana Lyons and I performed the song at the Friends of Lime Kiln State Park's one-hundred-year celebration for the park's lighthouse (see Ecosong.Net). We each performed it solo and both chose a more sedate, reflective style for the song. There were several members and leaders of local conservation organizations in attendance, and a petition was being circulated to deal with boat noise on a county level. A few weeks later Tim Gustafson and I performed the song for a group of local environmental stewards, the Friends of Moran (Pedelty 2019). Bryanna McCorkle and I sang it together at an Earth Day event for the Friends of Saltwater State Park, and Felicia Harding and I sang it in Vancouver. Dana Lyons is continuing to sing LOUD for diverse audiences on his Rise for the Orca Tour (2019). Those events and others like them have helped LOUD to more successfully fulfill its promise. Ultimately all of the talented singers who took part in the LOUD video are making it work. As we sat, sang, and looked out to sea at Lime Kiln, there was a palpable sense of hope that such efforts, even a song and music video about noise pollution, might actually matter.

REFERENCES

Almiron, Núria. 2016. "Beyond Anthropocentrism: Critical Animal Studies and the Political Economy of Communication." *The Political Economy of Communication* 4, no. 2: 54–72.

Beyoncé. 2016. *Formation.* Video: 4:47. https://youtu.be/WDZJPJV__bQ.

Bob, Clifford. 2005. *The Marketing of Rebellion: Insurgents, Media, and International Activism. Cambridge Studies in Contentious Politics.* New York: Cambridge University Press.

Buxton, Rachel T., Reina Galvan, Megan F. McKenna, Cecilia L. White, and Victoria Seher. 2017. "Visitor Noise at a Nesting Colony Alters the Behavior of a Coastal Seabird." *Marine Ecology Progress Series* 570: 233–46.

Firestein, Stuart. 2012. *Ignorance: How It Drives Science.* New York: Oxford University Press.

Fisherkeller, JoEllen. 1997. "Everyday Learning about Identities among Young Adolescents in Television Culture." *Anthropology & Education Quarterly* 28, no. 4: 467–92.

Freire, Paulo. 2000. *Pedagogy of the Oppressed, Thirtieth Anniversary Edition.* New York: Continuum. Ebrary, Inc.

Goffman, Erving. 1959. *The Presentation of Self in Everyday Life.* Garden City: Doubleday.

Hausmann, A., R. Slotow, I. Fraser, and E. Di Minin. 2017. "Ecotourism Marketing Alternative to Charismatic Megafauna Can Also Support Biodiversity Conservation." *Animal Conservation* 20, no. 1: 91–100.

Holt, Marla M., Dawn P. Noren, Val Veirs, Candice K. Emmons, and Scott Veirs. 2009. "Speaking Up: Killer Whales (Orcinus Orca) Increase Their Call Amplitude in Response to Vessel Noise." *The Journal of the Acoustical Society of America* 125, no. 1: EL27–L32.

Houghton, Juliana, Marla M. Holt, Deborah A. Giles, M. Bradley Hanson, Candice K. Emmons, Jeffrey T. Hogan, Trevor A. Branch, and Glenn R. VanBlaricom. 2015. "The Relationship between Vessel Traffic and Noise Levels Received by Killer Whales." *The Journal of the Acoustical Society of America* 134, no. 5: 4044. https://doi.org/10.1121/1.4830758.

Kivitso, Sharon. 2018. "Profits or Protection? Killer Whales in the San Juan Islands." *San Juan Islander*, September 16, 2016. https://sanjuanislander.com/news-articles/environment-science-whales/whales-2/27997/follow-the-money.

Klein, Naomi. 2009. *No Space, No Choice, No Jobs, No Logo: Taking Aim at the Brand Bullies.* Toronto: Vintage Canada.

Langer, Ellen J. 2007. *On Becoming an Artist: Reinventing Yourself through Mindful Creativity.* New York: Ballantine Books.

Latour, Bruno. 2018. *Down to Earth: Politics in the New Climate Regime*. Medford, MA: Polity.

Levitin, Daniel J. 2006. *This Is Your Brain on Music: The Science of a Human Obsession.* New York: Dutton.

Lyons, Dana. "Rise for the orca tour." Accessed July 3, 2019. https://cowswithguns.com/rise-orca-tour/.

Ma, Guancong, Xiying Fan, Ping Sheng, and Mathias Fink. 2018. "Shaping Reverberating Sound Fields with an Actively Tunable Metasurface." *Proceedings of the National Academy of Sciences* 115, no. 26: 6638–43.

Ma, Jing, Chunjiang Li, Mei-Po Kwan, and Yanwei Chai. 2018. "A Multilevel Analysis of Perceived Noise Pollution, Geographic Contexts and Mental Health in Beijing." *International Journal of Environmental Research and Public Health* 15, no. 7: 1479.

MacGillivray, Greg, Meryl Streep, and Sting. 1995. *The Living Sea*. Chatsworth, CA: Image Entertainment.

Mattern, Mark. 1998. *Acting in Concert: Music, Community, and Political Action*. New Brunswick, NJ: Rutgers University Press.

Mulvey, Laura. 1989. *Visual and Other Pleasures: Theories of Representation and Difference*. Bloomington: Indiana University Press.

Nisbet, Matthew C. 2010. "Knowledge into Action: Framing the Debates over Climate Change and Poverty." In *Doing News Framing Analysis: Empirical and Theoretical Perspectives*, edited by Paul D'Angelo and Jim A. Kuypers, 59–99. New York: Routledge.

Pailthorp, Bellamy, 2019. "Compromise on Orca Protection Removes Whale Watching Moratorium, Garners Criticism." *KNKX radio*, February 27, 2019. https://www.knkx.org/post/compromise-orca-protection-removes-whale-watching-moratorium-garners-criticism.

Pedelty, Mark. 2016. *A Song to Save the Salish Sea: Musical Performance as Environmental Activism*. Bloomington: Indiana University Press.

Pedelty, Mark. 2019. "July 23, 2019: Friends of Moran State Park," Public Lands Podcast. Posted July 23, 2019. https://podcasts.apple.com/us/podcast/public-lands-podcast/id1073287661.

Pedelty, Mark, and Morgan Kuecker. 2014. "Seen to Be Heard? Gender, Voice, and Body in Television Advertisements." *Communication and Critical/Cultural Studies* 11, no. 3: 1–20.

Radford, Andrew N., Emma Kerridge, and Stephen D. Simpson. 2014. "Acoustic Communication in a Noisy World: Can Fish Compete with Anthropogenic Noise?" *Behavioral Ecology* 25, no. 5: 1022–30.

Shannon, Graeme, Megan F. McKenna, Lisa M. Angeloni, Kevin R. Crooks, Kurt M. Fristrup, Emma Brown, Katy A. Warner, Misty D. Nelson, Cecilia White, Jessica Briggs, Scott McFarland, and George Wittemyer. 2016. "A Synthesis of Two Decades of Research Documenting the Effects of Noise on Wildlife." *Biological Reviews* 91: 982–1005.

Star, Susan and James Griesemer. 1989. "Ecology, 'Translations' and Boundary Objects: Amateurs and Professionals in Berkeley's Museum of Vertebrate Zoology, 1907–39." *Social Studies of Science* 19, no. 3: 387–420.

Von Glahn, Denise. 2013. *Music and the Skillful Listener: American Women Compose the Natural World*. Bloomington: Indiana University Press.

White, Dave D., Amber Wutich, Kelli L. Larson, Patricia Gober, Timothy Lant, and Clea Senneville. 2010. "Credibility, Salience, and Legitimacy of Boundary Objects: Water Managers' Assessment of a Simulation Model in an Immersive Decision Theater." *Science and Public Policy* 37, no. 3: 219–32.

Chapter Fifteen

The Human Rights of a River

Codifying the Posthuman

Chris Ingraham

In March 2017, a high court in the Indian state of Uttarakhand, the region home to the glacial source of the Ganges in the north Indian Himalayas, ruled that the Ganges and Yamuna rivers could be granted the legal status of "living human entities" and accorded the full set of rights thereby implied. This meant that, in the eyes of Indian law, to pollute either river would now be tantamount to harming a human being, and prosecutable on that basis. In the court's ruling opinion—bolstered by a similar ruling in New Zealand from just a week earlier—the presiding judges justified the decision as a matter of improving the "preservation and conservation" of these rivers, which are each terribly polluted yet essential to the Indian social imaginary—not to mention to the everyday lives of the country's 1.3 billion people, most of whom are directly or indirectly impacted by these rivers in innumerable material and practical ways. By that July, however, the case had been appealed on grounds of being too impractical, and it remains stayed in the Indian Supreme Court as of this chapter's publication, over two years later. Articulating the Indian case of a river's human rights within the recent history of a broader legal movement advocating the rights of nature at large, in this chapter I discuss how such movements can be understood as endeavors to codify the posthuman.

While the mantle of "the posthuman" can be used in many ways, recognizing a river as a living entity possessing the legal status of a person certainly challenges us to go beyond traditional conceptions of "the human" and our supposed exceptionalism. A number of questions follow: Who represents the river's interests? (In both New Zealand and India, a pair of state officials were appointed as "legal guardians" to represent the rivers.) What sorts of actions constitute a threat to the river's rights? (Pollution, presumably, but how to define that?) And what if the river seems complicit? (Is the river now culpable if a person drowns in it?) Are we now to look for ways that the river "speaks" for itself? And why is India giving human rights to some rivers when, socially and culturally, it has yet to recognize the immanent self-worth of women? The very enterprise of "giving" a river human rights, after all, is a deeply anthropocentric condescension, as if human laws or rights were even desirable to the river (from the river's standpoint), and not merely a paternalistic political act to bring the river within the jurisdiction of further human control and dominance over the environment. In a time when posthuman thinking is sweeping the academy, the legal assignation of human rights on a river can help us think more clearly about what is at stake in moving beyond anthropocentrism, particularly when it comes to environmental justice and those rhetorical measures undertaken to achieve it.

The case of the Ganges and Yamuna may be especially entangled with environmental justice because these rivers are so inseparable from the ecological, agricultural, economic, social, spiritual, and cultural well-being of the Indian people. The Ganges in particular poses challenging complexities. The historian Sudipta Sen (2019a, 2019b) has probably laid them out better than anyone. In northern India alone, he observes, the Ganges provides water for more than five hundred million people. Those with claims to the water as a resource are too vast to quantify, let alone monitor. From the thousands of towns and villages whose groundwater comes from the river, to those farmers, fishers, boaters, and others who live along its banks, millions of people hold traditional or customary rights to use the water in accordance with their long-standing rituals and ways of being, let alone the many (and often conflicting) claims made upon the river by those operating from commercial, governmental, or civic positions. Factor in the many engineering projects (some legal, others wholly DIY) to divert the river or to harness its energy, to clean it, to build tourist infrastructure around it, to regulate its flow or to dam it for agriculture and other purposes, along with the messy conflicts and corruption associated with a single river that flows through five separate Indian states, each with its own jurisdictional regulations, and it's easy to see how "environmental justice" in this case is itself a mighty stream with many, many tributaries.

Rather than attempt to analyze only a small brook of a case too complex to be done full justice in a chapter of this size, then, I wish to use the Ganges case as an anchor to hold in place a broader inquiry into how the attempt to give nature rights can be understood to operate rhetorically in service of environmental justice. In evoking "environmental justice," I mean something more than its standard meaning as a project of improving the uneven distribution of natural resources and human access to clean, thriving environments for all people. Legal avenues to pursue environmental rights and protections through the justice system, that is, also imply a commitment to justice *for the environment*: not merely justice for humans seeking a more equitable and sustainable relationship with "their" environment, but also a commitment to the immanent justice due to the environment itself, irrespective of what people might look to gain or take from a relationship with it. Of course, the effort to gain human rights for rivers is undeniably a practical, tactical, and thoroughly rhetorical endeavor to open more pathways toward their preservation and conservation, in no small part to make the rivers safer, cleaner, and more sustainably suitable *for* humans. But this is not to suggest that it is productive or tenable to partition concerns for "the environment" here and concerns for "the human" there, as if the two weren't interdependent.

To the contrary, a more inclusive notion of environmental justice affirms an observation that Timothy Morton (2007) has made about the flaws inherent in the all-too-human concept of "the environment" to begin with. As Morton points out, "The very word *environmentalism* is evidence of wishful thinking. Society would be so involved in taking care of 'it' that it would no longer be a case of some 'thing' that surrounds us, that environs us and differs from us" (141). Despite the strategic advantage of giving rivers and other natural entities human rights, in other words, the effort to do so can also be read as a belated corrective for the environmental *injustice* perpetrated by the long-standing anthropocentrism that has made us vain enough to invent "the environment" as something separate from the human on account of its not always working the way we want. "In a society that fully acknowledged the idea that we were always already involved in our world," Morton observes, "there would be no need to point it out" (141). What I modestly aim to point out in what follows are some ways that legal interventions like the ones in India, New Zealand, and elsewhere begin to enact the very principles of ecologically inflected posthuman thought, with great consequences for how people perceive their relationship with the sensuous world around them.

ECOLOGICAL POSTHUMANISM

There's an old joke about someone hesitating to swim across a river. What finally persuades him across is some specious reasoning: if 60 percent of the human body is water, only 40 percent of him will have to swim. It may be a bad joke, but it does a good job of illustrating what for some scholars is a core tenet of posthumanism: namely, that any notion of "the human" must account for the constitutive entanglement of humanity with the nonhuman. If what we think of as the human consists largely of things that we do not think of as human, then we are left to make sense of humanity in different ways. In the joke, of course, whatever humor it attains derives from the sheer outlandishness of imagining that a human body is somehow part of water because water is part of it. Humans may be mostly water, but we're *not only* water, and it's just that "not only" that authorizes a relationship to water predicated on our exceptional distinctiveness from it. To be human here is to be more than what is not, and that remainder is what our swimmer understands as the work that his very humanity mandates: precisely not "to be one with" the nonhuman, but to exert himself against it.

In the same way that explaining a joke can flatten all its humor, trying to account for the diverse constellation of concepts and commitments evoked by the term "posthuman" can deaden whatever feral vitality animates its multiplicity. Academic discussion of the posthuman today has become so broad that there's no guarantee that two scholars talking about it will be talking about the same thing. Generally, as I treat it here, posthumanism involves a project of challenging or thinking beyond normative (read: taken-for-granted) conceptions of the human and the human condition. In this sense, the "post-" in "posthuman" sometimes indicates a historical or temporal shift of some kind, perceived to mark an "after which" around how humans have come to understand themselves and our entangled relationship with what is not human. In contrast to conceiving of the posthuman's starting point as some pivotal event, technological singularity, or historical epoch that either already happened or may yet lie ahead, however, I instead follow Andy Miah's (2008) position that "the starting point should be an attempt to understand what has been omitted from an anthropocentric worldview" (72). In other words, in this chapter, the posthuman will designate the many-folded project of identifying ways to understand humanity that don't involve the assumption of human exceptionalism and anthropocentrism that can give license to appropriative relations toward the nonhuman "other."

Ecological thought widely acknowledges the basic view that "we are all in this together" even while acceding that those comprising such a "we" are nevertheless different from one another and implicated differently in ways of experiencing so-called common worlds. Cary Wolfe (2017), for example (extending Derrida's work on animals), has written that the view "that dominates ecological thought and environmental ethics" is that "we share the same world, we just experience it differently" (140). What ecological thought accordingly needs to do more of is to address how the different ways of experiencing and being are produced by legal systems, social practices, technological mediations, and cultural productions capable of fostering or destroying the nonhuman.

At risk of veering into bad stand-up, there's another joke that's useful here to illustrate the ways phenomenal experience is determined differently for different beings. Variations of it have been made by Aristotle, Marshall McLuhan, John Durham Peters, and others (for a summary, see Peters 2015, 55). The version I'll tell is from David Foster Wallace (2009): Two young fish are swimming along when an older fish swims by. "Nice water today, 'eh boys?" says the older one. When he's passed, the two younger fish turn to each other and say, "What the heck is water?" There are a lot of things the joke might illustrate, but its punchline pivots on the comical obliviousness of the younger fish to what, from a human standpoint, is so fundamental to a fish's existence as to go without saying. But the joke gets at something more substantive: namely, that some degree of this obliviousness is inevitable—and not just for fish, but for all living beings. In other words, there's a sense in which the environments in which all things are situated, and on which all living beings existentially depend, are nevertheless invisible to them. Disputations about human self-consciousness or symbol-use aside, humans are not unique in being epistemically and phenomenologically privy only to what the affordances of our bodies and technologies allow us to know and experience.

One productive way to think about the posthuman, then, is as a project that seeks to interrogate and negotiate the terms on which humans can, from the vantage of our inescapably embodied humanness, begin to imagine our own involvement in what exceeds the human. What is the form of thought for imagining the more-than-human from within the human? Is it hypothetical? Speculative? Science-fiction? Fantasy? Well, it can be all of these and more. But one of its forms is also *legal*. And it's the legal forms of thinking the posthuman—thinking beyond the free and rational human subject—that are particularly consequential because the law is a system, *the* system, that operates by taking an otherwise abstract notion of the "subject" and making it enforceable as a set of rules.

HUMAN RIGHTS AND THE ENVIRONMENT

The historical development of legal systems by and for humans has tended to privilege human interests at the expense of concerns for the nonhuman. Human laws, in other words, often justify and condone the exploitation or degradation of the natural world on the basis of an unexamined anthropocentrism whereby being human authorizes a dominating and imperialist relationship toward the environment. To the extent that legal environmental protections do exist, then, they tend not to protect the environment *per se*, but to protect the competing rights of humans to use or exploit natural resources and spaces without encroaching on the respective rights of other people and their mutual prerogatives to do so. As Edward Mussawir (2019) has argued, "There is no ecological perspective on justice that is broader than that defined by the narrow, private, even petty juridical interests of the claimants" (257). And these claimants are none other than humans, the only "subject" with rights authorized by law. In other words, the law protects human claims upon the environment from competing human claims thereupon, but it does not as deliberately protect the environment itself from these claims.

Such is the legal anthropocentrism that pervades jurisprudential thought and codifies an ethical system—to the extent the law regulates minimal standards of ethical behavior—premised upon the centrality of human purposes and prerogatives relative to all else. This unexamined anthropocentrism, which is already baked into the law, "is not just the fact that law is composed and constructed by humans for needs that they understand as uniquely their own. Nor is it the typical condition of recognizing and giving preference to the rights and interests and values of humans over those of other species. It is the projections of the very idea of 'man', the being and the ends and destinies of 'man', at the centre of the vision that law has of the world in general" (258). What Mussawir is pointing out here is how the law, when premised upon anthropocentrism, cannot help but replicate and codify the very social imaginary that, by conceiving of the human as the paramount protagonist of all existence, justifies the legal subordination of everything nonhuman in the human's service.

Though there is, of course, a longer history of environmental law and the many ways legal systems around the world have historically regulated the human use of natural environments and their resources, the anthropocentrism of much environmental law is particularly evident in the ways human rights rhetoric has, since the mid-twentieth century in particular, been used to promote legal attention to environmental protection and repair. By and large, lawmakers and citizens have been interested in improving the environment to the degree that humans have a basic right to enjoyment and benefit therefrom. That this point is rarely made in histories of legal approaches to human rights and the environment (at least those I have come across; see, e.g., Leib 2011; Boyle and Anderson 1998; Boer 2015; Knox 2018), attests to just how deeply rooted anthropocentric humanism is in the social imaginary that compels legal interventions pertaining to the environment.

In his history of legal approaches to human rights and the environment, for instance, John Knox (a professor serving in 2018 as UN Special Rapporteur on Human Rights and the Environment) presents a story whereby lawmakers have been interested in improving the environment to the degree that humans have a basic right to benefit from its affordances. In other words, without seemingly meaning to, Knox (2018) shows that the environment is important to the law because the environment is important for humans to use or enjoy. Consider the trajectory he charts. In 1968, the United Nations General Assembly held the first international conference on the environment to discuss "the continuing and accelerating impairment of the quality of the human environment." In 1971, Pennsylvania amended its constitution to include environmental rights, becoming the first government in the world to do so. The year after that, at another UN conference on the environment, the General Assembly adopted the Stockholm Declaration, maintaining that both human-made and natural environments are essential "to the enjoyment of basic human rights." By 1976, Portugal became the first country to implement the constitutional "right to a healthy and ecologically balanced human environment." And the first *international* agreement to include an environmental right was signed in 1981, with the African Charter on Human and Peoples' Rights (649–50).

All of these examples rhetorically frame the environment as something *for* humans. For instance, multiple references to "the human environment" imply not an environment *of* humans but an environment somehow *belonging* to humans, just as the notion of "environmental rights" clearly does not imply the *environment's* rights, but the rights of humans to it—and to a particularly human formulation of what counts as its well-being, accessibility, cleanliness, and so on. Although using human rights as an avenue to vouchsafe protections for the environment is more liable to entrench anthropocentrism than to move beyond it, such a tactic does have its expediencies, as the scaling up of such protections from local to international levels attests. It is largely on account of successful arguments about the fundamental right of humans to a clean environment that local and international laws alike have enacted regulations to ensure that this right be fulfilled. Human rights arguments for environmental protections have accordingly been a source of meaningful legal change, even as they haven't meddled with initiating the larger and more epistemic social and cultural change that would involve questioning anthropocentrism itself.

Water justice has been an especially integral component in human rights approaches to environmental justice at large. In November 2002, the UN Committee on Economic, Social and Cultural Rights adopted the principle of a "right to water," defined as "the right of everyone to sufficient, safe, acceptable and physically accessible and affordable water for personal and domestic uses" (General Comment No. 15 2002). Almost a decade later, on July 28, 2010, the UN General Assembly went a step further, adopting the landmark Resolution 64/292, which affirmed the position that clean water for drinking and sanitation are essential to the realization of all human rights. In other words, water was not just one of the many rights humans should be able to count on, it was so essential to human existence that other fundamental human rights could not be attained without it. Water came first—but, of course, only to the extent it came second to the people who depended on it.

RIGHTS OF NATURE

It's fitting that the Ganges should lately emerge as an exemplar of how the notion of environmental justice is beginning to push beyond these anthropocentric approaches to environmental law. The Ganges, as it turns out, has been associated with some of the earliest postindustrial attempts by people to clean unhealthy environments in their own self-interest. As Jon McNeill (2000) describes in his sprawling environmental history of the twentieth-century, "The Ganges' fetid condition gave rise to one of the world's first antipollution societies, in 1886" (129). Considering that these fetid conditions have only grown worse since then—despite whatever the antipollution societies accomplished while undoubtedly acting less in the river's imagined interests than in their own—it's no wonder that the 2017 attempt to give the Ganges itself human rights should become the next progression in the history of the river's complicated relationship with those humans who both need and destroy it.

As I have explored elsewhere, the Ganges is materially entwined with the ceaseless cycle of life and death (Ingraham 2017). It has always been a sacred river, and accordingly a site for Hindu cremation rituals in which human flesh literally becomes river as bodies are burned and the ashes dispersed to disintegrate in the water. Of course, this means that the river becomes more polluted. But, in the context of human rights, it also means that the river, we could say, *already is human*. In other words, at least in a bare material sense, that means that any environmental law actuated in service of human interest is also already environmental law actuated in service of the river. Seen accordingly, the attempt to attribute legal personhood on the Ganges might be taken no more as an activist strategy than as an ontological point of clarification, calling out that we have always been posthuman.

The broader turn to rights of nature (as opposed to human rights *to* nature) is sometimes called "Earth jurisprudence," "Earth law," or "wild law," and it has emerged as an alternative framework recently gaining some global momentum in cases beyond just those in India (see Cullinan 2011; Burden 2014). The grim realities of climate change have disclosed how damaging it can be to presume that humans have a "natural" right to subdue and dominate the environment. In place of the human exceptionalism at the heart of this presumption some have sought to decenter the human in the interest of identifying universal rights shared by the human and more-than-human alike. Earth jurisprudence is associated with a social movement in environmental law known as rights of nature. A review of the vast literature and intellectual origins of the "rights of nature" is not my aim presently, but I do wish to show how some core tenets behind rights of nature invite us to think differently about rhetoric's role in action directed toward environmental justice.

To begin to do so, it's important to be familiar with the work of Christopher Stone, whose 1972 article, "Should Trees Have Standing?" galvanized the rights of nature movement by raising the question of whether "natural objects" should be accorded the legal right to bring an action or to appear in court (see also, Stone 2010). Responding to the question, Stone concludes that they should indeed have such rights, even if they cannot speak any human language. After all, he argues, plenty of things that can't speak are already accorded legal standing, from corporations to fetuses. Moreover, legal provisions already exist allowing for guardians to speak on behalf of those who themselves can't, or those prevented from bringing action before the law for other reasons (e.g., being underage). As Stone saw it, if someone wishes to defend the rights of a tree—legally speaking, that means if someone wished to be a tree's "guardian"—there was no reason they couldn't bring legal action against those that might harm it. Stone's project, however, was not to suggest the law forbid the felling of trees or even more widescale deforestation. He rather wished to show that there was no tenable case, consistent with existing American law, for denying trees the right to standing.

What that right implied, as many have realized since, could apply to much more than just trees. Stone's argument began to decenter the human as the basis for environmental law, opening a path for valid legal arguments about the rights due to all parts of the natural environment, including Mother Earth herself. In 2008, Ecuador became the first of the world's nations to recognize the rights of nature in its constitution. According to Article 71:

> Nature, or *Pachamama*, where life is reproduced and occurs, has the right to integral respect for its existence and for the maintenance and regeneration of its life cycles, structure, functions and evolutionary processes. All persons, communities, peoples and nations can call upon public authorities to enforce the rights of nature. (Constitution of the Republic of Ecuador 2008)

By 2010, a "Universal Declaration of the Rights of Mother Earth" had been drafted at the World People's Conference on Climate Change and the Rights of Mother Earth, in Cochabamba, Bolivia. The declaration's preamble marked its signatories' conviction "that in an interdependent living community it is not possible to recognize the rights of only human beings without causing an imbalance in Mother Earth" ("Preamble" 2010). Article 1.6 elaborated: "Just as human beings have human rights, all other beings also have rights which are specific to their species or kind and appropriate for their role and function within the communities within which they exist" ("Preamble" 2010). The Bolivian Plurinational Legislative Assembly passed the fundamentals of the declaration into law in December 2010, and a longer revised version in 2012.

The conferral of human rights on things widely agreed to be nonhuman does not mean suddenly saying they *are* human. It means extending certain legal benefits previously associated only with humans to that which is not. Conceptually, at least, the principle is not much different from extending the bedtime of one child to match that of an older sibling. Doing so doesn't suddenly make both siblings the same; it merely acknowledges that the younger should have some of the same privileges as the older. The challenge is that we don't have a language, let alone a legal framework, for a notion of "rights" or liberal justice that does not presume a human basis for the justification of its existence. That basis is not just "human," full-stop; it rather figures the human in a particular way—that is, as a "subject" or legal person whose presumably exceptional capacities compared with the nonhuman, serve as the benchmark for our ethical rights. As Deleuze (2006) once said, though, "A conception of law as founded on jurisprudence can do without any 'subject' of rights" (350). The rights of nature conceived by Earth jurisprudence movements accordingly codify the posthuman in a Deleuzian way by challenging us to think about a subjectless, more-than-human basis for ethical rights.

In this sense, what Cary Wolfe (2009) has argued about both animal studies and disability studies could as well apply to rights of nature research: it fundamentally challenges "a model of subjectivity and experience drawn from the liberal justice tradition and its central concept of rights, in which ethical standing and civic inclusion are predicated on rationality, autonomy, and agency" (127). As Wolfe points out, the sense of agency taken for granted in the tradition of liberal humanism is one believed to express the intentionality of a willfully acting subject. And, I'd add, the quintessential expression of liberalism's rational, intentional, and freely acting subject is, of course, speech—a belief tied up with the historical invention of "communication" as such, that speech is the conveyance of private property into a commons (see Locke 1980, Chapter V, especially Section 27; also see Peters 1989). To acknowledge the rights of nature by law does not therefore just codify nonanthropocentric posthumanism. In doing so, it poses a radical challenge to traditional conceptions of rhetoric, and of rhetoric as a *human* art of speech in accordance with long-standing tendencies to think the human, à la Kenneth Burke (1966), as "the symbol-using animal" (3–9).

THE RHETORIC OF ANIMACY

When Christopher Stone (1972) made his important case for giving trees standing, their *inability to speak* was the tacit point of contention in his argument. Without much addressing speech or language directly, Stone essentially asked, *Why should speech matter?* The question was not just about the ability to use language, but about animacy itself. As Stone put it, "The world of the lawyer is peopled with inanimate right-holders: trusts, corporations, joint ventures, municipalities, Subchapter R partnerships, and nation-states, to mention just a few" (452). If these things have rights despite being inanimate and "incompetent" insofar as they can't speak except through humans who represent their interests as custodians or guardians, then why should animacy or speech matter for the rights of trees? Without perhaps meaning to, Stone had struck upon a connection between the animate and language that has, in the study of linguistics as well as in political, cultural, and other spheres, long been used as a way to justify kinds of exceptionalism that are not just anthropocentric, but also predominantly male, straight, white, Western, and privileged.

As Mel Chen (2012) has shown in their book *Animacies* the very means of language is its animacy, a quality of liveness or sentience in how language operates. "Animacy" is accordingly a term used among cognitive linguistics to describe this quality and phenomenon of language. But Chen goes further, arguing that "animacy is political, shaped by what or who counts as human, and what or who does not" (30). From the great taxonomer Aristotle onward, animacy has been something meted out in hierarchies with political consequences (it's not by chance that most of Aristotle's writing about plants and animals—about the nonhuman—appears in his *Politics*). While many cultures and cosmologies have animacy hierarchies, however, their specific ordering is far from universal. Indigenous peoples the world over, in particular, have tended to treat the nonhuman as more animate, including the quality of being more communicative, than cultures from a dominant Western tradition.

The case of giving human rights to the Ganges again illustrates the stakes that are accordingly at hand. In a country with a majority Hindu population, many claims upon the river are entangled with religious beliefs that the water is holy, and hence immune to pollution. Yet when sewage, effluvia, and overall wastewater amount to over three billion new liters of unsanitary discharge pouring into downstream portions of the river *per day* (Scarr et al. 2019), the risk to public health is immense. This has only compounded concern by legal and environmental advocacy groups that the millions of Indians who aren't Hindu, yet are still connected to the Ganges in other ways, might be underrepresented by those appointed as the river's custodians and inclined to privilege regnant interests over others. Practically, that is, even the broad inclusivity of legally acknowledging a certain animacy of a river comes with its own threats of *exclusivity* by potentially favoring a dominant hierarchy of interests.

Nevertheless, as a historical matter, it has been the cosmological perspectives of religious and indigenous groups that have given rise to the first and most prominent adaptations of rights of nature around the planet. Ecuador would not have been the first country constitutionally to adopt such laws without the widespread belief of its indigenous people in *Pachamama*, or nature as a living being (for more on *Pachamama*, see Mamani-Bernabe 2015; for more on Ecuador's adoption of rights of *Pachamama*, see Humphreys 2017). The same is true of Bolivia, where the majority of its people are indigenous. Similarly, New Zealand's 2017 decision to grant the Whanganui River the status of a living entity would not have happened if not for more than a century of negotiating by the indigenous Māori iwi who live along the river and see it as an ancestor. At issue is not just widespread social imaginaries or cosmologies in certain cultures, but the rhetoric of animacy that they abide and promote. Though indigenous cosmologies of course differ greatly, many of them don't abide the same Western separation of human and nonhuman into hierarchies whereby animacy is a matter of autonomy, rationality, and speech. Indeed, often just the opposite: in place of autonomy, interconnection; in place of rationality, intuition; in place of speech, sensation.

For Chen (2012), "If animacy not only works in different ways for different cultures but indicates different hierarchalizations of matter, then it is critical to distinguish between relatively dominant formulations of animacy hierarchies and relatively subordinated ones, a project that seems all too vital for studies that reify the place in 'nature' of non-Western or subordinated cosmologies" (29). The stakes here, in other words, are not just about how humans subjugate rivers or "nature" in general, but about how that subjugation is often carried out in ways that also subordinate nondominant groups, whether racialized, genderized, sexualized, or colonialized. This subjugation, moreover, stains a given culture's or society's ways of imagining how they can, should, or actually *do* get on together—or apart.

The great "we" now made evident by climate change may be marked by radical difference, but it includes humans, animals, and all biotic and abiotic "existents"—from water to bacterium to trees—insofar as we are all implicated in and impacted by a shared planetary crisis. "Existents" is the term that Elizabeth Povinelli (2016) uses in her book, *Geontologies*, as well as elsewhere, in lieu of less animate alternatives like "entities" or "things," to refer to more-than-human phenomena and bodies that have ontological weight. As I read Povinelli, "existents" is a sort of ontologically oriented analog to the agency-oriented "actant" we get from Latour. Existents can be actants and actants are existents. The mattering of all existents we can take as given. What merits ungivenness is the difficult question of how existents are given to matter or not, and to matter or not in particular ways, based upon their mattering to or for humans who have the power to nurture or extinguish them.

That power is so often justified and executed through a legal apparatus that condones it. For this reason, Povinelli has wondered if more-than-human existents, from fog to rocks to creeks, "should have an equal say in legal, political and ethical debates in late liberalism" (122). But is there a viable way for humans to envision the rights of nature *from nature's standpoint*? We are, after all, like the fish swimming in water without realizing it, unable to get outside our own condition except by speculation. It bears mentioning that there's an ethical division in play here that's roughly analogous to the Christian golden rule, as opposed to a Levinasian ethic of responsibility to the Other. Should we treat others as *we* would like to be treated, or treat others as *they* would like to be treated? If we even provisionally operate according to the latter code, the challenge then becomes a matter of how to ascertain the ways that the Other would like to be treated, when language is not a viable means of communicating as much between species and existents. Might a river not "want" *river rights* instead of *human rights*? How would we know, and what might river rights involve?

THE ANIMACY OF RHETORIC

To think the posthuman from outside the human means acknowledging that if there is a rhetoric of animacy it is inseparable from the animacy of rhetoric. The recognition of rivers as living entities is one way to fulfill Latour's position that a parliament of nonhuman things needs to be better acknowledged and represented in human politics (1993, 144–145; 2005). When the legal infrastructure that establishes the legitimate terrain and rules for the political begins to open that terrain to include the nonhuman, a parliament of things begins to take shape. Access to water—and the use, pollution, and diversion of it—will always be "matters of concern" so long as water remains fundamental for the existence and sustenance of life. But that need not mean the concern should be limited to its human ramifications. If global warming teaches humans anything, it should be the power of nature to kick us out. That's a humbling power, one that persuades not through words, but through material animacies and obduracies that disclose the wisdom in recognizing that rhetoric is an animate energy, not limited by language but immanent to the being together of all existents. To cede such a recognition leads to an ethical imperative to organize this being together jurisprudentially around something other than a speaking "subject" of rights.

We should not forget that "nature," which sometimes goes by the name of physics or mathematics, has its own legal system: the laws of thermodynamics, the laws of classical mechanics, the laws of motion, and so on. These are not just rhetorically discussed as laws. The so-called laws of nature or physics are taken to be universal and obdurate, not symbolic or discursive, because they are believed to be permanent and unalterable by human design or engineering. As consistencies that organize the universe, nature's laws supersede human laws, providing the meta-framework in which all systems operate, legal or otherwise. But, as Foucault has shown, the very notion of the human has not been subject to such consistent understanding over time. In one of his more ominous passages, Foucault (1973) writes that despite the supposedly "luminous consciousness" of humanity, "Man [*sic*] is an invention of recent date. And one perhaps nearing its end" (387). The imminent end of "man" is not, however, an impending apocalypse, at least not as Foucault seems to conceive it, so much as an inevitable and forthcoming shift in discourse about what it means to be human. If that shift has begun to occur, it is part of a long and ongoing project of creating hierarchies to delineate the order of things. That project is not just ongoing in places that give indigenous people more credence. It is also happening in the United States, where indigenous beliefs that long predate America as such nevertheless remain largely unheeded.

In February 2019, voters in Toledo, Ohio, approved a "Lake Erie Bill of Rights" ballot measure to legally protect the heavily polluted Great Lake by giving it rights normally associated with a person. This was not the first measure passed in the United States acknowledging the rights of nature as such, that is, as distinct from the rights of humans to a certain standard of environmental well-being. As David Humphreys (2017) has noted, the first was in 2006, when a local ordinance passed in Tamaqua Borough, Pennsylvania, recognizing ecosystems as "legal persons" in order to prevent the dumping of sewage sludge on wild land (463). In 2010, a similar ordinance passed in Pittsburgh, stating that "Natural communities and ecosystems, including, but not limited to, wetlands, streams, rivers, aquifers, and other water systems, possess inalienable and fundamental rights to exist and flourish within the City of Pittsburgh." The ordinance goes on to specify essentially what Stone had argued in his 1972 paper: that people "shall possess legal standing to enforce those rights on behalf of those natural communities and ecosystems" (Pittsburgh Pennsylvania Code of Ordinances 2013).

It's no accident, I think, that many efforts to achieve Earth jurisprudence through rights of nature have focused on water. What makes water an especially powerful locus of these movements is its archetypal and biological association with the sustenance of life itself. All life. Just think of media reports about the discovery that liquid water once flowed on Mars, or that a lake of water persists near its south pole even now, and how such reports are immediately followed by speculative inferences that the planet may have, or once have had, the conditions necessary to foster life. Liquid water is often hailed as *the* essential ingredient for all known forms of life, human or otherwise, on Earth or any other planet. Without water, no life. This is as good as law. If we are looking for a replacement to anthropocentrism, aquapocentrism would make a decent alternative. But the point is not to replace one center with another (nor to loose anarchy upon the world when the center will not hold). Instead, the aim should be to continue seeking ways to cultivate a new social imaginary that does not take as its for granted horizon the supposition of human superiority and centrality as a way to justify environmental justice. To do so will require greater attunement to the sensorial and animate more-than-human world all around, but also a daily practice of "serving" that world with deliberate attention and care.

Such was the aim in March 2017, when India's Uttarakhand High Court gave the Ganges and Yamuna rivers (and the glaciers that feed them) their own rights, each as a "juristic/legal person/living entity." As it played out, though, political motives pushed the other way, too, and the case was soon brought to the Indian Supreme Court in an effort to overturn the High Court decision. When, in July 2017, the country's Supreme Court stayed the Uttarakhand High Court's order, it put the status of the rivers in limbo while awaiting further legal procedure that appears indefinitely delayed. In a way, this is appropriate: the river subsists in an ontological limbo of its own, always already human, always already not. The Supreme Court's preliminary basis for pausing the Uttarakhand ruling was that neither the Ganges nor Yamuna could be recognized as "living entities" because to do so would be far too impractical. Too many legal complications would follow, enforcement would be impossible, and besides, laws already existed to curtail pollution and to conserve the health of these rivers and their ecosystems. The trouble in India has been that existing laws have largely been insufficient given the rampant corruption around the regulation of the country's holy rivers, to say nothing of their overengineering by competing sources (privatized, state, and homemade alike; for more on the river's engineering problems, see Acciavetti 2015).

The limitations of the law as a solution to environmental justice, despite its many promises, comes into view when considering any legal apparatus's finite jurisdictional reach. The Ganges River may originate in the Indian Himalaya, after all, but it ends in Bangladesh's Bay of Bengal. Though geographers recognize the Ganges as one continuous river, that is, its legal recognition falls into two discontinuous systems. And in July 2019, Bangladesh one-upped the indecisiveness of Indian law when it comes to the Ganges's rights to personhood. Indeed, with its Supreme Court's landmark decision, Bangladesh became the world's first country to treat *all* of its rivers as living entities, granting each one, including the Ganges, the same legal status as humans. Though Bangladeshi law, of course, has purchase only in Bangladesh, the river doesn't know such geopolitical boundaries, and maybe that in itself is reason to think beyond restrictive rights and their codification when it comes to recognizing their worth and importance.

What it's important to acknowledge, in other words, is that the impracticality of more "posthuman" laws aligned with Earth jurisprudence is not sufficient to invalidate them. These laws are important not primarily because they help to ensure limited legal protections for important rivers, vouchsafing their preservation and conservation when other laws have failed, though these protections are real and desperately needed. Instead, rhetorics around the rights of nature are imperative even without their successful implementation, because only through the cultivation of such nonanthropocentric thought can we begin to live in more sympathy with the more-than-human with which we are invariably entangled and to which we have an immanent responsibility. Though it is not always wrong to act in human self-interest, it is almost always not wrong to support the immanent value of others, human or not.

REFERENCES

Acciavetti, Anthony. 2015. *Ganges Water Machine: Designing India's Ancient River*. New York: Applied Research + Design Publishing.

Constitution of the Republic of Ecuador. 2008. Article 71.

Boer, Ben, ed. 2015. *Environmental Law Dimensions of Human Rights*. Oxford: Oxford University Press.

Boyle, Alan E., and Michael R. Anderson, eds. 1998. *Human Rights Approaches to Environmental Protections*. London: Clarendon Press.

Burden, Peter. 2014. *Earth Jurisprudence: Private Property and the Environment*. London: Routledge.

Burke, Kenneth. 1966. *Language as Symbolic Action: Essays on Life, Literature, and Method*. Berkeley: University of California Press.

Chen, Mel. 2012. *Animacies*. Durham, NC: Duke University Press.

Cullinan, Cormac. 2011. *Wild Law: A Manifesto for Earth Justice*, 2nd ed. Devon, UK: Green Books.

Deleuze, Gilles. 2006. *Two Regimes of Madness: Texts and Interviews 1975–1995*. Translated by Ames Hodges and Mike Taormina. Cambridge: The MIT Press.

Foucault, Michel. 1973. *The Order of Things: An Archaeology of the Human Sciences*. New York: Vintage.

General Comment No. 15. 2002. "The Right to Water." UN Committee on Economic, Social and Cultural Rights, November 2002.

Hayles, Katherine. 1999. *How We Became Posthuman: Virtual Bodies in Cybernetics, Literature, and Informatics*. Chicago: University of Chicago Press.

Humphreys, David. 2017. "Rights of Pachamama: The Emergence of an Earth Jurisprudence in the Americas." *Journal of International Relations and Development* 20: 459–84.

Ingraham, Chris. 2017. "The Suddener World: Photography and Ineffable Rhetoric." *Philosophy and Rhetoric* 50, no. 2: 129–52.

Knox, John H. 2018. "The Past, Present, and Future of Human Rights and the Environment." *Wake Forest Law Review* 53: 649–65.

Latour, Bruno. 1993. *We Have Never Been Modern*. Cambridge: Harvard University Press.

Latour, Bruno. 2005. "From Realpolitik to Dingpolitik or How to Make Things Public." In *Making Things Public: Atmospheres of Democracy*, edited by Bruno Latour and Peter Weibel, 14–41. Cambridge, MA: The MIT Press.

Leib, Linda Hajjar. 2011. *Human Rights and the Environment: Philosophical, Theoretical and Legal Perspectives*. London: Martinus Nijhoff Publishers.

Locke, John. 1980. *Second Treatise of Government*. Indianapolis: Hackett Publishing.

McNeill, Jon. 2000. *Something New under the Sun: An Environmental History of the Twentieth-Century World*. New York: W. W. Norton & Company.

Mamani-Bernabé, Vicenta. 2015. "Spirituality and the Pachamama in the Andean Aymara Worldview." In *Earth Stewardship: Linking Ecology and Ethics in Theory and Practice*, edited by Ricardo Rozzi, F. Stuart Chapin III, J. Baird Callicott, S. T. A. Pickett, Mary E. Power, Juan J. Armesto, and Roy H. May, 65–76. Heidelberg: Springer.

Miah, Andy. 2008. "A Critical History of Posthumanism." In *Medical Enhancement and Posthumanity*, edited by Burt Gordijn and Ruth Chadwick, 71–94. London: Springer.

Morton, Timothy. 2007. *Ecology without Nature: Rethinking Environmental Aesthetics*. Cambridge, MA: Harvard University Press.

Mussawir, Edward. 2019. "A Modification in the Subject of Right: Deleuze, Jurisprudence and the Diagram of Bees in Roman Law." In *Posthuman Ecologies*, edited by Rosi Braidotti and Simone Bignall, 243–64. Lanham, MD: Rowman & Littlefield.

Peters, John Durham. 1989. "John Locke, the Individual, and the Origin of Communication." *The Quarterly Journal of Speech* 75, no. 4: 387–99.

Peters, John Durham. 2015. *The Marvelous Clouds*. Chicago: University of Chicago Press.

Pittsburgh Pennsylvania Code of Ordinances. 2013 "Code of Ordinances City of Pittsburgh Pennsylvania. Codified through Ordinance No. 12-2013." http://library.municode.com/index.aspx?clientID=13525&stateID=38&statename=Pennsylvania.

Povinelli, Elizabeth. 2016. *Geontologies: A Requiem to Late Liberalism*. Durham, NC: Duke University Press.

"Preamble: Universal Declaration of Rights of Mother Earth." 2010. World People's Conference on Climate Change and the Rights of Mother Earth in Cochabamba, Bolivia. April 22, 2010. https://therightsofnature.org/universal-declaration//.

Scarr, Simon, Weiyi Cai, Vinod Kumar, and Alasdair Pal. 2019. "The Race to Save the River Ganges." *Reuters Graphics*, January 18, 2019. https://graphics.reuters.com/INDIA-RIVER/010081TW39P/index.html.

Sen, Sudipta. 2019a. *Ganges: The Many Pasts of an Indian River*. New Haven, CT: Yale University Press.

Sen, Sudipta. 2019b. "Of Holy Rivers and Human Rights: Protecting the Ganges by Law." *Yale University Press Blog,* April 25, 2019. http://blog.yalebooks.com/2019/04/25/of-holy-rivers-and-human-rights-protecting-the-ganges-by-law/.

Stone, Christopher D. 1972. "Should Trees Have Standing? Towards Legal Rights for Natural Objects." *Southern California Law Review* 45: 450–501.

Stone, Christopher D. 2010. *Should Trees Have Standing? Law, Morality and the Environment*. Oxford: Oxford University Press.

Wallace, David Foster. 2009. *This Is Water*. New York: Little, Brown and Company.

Wolfe, Cary. 2009. *What Is Posthumanism?* Minneapolis: University of Minnesota Press.

Wolfe, Cary. 2017. "Of Ecology, Immunity, and Islands." In *Posthumous Life: Theorizing Beyond the Posthuman,* edited by Jami Weinstein and Claire Colebrook, 137–52. New York: Columbia University Press.

Chapter Sixteen

Preventing Another Great Garbage Patch

Attuning to an Ecospheric Rhetoric

Jacob A. Miller

Every year, we pump out a mass of plastic equal to the amount of all humans on Earth (Hanson 2016). Of the 260 million tons of plastic the world produces each year, roughly 10 percent ends up in the oceans and inevitably circulates in ocean gyres. The largest and perhaps most well-known aggregate of plastic is the Great Pacific Garbage Patch (GPGP). The GPGP is located approximately halfway between China and the United States, the world's two biggest plastic polluters. The Patch was discovered in 1997 by oceanographer Charles Moore and named by Curtis Ebbesmeyer (Maloney 2018). It is now a whopping 994,194 square miles in size (Lebreton et al. 2018), or 3.7 times the size of Texas.

A United Kingdom social media and entertainment company called LADbible teamed up with the Plastic Oceans Foundation to turn this country-sized trash patch into the world's 196th country. They named the patch "Trash Isles" ("Trash Isles" 2019). On June 8, 2017, World Oceans Day, they submitted a Declaration of Independence to the United Nations to seek its recognition as an official country. According to the Montevideo Convention on the Rights and Duties of States, an official state must meet four requirements. Journalist Brittaney Kiefer (2018) explains how Trash Isles met them all:

> First, define a territory. The campaign claimed all the plastic in the North Pacific Ocean as theirs. Second, form a government. Trash Isles established a monarchy and appointed Dame Judi Dench as queen. Third, be able to com-

> municate with other states. The PR, social media, advertising and online content of the campaign covered this off. Finally, establish citizens. This is how the campaign brought in members of the public, by asking people to become citizens and sign a Change.org petition urging the UN to recognize Trash Isles as an official country.

When the petition closed on January 24, 2018, it had 240,906 supporters signing up to be symbolic citizens (Roberts and Nunn 2018). Celebrities soon joined the campaign. For instance, although Al Gore did not become the U.S. president in 2000, he did become president of Trash Isles (Bengle 2017). John Cena followed as treasury secretary, Judi Dench the queen, and Mark Ruffalo and Chris Hemsworth as honorary citizens (Bengle 2017). New citizens received Trash Isles passports, one of the unique design elements of the campaign. LADbible also designed stamps, currency, and a flag.

Some rhetorical analyses have been conducted to make sense of Trash Isles visuals, discourses, and implications for environmentalism. For instance, Jordan Wong's 2018 analysis, entitled "Trash Isles: The Rhetoric of Statehood," uses Castelló's (2009) model, which considers how fictional media narratives unify nationalist movements. Wong concludes that "Trash Isles forces us to reconsider our conceptions of statehood. While other states are based on a combination of territory and cultural unity, Trash Isles primarily uses discourse to establish a globalized national identity." An implication of this, Wong argues, is a potential new trend in geopolitics. Social discourse could be more important than physical proximity in establishing statehood. Moreover, if the UN recognizes Trash Isles as a nation, the international body will be required by its own policies to clean it up. Wong furthers that this raises some questions on the rhetoric of dissolution. If the main goal of establishing nationality is to dissolve itself, other environmental activist groups could follow suit.

Another example is an ethnographic analysis of GPGP written by Alli Maloney (2018). Maloney took a ship to the middle of the GPGP to see how bad the waste was compared to Trash Isles' description. Her primary claim is that the "rhetoric of Trash Isles" mischaracterizes its reality. She wrote that she could not see the "landmasses" of trash portrayed by LADBible. This is understandable given plastic only aggregates in some places and can disperse given the gyre's flux. Nevertheless, the fact still remains that five hundred sunken jets' worth of plastic tonnage floats within the GPGP (Lebreton et al. 2018).

Wong and Maloney exemplify analyses of Trash Isles discourses *placed upon* the Great Pacific Garbage Patch. However, the focus of their analyses is not what the GPGP's existence *itself* tells us about the materiality of past and current discourses. Analyzing the rhetoric of GPGP tell us more about ourselves than can the GPGP headlines. Furthermore, a materialist analysis of GPGP is appropriate given rhetoric's recent "ecological turn" (Wells et al. 2018). As Wells, McGreavy, Senda-Cook, and McHendry Jr. (2018), ecological turn scholars study matter's agential, immanent, and mutual qualities within rhetorical situations. For example, Barnett and Boyle's *Rhetoric, through Everyday Things* (2016) expands the scope of rhetorical situations beyond the humanist tradition of purpose, audience, and speaker. It seeks to include the study of assemblages of nonhuman actors, inanimate objects, and humans. Examples of assemblages include cookery, bicycles, dragonflies, prairie, trees, turtles, tidal waves, dammed rivers, sand footprints, mimic octopi, urban neighborhoods, fuel, bicycles, GPS, and mudflats (McGreavey et al. 2018). The ecological turn has successfully argued that, in the words of Jane Bennett (2010), "materiality matters" (xvi).

The turn distinguishes itself in other ways. It moves from a concept of "environment" to "ecology" (McGreavey et al. 2018), because ecology de-objectifies its components and sees humans as existing within the natural world without controlling it. The turn shifts emphasis from the meaning-making of human subject to that of nonhuman matter, thereby treating "objects" as sacred representatives of ecological limits. Even so, to avoid the subject-object dualism, these objects are imbued with a Whiteheadian process-relational metaphysic that is designated by the term "assemblages" (Bennett 2005). By doing so, the turn does not wish to objectify matter. It wishes to acknowledge materials' steadfast limits.

Rhetoric's ecological turn is necessary to attune the cascading ecological crises of our time. However, the ecological turn lacks enough description about the nature of the materiality that comprises the rhetorical situation at the scale of Earth systems. For instance, GPGP's materiality is more pervasive, expansive, and quantitatively larger than cookery, mimic octopi, or bicycles. The middle of the Pacific Ocean is one of the least-explored places by humans, and yet we have managed to litter it with tons of trash. Plastics were even found in the Mariana Trench (Gibbens 2019). The ubiquity of objects like plastics, nuclear radiation, or GHGs requires us to consider a rhetoric of things on a planetary scale, to expand the rhetorical situation to fit the magnitude of our impact. Taking the next step requires addressing the Anthropocene's temporality, scale, systems, and totalization.

Working from the ecological turn's foundational assumptions, and to best analyze GPGP, I foreground the previously imagined theory of "Ecospheric rhetoric" (Miller 2019). Ecospheric rhetoric follows in the footsteps of previous trailblazers who write within rhetoric's ecological turn because it uses similar interdisciplinary form. Case in point, just as Barad (2007) develops her "intra-action" theory by drawing on insights from physics, or as Rickert (2013) arrives at "ambient" rhetoric by looking to ecology, this chapter pulls from "Ecospheric" rhetorical theory that consults theoretical ecologists and Earth systems scientists. I treat "Ecospheric" rhetoric as an ontological distinction, because as Latour (1993) writes, "the discourse of the Ecosphere is too real and too social to boil down to meaning effects" (6). Ultimately, my Ecospheric analysis shows of the Great Pacific Garbage Patch reveals the deeper systemic issues Trash Isles represents. It uncovers the effects we have on the ocean, ourselves, and on the entire Ecosphere.

ECOSPHERIC ANALYSIS OF THE GREAT PACIFIC GARBAGE PATCH

"Ecosphere" was coined in 1953 by physician and space medicine pioneer Hubertus Strughold. The term was picked up in 1961 by ecologist J. Stan Rowe and became the focus of his academic career. Rowe (1989) defined it as the global sum of all ecosystems without the living/nonliving distinction. Although Rowe was an ecologist, he was really an Earth systems scientist before there was such a thing as one. In scholarly disciplines, "Ecosphere" remains mostly unknown, misunderstood, or is used incorrectly in a synonymous manner. However, Rowe emphasizes that it represents a conceptual shift. The ecosphere includes all the Earth's living organisms and their physical environments within the planetary ecosystem. The Ecosphere's parameters are comprised of three limit-bound spheres: atmo-, litho-, and hydro- (Rowe 2006). In contrast, an environment's parameters can shift depending on its holder. The assumption of someone who has an environment can make it *better than* what it was prior to them establishing it as *their* environment (Rowe 2006).

When considering the rhetoric of the Ecosphere, it becomes totalizing, subsumes local discourses, and prohibits individual rhetors from "wielding rhetoric as a tool" (Lind 2017). If we attune to an Ecospheric rhetoric we can begin to reconfigure our relationship to our Earthly surroundings and turn from humanity's crash course, or at least recognize the severity of the crash. To achieve this end, I define the parameters of the GPGP artifact, then highlight three properties of the material forming the rhetorical situation as designated by Ecospheric theory: (1) *Encapsulation*, (2) *Earth Alive!*, and (3) *Emergence.* After immediately applying each to the GPGP, I offer two implications: (a) a way to see reflected in our language a disconnect from the material world, and (b) the potential for Ecospheric rhetoric to unify a fractured environmental movement.

ARTIFACT PARAMETERS

The GPGP represents the perfect convergence of human and nonhuman discourses in the Ecosphere's three spheres, which makes it difficult to designate specific parameters. The GPGP's images, texts, objects, and their genealogy generally comprise the boundaries of the artifact. Since fragmentary pictures cannot provide a composite sketch of GPGP, the artifact is not limited to particular visual images. Neither is it limited to popular texts or LADbible reports. Defining the objecthood of GPGP is even more difficult. Lebreton et al. (2018) estimate seventy-nine thousand tons of ocean plastic are floating inside an area of 994,194 miles^2 (1.6 million km^2); a figure four to sixteen times higher than previously reported. Most of the dense debris sinks several feet or miles beneath the surface, which makes the vortex's depth nearly impossible to measure. Most intuitively, the ocean represents the hydrosphere, the petroleum-based plastics represent the lithosphere, and the plastics slowly break down in part thanks to their interaction with the atmosphere. Therefore, GPGP, as an artifact, includes the assemblages of materialist constructions, their accompanying discourses, and localized manifestations of Ecospheric properties.

Encapsulation

In contrast to an environment, whose parameters can shift depending on its holder, the Ecosphere's parameters are always comprised of three spheres: atmo-, litho-, and hydrospheres. The Ecosphere swallows all properties of life, including us. And insofar as it subsumes us, it also subsumes all of our signs, symbols, and discourses. Wells et al. (2018) place humans within the interaction of the hydrosphere by highlighting its totalizing presence:

> A material fact about human experience on Earth is that we are, mostly, confined to a terrestrial existence. . . . Yet, with water covering three quarters of the Earth's surface and with our own bodies mostly composed of this material, our very existence is much more fluid than we might, at first, perceive. As environmentalist Rachel Carson once remarked, "this fluidity connects us to ecologies across time and space: To stand at the edge of the sea, to sense the ebb and flow of the tides, to feel the breath of a mist moving over a great salt marsh, to watch the flight of shore birds that have swept up and down the surf lines of the continents for untold thousands of years, to see the running of the old eels and the young shad to the sea, is to have knowledge of things that are as nearly eternal as any earthly life can be."

This vivid, affective, and embodied description of the ocean's grandness and inescapability is confirmed by an empirical assessment. The ocean takes up "71 percent of the Earth's surface" and houses "99 percent of the biosphere" (Gass 2013). That means only *1 percent* of the biosphere takes up residence on dry land. Humanity has long gazed upon the ocean horizon and felt enmeshed in its grandeur.

Just as 95 percent of the ocean is unexplored and filled with mystery, so is the GPGP mysterious. We know little about the best ways to clean it up, not to mention the extent of its properties, depth, or boundaries due to ocean currents. Lebreton et al. (2018) estimate it to be 994,194 square miles, although they say their estimation is "conservative" (13). The Ecosphere is both apparent in, and affected by, the severity of the GPGP's material accumulation. Lebreton et al. found 99.9 percent of all marine litter to be plastics. Plastics are polymers that have various manifestations, including polyethylene, polystyrene, and polypropylene. They are primarily made of hydrogen and carbon but sometimes oxygen, sulfur, chlorine, fluorine, silicon, or nitrogen. Most are manufactured from petroleum, which was first assembled billions of years ago in the cells of organisms. Trace the source of the plastic, and it comes from the same biotic material as you or I. The Ecosphere encompasses us both.

The egregiousness of the GPGP's effect is as a result of plastics being a combination of these materials that severely deviates from their natural state. As Wolchover (2011) says of plastics, "Nature doesn't make things like that." The organisms that quickly degrade a banana from green to yellow to brown have evolved over billions of years to attack bonds commonly found in nature. But when faced with polypropylene, the critters don't have the metabolism to break them down. While plastics' constituent molecules may be found naturally in the Ecosphere, their chains are not. Trace amounts of hydrogen in the atmosphere interact with carbon in the hydrosphere, and intra-action of hydrogen occurs within the hydrosphere itself. When introducing plastics with the same elemental components rearranged through rapid manufacturing, the nature of the interaction changes. The Ecospheric properties are apparent in both manifestations, and it is inescapable in both cases. Ecosphere encapsulates all, but humans and our precious plastic are out of spatiotemporal sync.

Earth Alive!

Rowe (1961; 2006) takes issue with equating life to the biosphere because doing so privileges organisms, especially humans, over their abiotic surroundings. Treating the three spheres as dead diminishes their importance and ignores their limits, Rowe argues. Following this line of thinking, populations could spread unchecked by their ecosystems. Rowe emphasizes how this fundamental mismatch perpetuates in our linguistic, social, and political systems. Rowe's primary assertion is that to reattune to the Ecosphere, we should treat it, and everything that makes it up, as Alive! The capital "A" and exclamation mark emphasize his point.

Alive! are not just the turtles, fishing populations, fishing experts, scientists, LADbible activists, Trash Isles citizens, or United Nations members. The ocean is Alive! The plastic is Alive! Yes, they do not meet biologist's definition for the necessary conditions of life because they do not maintain homeostasis, are not composed of cells, nor do they have a life cycle, undergo metabolism, grow, adapt to their environment, respond to stimuli, or reproduce and evolve. But what allows all these generative processes in the first place is the interobjectivity and encapsulation of the Ecosphere. This can be understood through a quirky example. In the GPGP there is likely to be one of the twenty-eight thousand rubber duckies that was spilled out of a shipping container in 1992. If one rubs the rubber ducky with sodium polyacrylate, the same waterlock material used in diapers, it seems to spring to life. The individual ducky does not meet biology's definition of life, but it does contain the property of responding to the stimuli of a natural element. Although not considered "life," mimicking movement can be a satisfying simulacrum of life. And even though plastic, rubber ducks, or water are considered abiotic and inanimate, these materials present themselves to us Alive! if we reimagine the definition of "life."

The ontological status of those affecting should be the same as the affected for the sake of increasing the importance of our earthly surroundings. For another example of the dead affecting the living, take microplastics, which we currently consider abiotic. Scientists estimate there are "51 trillion microplastic particles in the oceans," which is "500 times more than stars in our galaxy" (Villemain 2017). If that is not alarming enough, a 2018 study found microplastics in 93 percent of bottled water. It involved "259 bottles of water from 11 brands across nine countries" (McCarthy 2018). These results make it safe to assume that humans have ingested residual microplastics. Seemingly out-of-sight waste, its mismanagement—and all the industrial, political, and social discourses that created it—are coming back to us. We are *literally* eating our words in fish and drinking them in water. Scientists warn that microplastic accumulation in the bloodstream could cause blood clots, that microplastics swirling in the GPGP can make their way into us. This highlights that they are Alive! And insofar as microplastics affect our food, water, and bloodstream, they also affect speech acts, nonverbals, and meaning-making in general. If nothing else, "dead" plastics are Alive! because of their tremendous effect on the living.

Although the Ecosphere is limit-bound, it is constantly changing. The ebb and flow of the gyre especially highlights the processual nature of oceans. Formed by global wind patterns and forces created by Earth's rotation, its movements are akin to those of a conveyor belt. The enclosed variation is also apparent when watching the flow of carbon dioxide (CO_2) over the course of twelve months (Krulwich 2016). In June, oceans and tree leaves vacuum CO_2. In the winter months, when oceans turn to ice and trees bare leaves, there is a great exhale of carbon. Play the tape on repeat and watch the Ecosphere's lungs breathe like the lungs of any human. Humans are straining the Ecosphere's breath. Scientists warn that the Ecosphere cannot "keep up" with the buildup of CO_2 (Krulwich 2016). As a result, more CO_2 is absorbed into the ocean and the PH level decreases, causing ocean acidification, food and habitat loss that damages the marine life chain, and hypoxia (oxygen deficient) areas within the North Pacific oceanic gyre (Valdes et al. 2009). Moreover, it speeds up the breakdown of plastic that isn't going anywhere (Kolbert 2014).

This recalls a joke: Why does the person afraid of death want to be reborn as plastic? Because they will last forever. The punchline should punch us in the gut. Plastics are no joke. Because polystyrene materials will last tens of thousands of years, beyond biological forms and social entities, they are beyond the scope of our comprehension (Rickert 2013). Considering the plastic, marine life, and ocean are Alive!, so too is the Great Pacific Garbage Patch Alive! The GPGP is an Ecospheric lifeworld of swirling objects, effects, and discourses making itself known to us in startling ways.

Emergence

Alive! things do not stay stationary. They emerge. Emergence is when "something else emerges from nothing but" (Abrams 2015, 43). Distilled, emergence means "more is different" (38). Emergence exists at every scale. It existed in the expanding cosmos, when microorganisms became the first eukaryotic cells 1.5 billion years ago, in E. O. Wilson's famous ant colonies, or in basic biology. We learn parts—nucleus, mitochondria, ribosomes, cell membrane—that interact together and make up an entirely new whole that our language labels as "life." The GPGP perfectly exemplifies emergence. A water molecule has emergent properties that arise from the properties of oxygen and hydrogen atoms. Many water molecules form together as ocean waves. An emergent approach to a rhetorical situation considers not only plastic, water systems, or the ocean environment, but also the complex "relationship among them" (Bar-Yam 2011).

GPGP's plastic itself is emergent property. It is an assemblage of the material and the historical, human discourses that produced the material. The plastics imbue recent discourses—like LADbible's attempt to eradicate the GPGP by making it its own country—and the past discourses of its origin. The history of plastics helps explain how plastics emerged to be something completely different than originally conceived. Celluloids, the first industrial plastics, were patented and made commercially viable in 1869 by John Wesley Hyatt, "who was inspired by a New York firm's offer of $10,000 for anyone who could provide a substitute for ivory" (Science History Institute 2016). Hyatt's thinking was that nature only supplied so much wood, coal, bone, tusk, horn, and so forth, but plastic was unconstrained by nature's limits and could replace the use of those natural materials. At the time, plastic's proponents touted it not just as revolutionary but *environmentally friendly*. Advertisements even praised plastic for being able to protect the planet from the destructive forces of humans. It would "save the elephant and the tortoise" (Science History Institute 2016). Today, humans have produced the plastic weight equivalent of one billion elephants, and regularly see turtles choked by six-pack plastic rings. If alive today, Hyatt would be shocked that plastic has had such a deleterious emergent effect.

The GPGP has grown to a large size in a short period of time. It was only discovered in 1997, and just twenty-one years later Lebreton et al. (2018) conclude that GPGP's area is "four to sixteen times bigger than previously reported" (1). As it is, plastic is accumulating too rapidly, beyond the natural capacity for time to heal our wounds. As Rachel Carson (1962) observed, "Time is the essential ingredient, but in the modern world there is no time" (6). Humans have accelerated the emergence of plastic far beyond what the Ecosphere is comfortable with. We have done so unwittingly, which is, as I will argue in the next section, inevitable given the language we use every day.

ECOSPHERIC DISCOURSES REVEAL THE DISCONNECT OF LANGUAGE FROM MATTER

Most humans are not able to recognize the GPGP's severity, pay it the attention it deserves, or even recognize that it exists in the first place. This Ecospheric analysis provides a unique understanding of how the internal mechanisms of language operate and how language drives humanity's degradation of the material Ecosphere. Its structured production of meaning cannot truthfully represent the material that begets that meaning. There is an inevitable and persistent gap in naming things and understanding the nature of the things themselves. Better identifying this gap can potentially make us aware of the discourses and actions that create monstrous rhetorical situations like the GPGP.

Interaction of the Ecosphere's three spheres bears all the meaning of our existence even if we are unable to recognize it. The GPGP exemplifies that the reality of our Ecosphere is out of sync with the reality constructed by our symbolic code and its material effects. The associated symbols that comprise the GPGP—"plastic," "trash," and "debris"—reference the materiality of the thing-in-itself. These objects are inescapably Ecospheric, meaning the materials necessary to build them came from what was already present on Earth. This disrupts the standard semiotic triangle's relationship of object (sign), signifier (sound-image), and signified (idea/concept) (de Saussure 1966, 66–67). These are staple concepts in both the French (de Saussure and Levi-Strauss) and American (Pierce and Chomsky) semiotic and linguistic traditions (Lodge and Wood 2008). Insofar as the object is Ecospheric, so too are the signifier and its signified. The symbolic code is circulating and dynamic, but the signifier and signified are not ethereal abstractions. All involved members of the process are at once interpreters, encoders, and decoders (Irvine 2012).

Given a bound Ecosphere, objects can only support a limited quantity of meaning. All the meaning tied to the objects is bound by what the Ecosphere grants, so there are limits to the generative properties of language. This claim would seem obvious. Words cannot procure meaning outside of this earth; even the imaginary register is inescapably tied to our world. However, semiotic theory within the poststructuralist tradition has us believe that meaning is not anchored to the material. For instance, Lacan claims a "chain of signifiers" infinitely defers meaning, and Derrida's *différance* says there is no origin or center of meaning (Lodge and Wood 2008, 210). These build from the assumption that the object-signifier-signified relationship is arbitrary and that objects here on Earth will always exist in more-or-less stagnant form. Because the relationship is considered arbitrary, signs are thought fundamental to how "societies construct, produce, and circulate meanings and values" (Irvine 2012). But this circulation is still entirely dependent on the persistence of objects that grant life and its encircling conditions for the signifier and signified. In other words, the triadic relationship is not quantitatively arbitrary because the triangle's equidistant lines are not the same width. Harkening a popular process philosopher phrase, the "map not the landscape" (Cobb Jr. 2015, 18). The map represents the material landscape but should not be confused as ontologically equal to the signifier "map." Without the oceanic landscape we would have no conception of "map" or "ocean."

What does a limit-bound process of signification mean for meaning? It means that every translation carries with it traces of material effects. For example, consider signifiers like "water" or "air." These stand in for their objects, which underwent a quantitatively entropic physical transformation over a certain amount of time relative to the physical transformation of the Ecosphere. Now consider "plastic" or "trash." These, too, stand in for their objects, which also underwent a quantitatively entropic physical transformation over a certain amount of time relative to the Ecosphere's transformation. The difference is the speed at which the latter scaled up in size and impact—hundreds of years compared to billions. In sum: *automatic systemic translations of difference are dependent on the temporal contingencies of the ordering relations of each object.*

This remains the case even as signifiers beget signifieds, phrases, sentences, and become bound by syntagmatic relations, myths, codes, and ideologies (Irvine 2012). Letters, words, and sentences may serve as the building blocks to produce an infinite number of meanings, but the objects they signify are finite. Due to each word's individual effect on a finite Ecosphere, each carries a distinct weight. The object weights the signifier and signified. Thus, they should be held accountable.

It stands to follow that objects that are nonhuman and not altered by humans are more responsible for meaning-making than often given credit for. They make up a majority of rhetorical situations. Again, let us consider the word "water." As I say the word "water" and then drink it, it becomes part of my body, adding to the 60 percent of water that makes up the material me. But it is also not water as we know "it." Water has four emergent properties—hydrophilic and -phobic substances, evaporation, cohesion, and temperature moderation—that disrupt our signifying name of "water." It is more or less all these things, but the word "water" does not grant the substance of these properties. Its many ordering relations grant me life and prepare it for signifying; it is not solely I who does so. Greek philosophers thought water to be the original substance and that all things were made from it, and a common English translation of verse 21:30 of the Qur'an says, "From water we have made all things." But *we* haven't made all things from water. Rather, water, and its processual interactions, made me from it when evolving us out of the oceans and later called me to call it so.

Now let us consider the signifier "plastic." It imbues a genealogy that incorporates an assemblage of effects from hundreds of previous years, including the discovery of petroleum to produce plastic, its initial advertising as a viable alternative, and the businesses' willingness to mass produce it. Taken at face value, "plastic" merely represents but a material referent made up of an assemblage of polymers; in the semiotic triad this makes sense. However, because plastic's emergence is so completely different from the natural emergence of anything in the Ecosphere, the material effects produced by the proliferation of the "plastic" signifier and associated concepts far outpace the material effects produced by "ocean" within a relatively short time period.

The stark juxtaposition of two wildly different Ecospheric assemblages—plastic and ocean—reminds us of our common Ecospheric origins. The Ecosphere that gave rise to our ancestors and sustains our life consumes everything we do; the plastics we make and the matter we rearrange. The GPGP reminds of time united. A mass of floating assemblages exists due to the multiple, previous discourses that went into the conceptualization, development, manufacturing, distribution, use, and discarding of the plastics. The 1970s Pepsi bottles, 1990s plastic tubs, and 2000s *Suave* microbeads converge in the GPGP. The discourses of the past connect with the present in one swirling gyre. A temporal flow occurs. Plastic bobs in and out, juxtaposing the 200-million-year-old ocean.

Consider what I have laid out in this section as a heuristic, a form of semiotic and rhetorical accounting that can be used to understand a direct connection between our language and its material effects. Historians, geologists, anthropologists, and those in other fields have taught us that we have been on this Earth for such a short amount of time relative to most other objects, yet we altered it tremendously in that time. This section is making the case that there is a way to see that fact reflected in our everyday language, that we can measure anthropogenic impacts in an ordinary sentence. Every modern word and modern sentence—for example, "progress," "innovation," or "technology"—is densely packed and speaks volumes about modernity's pernicious effects (Wallace-Wells 2019). Language is just not tuned to hear it.

ENVIRONMENTALISM'S UNIFYING VOICE

More practically, Ecospheric discourses could aid in healing the environmental movement that is fractured over how to solve crises. As Danielle Endres (2014) asserts, environmentalism faces a "voice problem." Endres argues the "collective voice(s) of environmental advocacy have not yet achieved widespread success in persuading publics to adopt the large-scale behavioral and political changes" that are needed to mitigate complex problems (111–12). Endres further notes that these problems may stem from "internal tensions about the many causes, solutions, and definitions of the environmental crisis" (116). For instance, there is an ongoing debate about how to best capture GHGs, with some environmentalists arguing for carbon capture and storage devices while others promote more natural carbon sequestration through sustainable farming practices. Endres has us ask, "Is environmental advocacy better suited to a multiplicity of voice or to being constructed univocally?" (119).

Multiplicity of voice is needed to ensure a just transition; we must include in the conversation indigenous, minority, underrepresented, and marginalized voices. Nevertheless, the climate changes happening on an earthly scale will affect us all, so we should all agree to reattune to the same *thing*—the Ecosphere. Once we are clear that our efforts fully help the Ecosphere, and not just our environments, symbolically constructed bottom lines, or quarterly profits, then and only then can we deliberate the best solutions. Otherwise, the underlying forces of our problems—neoliberalism, progress, capitalism, and consumerism (Endres 2014, 117)—will continue to wreak havoc because they are the Ecosphere's adversaries. To support this claim, let's return to GPGP cleanup efforts focused on solving surface symptoms.

LADbible is pressuring the United Nations to gain sovereign status and mandatory cleanup, but another group called The Ocean Cleanup (TOC) is done waiting for a governing body to take action. They are cleaning it themselves. The San Francisco-based environmental organization was founded by Dutch inventor Boyan Slat in 2013 when he was just eighteen. Slat invented a 600-meter-long floater system that is estimated to capture "50% of the Great Pacific Garbage Patch every 5 years" (Ocean Cleanup, n.d.). It mimics a coastline, collects the trash, and then a boat collects the garbage and disposes of it onshore. However, the technology is already facing setbacks, behind its goals (Martin and Morris 2019), and the collected garbage still needs disposal on land.

Plastic straws have become another culprit. They have come under fire for their contribution to the GPGP's plastic waste, and individuals caught using them have been chastised. However, plastic straws account for only .03 percent of the eight million metric tons of plastics (Minter 2018). Comparatively, 47 percent of the GPGP is old or discarded industrial fishing equipment (Lebreton et al. 2018). Lebreton et al. (2018), who work for TOC, discuss the need to change not just individual behavior but entire industrial systems. Casey R. Schmitt (2019) refers to this as "scapegoat ecology," wherein a few actors are blamed for significant environmental crises even though the system is at fault. Individual blame shifts the focus away from the industries that should be held responsible.

These two approaches are representative of standard environmentalist approaches: innovate our way out of the problem without acknowledging that innovation got us into the problem in the first place; or, scapegoat a few individuals while assuaging industrial guilt or the underlying societal forces. The Ecospheric movement could have the potential to not just rely on LADbible or TOC but also to unify a movement that does not carry the baggage of previous environmental discourses.

A unified movement arguing for Ecospheric justice is needed. It can save legislators from hearing the same worn-out environmentalist debates, like jobs versus the environment, and can move us past a cycle of blaming parties. Why not pressure the United Nations to ban plastics altogether because they are simply bad for the Ecosphere? One hundred and twenty-seven countries already regulate plastic bags, while some ban them (Excell 2019). Yes, few will be open to rethinking our very role on Earth through a new concept. And yes, I recognize it is idealistic. Ecospheric rhetoric is a radical, difficult, and profound departure from our current way of being. But it is not like we have a choice. In the age of the Anthropocene, it is necessary in every sense of the word. We no longer have the privilege to apply Band-Aids to broken systems.

The trickiest part about unifying around the "Ecosphere" concept, or any concept for that matter, is that language is slippery. As Anthony Leiserowitz, Director of Yale's Program on Climate Change Communication, jokes, "the key thing about terms like [global warming or climate change], they are plastic. Or, well, maybe since we are talking about the environment, we should say words are renewable organic latex or something. . . . Essentially, meaning changes" (Thompson 2018). The plasticity of language makes it that much more difficult to identify what modern terms like "plastic" actually mean. Since starting with a revamp in language is a bottom-up approach, environmental activists of a new generation can embrace this conceptual shift. This can happen through education. It already is happening through education (Land Institute, n.d.).

CONCLUSION

The Great Pacific Garbage Patch is now visible from Google Earth and Google Maps. This is an absurd statement, especially when one considers the first viable plastic synthesis was not mass produced until the 1950s. An estimated 269,000 tons is circulating the oceans in 2019. This is but a fraction of the 9.1 billion tons of plastic produced each year ("Earth Day 2018" 2018). This annual plastic weight equals the total number of humans that will be on Earth in 2050. Humans managed to go from not having a word for a nonexistent thing to producing 9.1 billion tons of it annually *in just over a hundred years*. Although it could be argued that plastics are just an emergent rearrangement of what has been on Earth for centuries, its effects highlight humanity's irreparable destruction.

We need nothing less than a new way of talking about the ways in which humanity's language impacts the Earth. We may have different environments, but we all share the same Ecosphere. Using new language begins by understanding the Ecosphere's latent discourses, which currently communicate that our actions are out of spatiotemporal sync. The ecological turn has proven that matter matters. If we begin to reconceptualize our signification process and discourses to attune to this one Ecosphere, perhaps matter can begin to matter more.

REFERENCES

Abrams, Nancy. 2015. *A God That Could Be Real*. Boston: Beacon Press.

Bar-Yam, Yaneer. 2011. "Concepts: Emergence." New England Complex Systems Institute, Concepts: Emergence. https://necsi.edu/emergence.

Barad, Karen. 2007. *Meeting the Universe Halfway*. Durham, NC: Duke University Press.

Barnett, Scott, and Casey Boyle. 2016. *Rhetoric, through Everyday Things*. Birmingham: The University of Alabama Press.

Bengle, Josie. 2017. "Al Gore Joins Campaign to Turn Ocean Plastic Landmass into Official Country," *Resource*, September 7, 2017. https://resource.co/article/al-gore-joins-campaign-turn-ocean-plastic-landmass-official-country-12089.

Bennett, Jane. 2005. "The Agency of Assemblages and the North American Blackout." *Public Culture* 17, no. 3: 445–66.

Bennett, Jane. 2010. *Vibrant Matter: A Political Ecology of Things*. Durham, NC: Duke University Press, 2010.

Carson, Rachel. 1962. *Silent Spring*. Cambridge: Riverside Press.

Castelló, Enric. 2009. "The Nation as a Political Stage: A Theoretical Approach to Television Fiction and National Identities." *International Communication Gazette* 71, no. 4: 303–20.

Cobb Jr., John B. 2015. *Whitehead Word Book: A Glossary with Alphabetical Index to Technical Terms in Process and Reality*. Anoka: Process Century Press.

de Saussure, Ferdinand. Charles Bally and Albert Sechehaye, eds. 1966. *A Course in General Linguistics*. New York: McGraw-Hill Book Co.

"Earth Day 2018: End Plastic Pollution." 2018. Earth Day Network, March 7, 2018, https://www.earthday.org/2018/03/07/fact-sheet-end-plastic-pollution/.

Endres, Danielle. 2014. "Response Essay: The (Im)Possibility of Voice in Environmental Advocacy." In *Voice and Environmental Communication*, edited by Jennifer Peeples and Stephen Depoe, 110–23. London: Palgrave Macmillan.

Excell, Carole. 2019. "127 Countries Now Regulate Plastic Bags. Why Aren't We Seeing Less Pollution?" *World Resources Institute*, March 11, 2019. https://www.wri.org/blog/2019/03/127-countries-now-regulate-plastic-bags-why-arent-we-seeing-less-pollution.

Gass, Scott. 2013. "How Big Is the Ocean?" TED-Ed. Video, 5:25. https://www.youtube.com/watch?v=QUW_Zv_jJb8&vl=en.

Gibbens, Sarah. 2019. "Another Plastic Bag Found at the Bottom of World's Deepest Ocean Trench." *National Geographic News*, May 13, 2019. https://news.nationalgeographic.com/2018/05/plastic-bag-mariana-trench-pollution-science-spd/.

Hanson, Joe. 2016. "100,000,000 Years from Now." It's Okay to Be Smart. Video, 6:26. https://www.youtube.com/watch?v=oBcHf_eeYt4.

Irvine, Martin. 2012. "Structural Linguistics, Semiotics, and Communication Theory: Basic Outlines and Assumptions." Georgetown University Creative Commons. Accessed August 1, 2019. http://faculty.georgetown.edu/irvinem/theory/Semiotics_and_Communication.html.

Kiefer, Brittaney. 2018. "How Trash Isles' Creators Founded a Country of Rubbish to Tackle a Global Crisis." *Campaign*, June 27, 2018. https://www.campaignlive.co.uk/article/trash-isles-creators-founded-country-rubbish-tackle-global-crisis/1486119?utm_source=website&utm_medium=social.

Kolbert, Elizabeth. 2014. *The Sixth Extinction: An Unnatural History*. London: Picador.

Krulwich, Robert. 2016. "The Earth Has Lungs. Watch Them Breathe." *National Geographic*, March 9, 2016. https://www.nationalgeographic.com/science/phenomena/2016/03/09/the-earth-has-lungs-watch-them-breathe/.

Land Institute, n.d. "Ecosphere Studies." Accessed August 3, 2019. https://landinstitute.org/our-work/ecosphere-studies/.

Latour, Bruno. 1993. *We Have Never Been Modern*. Cambridge, MA: Harvard University Press.

Lebreton, Laurent, Boyan Slat, Francesco Ferrari, Bruno Sainte-Rose, Jen Aitken, Robert Marthouse, Sara Hajbane et al. 2018. "Evidence That the Great Pacific Garbage Patch Is Rapidly Accumulating Plastic." *Scientific Reports* 8, no. 1: 1–16.

Lind, Colene. 2017. "Ethos, Pathos, Logos, Wholeness." *The Land Report* 1, no. 119: 12–20.

Lodge, David, and Nigel Wood. 2008. *Modern Criticism and Theory*, 3rd ed. London: Pearson Education Limited.

Maloney, Alli. 2018. "I Went to the Great Pacific Garbage Patch. This Is What I Saw." *Teen Vogue*, December 22, 2018. https://www.teenvogue.com/story/i-went-to-the-great-pacific-garbage-patch?verso=true.

Martin, Michel, and Amanda Morris, 2019. "An Engineering Wunderkind's Ocean Plastics Cleanup Device Hits a Setback." *NPR*, January 5, 2019. https://www.npr.org/2019/01/05/682532583/an-engineering-wunderkinds-ocean-plastics-cleanup-device-hits-a-setback.

McCarthy, Niall. 2018. "Study Finds Microplastics in 93% of Bottled Water [Infographic]." *Forbes*, March 16, 2018. https://www.forbes.com/sites/niallmccarthy/2018/03/16/study-finds-microplastics-in-93-percent-of-bottled-water-infographic/#15a90a2a73fa.

McGreavy, Bridie, Justine Wells, George F. McHendry, Jr., and Samantha Senda-Cook. 2018. *Tracing Rhetoric and Material Life: Ecological Approaches*. London: Palgrave Macmillan.

Miller, Jacob. 2019. "Toward an Ecospheric Rhetoric: Cleaning Up 'Trash Isles.'" http://krex.kstate.edu/dspace/handle/2097/39598?fbclid=IwAR1XcXp8sGqimy3LZJSGFUXXVpPnmx_CEsbDnENmveBokHQonbhOoICqVJw.

Minter, Adam. 2018. "Plastic Straws Aren't the Problem." *Bloomberg*, June 7, 2018. https://www.bloomberg.com/opinion/articles/2018-06-07/plastic-straws-aren-t-the-problem.

Ocean Cleanup. n.d. "The Largest Cleanup in History." Accessed August 3, 2019. https://theoceancleanup.com/.

Rickert, Thomas. 2013. *Ambient Rhetoric: The Attunements of Rhetoric.* Pittsburgh: University of Pittsburgh Press.

Roberts, Sarah, and Tim Nunn. 2018. "Accept the Trash Isles as an Official Country & Help Protect Our Oceans." *Change.org*, January 24, 2018. https://www.change.org/p/un-secretary-general-ant%C3%B3nio-guterres-accept-the-trash-isles-as-an-official-country-help-protect-our-oceans.

Rowe, Stan. 1961. "The Level-of-Integration Concept and Ecology." *Ecology* 42, no. 2: 420–27.

Rowe, Stan. 1989. "What on Earth Is Environment?" *Trumpeter* 6: 1–6.

Rowe, Stan. 2006. *Earth Alive: Essays on Ecology.* Edmonton: NeWest Press.

Schmitt, Casey R. 2019. "Scapegoat Ecology." Blame, Exoneration, and an Emergent Genre in Environmentalist Discourse." *Environmental Communication* 13, no. 2: 152-64.

Science History Institute. 2016. "The History and Future of Plastics." Accessed August 31, 2019. https://www.sciencehistory.org/the-history-and-future-of-plastics.

Thompson, Andrea. 2018. "From Fish to Humans, A Microplastic Invasion May Be Taking a Toll." *Scientific American*, September 4, 2018. https://www.scientificamerican.com/article/from-fish-to-humans-a-microplastic-invasion-may-be-taking-a-toll/.

"Trash Isles." 2019. LADbible, accessed August 3, 2019. http://www.ladbible.com/trashisles.

Valdés, Luis, William Peterson, John Church, Keith Brander, and Marta Marcos. 2009. "Our Changing Oceans: Conclusions of the First International Symposium on the Effects of Climate Change on the World's Oceans." *ICES Journal of Marine Science* 66, no. 7: 1435–38.

Villemain, Cyril. 2017. "'Turn the Tide on Plastic' Urges UN, as Microplastics in the Seas Now Outnumber Stars in Our Galaxy." *UN News*, February 23, 2017. https://news.un.org/en/story/2017/02/552052-turn-tide-plastic-urges-un-microplastics-seas-now-outnumber-stars-our-galaxy.

Wallace-Wells, David. 2019. *The Uninhabitable Earth.* New York: Penguin Random House.

Wells, Justine, Bridie McGreavy, Samantha Senda-Cook, and George F. McHendry Jr.. 2018. "Introduction: Rhetoric's Ecologies." In *Tracing Rhetoric and Material Life: Ecological Approaches,* edited by Bridie McGreavy, Justine Wells, George F. McHendry Jr., and Samantha Senda-Cook, 1–36. London: Palgrave Macmillan.

Wolchover, Natalie. 2011. "Why Doesn't Plastic Biodegrade?" *Live Science*, March 2, 2011. https://www.livescience.com/33085-petroleum-derived-plastic-non-biodegradable.html.

Wong, Jordan. 2017. "Trash Isles: The Rhetoric of Statehood." Paper presented at American Forensics Association National Individual Events Tournament, Peoria, IL, April 3, 2017.

Concluding Remarks

Aimee Kendall Roundtree

This ambitious anthology makes a valuable contribution to our understanding of water rights and justice in the rhetoric of environmental science and health. In comparison to other books published in the past ten years on the topic of rhetorical interventions in water rights and justice, it offers a broad focus on community and governmental action on global water justice issues.

Over the past decade, three other books have undertaken the task of chronicling community and governmental rhetoric on water justice. *A Political Ecology of Women, Water, and Global Environmental Change* (Buechler and Hanson 2015) is a collection of feminist political ecology framework approaches to investigating rural and urban dependence on watersheds, rivers, lakes, and other natural water sources. The authors use ethnographic research, narrative, rhetorical analysis, and case studies to illustrate how gender is an important premise upon which socioecological policy and change has been based, and to document how gender and other intersectional identities affect access to water, response to water crises, and petitions for water justice in the United States and abroad. In *Water Policy Process in India: Discourses of Power and Resistance* (Asthana 2012), the author enlists discourse analysis to interrogate the process of policy production in urban water supply in India for the past two decades. The book charts the complex and evolving network of actors that shifted the terms of the discourse toward reforms meant to sustain domestic interests versus resistance to backdoor privatization of water access and safety. Finally, although water rights and justice are not exclusively the focus of *Topic-Driven Environmental Rhetoric* (Ross 2017), the book does provide a chapter that identifies the commonplaces, or statements of knowledge presumably shared by audiences and orators, that emerged in water hearings about aquifer conservation in

Texas. The chapter advances the book's overall goal of cataloging common communication tactics used in debates and arguments about water justice.

Each of these books takes a different tack on the subject matter—one essentializing gender as the optimal vantage from which to understand water rights, the second privileging policy as a key focal point, and the third indexing topics and tactics common to the debate. What differentiates the current collection from these efforts is the variety of cases, theories, and methods that it represents and, thereby, teaches readers and scholars in the field. Another important distinguishing feature is this collection's attention to national, international, and intercultural representation. It documents the water rights efforts of multiple marginalized identities. Finally, this collection shares with its predecessors and other scholarship on the topic of water rights and justice an attention to the rhetorical dynamics emergent in specific cases that inscribe and expose the working and reworking of power, representation, and action in water rights and justice on a local community and governmental level. These intimate revelations, in turn, help scholars and readers contextualize macroscopic activities pertaining to water justice and rights.

At the macroscopic level, the national and international history of water rights is rife with noteworthy, historic events. As Joshua Getzler (2004) describes in his book *A History of Water Rights at Common Law*, judges, lawyers, and authors as early as the 1800s framed the regulation of water in terms of Roman law and basic principles of common goods and natural rights of ownership. Through the years, technological developments such as the invention of the water wheel and irrigation systems added complexities to the framing of uses of water as an extension of servitude or the right to encumber land and natural resources. Even earlier, in the fifteenth and sixteenth centuries, courts established the idea of reasonableness in natural rights, or the right to use resources in ways that do not harm the resources of others. In proceedings and rulings that followed, courts grappled with interpreting and managing the tensions between standards of reasonableness, ownership, and servitude pertaining to water use.

In more recent years, international courts moved to recognize the universal human right to water. The Geneva Convention of 1949 acknowledged that water was a part of humanitarian law (Willaarts et al. 2014). In 1966, the International Covenant on Economic, Social, and Cultural Rights, part of the United Nation's bill of rights, included a right to an adequate standard of living and safe drinking water. The 1992 Convention on the Protection and Use of Transboundary Watercourses and International Lakes included a Protocol on Water and Health that underscored the importance of an adequate supply of safe drinking water and the protection of drinking water. In 2002, the U.N.'s International Covenant on Economic, Social and Cultural Rights made a general comment that the "human right to water entitles everyone to sufficient, safe, acceptable, physically accessible and affordable water for personal and domestic uses." Soon after, in 2006, the Convention on the Rights of Persons with Disabilities required equal access to clean water services for persons with disabilities. The U.N. High Commissioner for Human Rights issued a report in 2007 on human rights to safe drinking water and sanitation, and an independent expert issued a report on human rights obligations related to access to safe drinking water and sanitation in 2009. In 2010, the U.N. General Assembly issued a resolution on the right to water and sanitation and the U.N. Human Rights Council issued another on human rights and access to safe drinking water and sanitation.

While these landmarks in the timeline of water justice are evidence of some movement and commitment at the intergovernmental level to water rights, they provide a long view that overpasses many dimensions of the struggle over water rights. These high-level events do not, for example, recount the numerous tragedies and water crises that occurred in communities and compelled governmental action in the first place. They leave unseen the many local and regional tensions and contradictions of smaller governmental bodies and agencies that evade implementation of these principles in order to stimulate local economies or appease business and others with competing interests. They do not account for the public protests and outrage that compelled and, perhaps, shamed governmental leaders to reevaluate policies for the purpose of protecting safe and sanitary water for drinking and living. An even deeper criticism highlights how the framework of these federal and intergovernmental actions, themselves, are steeped in their own problematic histories of disenfranchisement, xenophobia, and subjugation. Consider that the Roman legal principles upon which we have built the discourse of water rights were complicit in centuries of colonialism for several generations of empire building by not only the Romans but also England and the United States. Collections such as this one that gather rhetorical analysis of water rights cases from the perspective of communities, protesters, and other actors and stakeholders on the ground, in the middle of the action, can help contextualize, agitate, and emancipate our understanding of water rights from traditional schemas born of dubious political, ethical, or moral origins.

In particular, three ontological postures emerge from this anthology—disruption, agency, and identity. By ontological posture, I mean that the chapters recount dealings (moments and responses) that establish and challenge the status and reality of official pronouncements and common knowledge about water use and safety. First, disruption presents itself in the chapters as a way of being and establishing the existence of a position on water rights, and as a means of setting terms and evaluating governmental environmental action pertaining to waterways and water safety. In rhetoric, disruption targets assumptions and stereotypes that invisibly anchor public and personal decisions by inserting a break from dominant opinion (Murray

2003; Murray 2015; Balzotti and Crosby 2014). Technology, changes in codes, ethics, practices, and standards trail behind media and technological disruption (Christensen 2013). In psychology, disruptions violate expectations and throw off equilibrium (Moriya and Tanno 2009; Rosenberg and Siegel 2016). The concept of disruption across these disciplines demonstrates and effects ontological shifts in the status of phenomena—in the case of water justice, the current state of affairs—from being determined (i.e., accepted and settled) to being indeterminant (i.e., unsettled and fractured). In this anthology, rhetorical disruptions undermine the warrants that uphold hegemonic structures in public thought about water use. For example, the first section presents demonstrations at Standing Rock and an annual flotilla demonstration that both interrupt the aura of technological progress that access pipelines and dams present. In the second section, discontinuation of water services to the communities who need it the most in Detroit taints public health and rends cracks between different factions within the community. In the third section, communities in China used social media to interrupt official denials of harm and danger, while in Australia an author unsuccessfully used an unconventional genre—satirical obituary—to advocate for the Great Barrier Reef. In all cases, rhetorical interventions serve as a wedge rupturing accepted or official narratives.

Second, these chapters present agency as evidence of water rights and justice realities. Agency theory in rhetoric posits orators as both leading an audience and being shaped and constrained by the audience's demands (Leff 2012). It has also been defined as, on the one hand, individuals authoritatively occupying a position and effecting change, and, on the other, a sequence of rhetorical events that, over time, evoke changes, often accomplished via authoritative force. Other theories of agency in rhetoric define it as the freedom to make choices and take beneficial actions; it involves intention on the part of the agent and recognition on the part of other agents. Recently, actor-network theory has interrogated nonhuman agency. In particular, the theory ascribes agency to a distributed constellation of actions and actors, thereby decoupling agency from intention, volition, and subjectivity, opening the prospect of different types of agency, and acknowledging that even nonhuman agents are imperative to society and mediators of meaning with moral and political associations (Sayes 2014). In all regards, agency is about power to act and influence. This collection illustrates several demonstrations of how rhetoric creates power to influence perception and approaches to water justice and policy. In the first section, public discourse in media coverage exposed the rhetorical battle between protestors and pro-hydroelectric-dam interests over false dichotomies that would dictate public perception about the dam's risk and its connection to climate change. In the second section, a governor promises seemingly contradictory potential for aquifer depletion to simultaneously uphold environmental responsibility and achieve economic growth, and a religious leader anthropomorphized natural water sources as a non-human agent with political and economic power and vulnerability. In the third section, an environmental performance serves as political action and empowers audiences to act in the present and envision a future that conserves the coastline. In the fourth section, two Asian and South Asian rivers are anthropomorphized as human entities by law, and a floating garbage patch in the Pacific Ocean might receive sovereign state status, which evokes its capacity for responsibility, liability, and enfranchisement. In all cases, discursive formations inscribe, create, imbue, or deprive power of action and influence in water rights and justice.

Third, these chapters reveal how communities and individuals define themselves in water justice and rights action. Rhetoric instantiates identity through performing and doing. We are formed and constituted by our utterances, and our utterances pronounce who and what we are; furthermore, rhetorical interactions delineate ourselves from others and lay stakes for ourselves inside (or outside) of the common experience (Grabill and Pigg 2012). Identity can be created and managed individually or corporately, as in the case of an organization, nation, or community with shared interests, stakes, values, and threats (Frandsen and Johansen 2011; Gross 2010). This collection illustrates how rhetorical justifications for water use and rights inform individual and community identities, and vice versa. For example, in section I, a chapter stresses how river restoration processes must earnestly incorporate marginalized voices from the community in order to reflect the most authentic representation of the community, one not entrenched in colonial justifications. In section II, water diversion requests and public hearings call into question the scope of stakeholders that comprise the public and community. In section III, diversity of local communities in Flint, Michigan, lend gendered authority by privileging the role of communities of women in environmental justice. Finally, in section IV, a community mourns the loss and shares in the process of grief when mine drainage damages a community river, and those bordering the Salish Sea produce an unsuccessful and disintegrous advocacy video, given the subtleties of multiple intersections of nationality, race, and gender. In total, these cases demonstrate how collective and individual identities impact the efficacy, composition, and reception of water rights and justice campaigns.

These chapters cover a lot of ground, and they provide intimate, in-depth looks at individual cases where these ontological strategies play out. However, the chapters also light the way for future research and scholarship on water justice and rights. The chapters lay a solid foundation from which to build upon, and they invite additional lines of inquiry. Given the attention to disruption, the anthology invites more theoretical work developing a specific model of disruption particular to water justice. What characteristics do all effective disruptions share? What entrenched discursive formations are common to water justice disruptions? The question also remains as to the nature and scope of agency as it directly pertains to water justice. From an artifactual perspective, these chapters describe agency as articulated through myriad tools including petition signs, speeches, media reporting, demonstrations, official documents, and social media. The question remains whether some performances of agency are more effective than others and to what extent concerted efforts maximize or diffuse the impact of agency. This collection also invites more scholarship and work in reception studies and the impact of these short-term cases on long-term trends toward water rights.

These next two points for further inquiry—calls for more work on reception and longitudinal studies—are important enough to illustrate with cases, themselves. Indeed, Danette Paul, Davida Charney, and I made the case that, to gain a fuller picture of the rhetoric of science generally—and, by extension, environmental and water science—studies of responses (including real-time responses) and studies that track the course of acceptance or rejection over time are as important as studies that pay attention to the disasters, spectacles, and other defining moments (see Paul, Charney, and Kendall 2001).

Let us turn to the first of our invitations—namely, to study reception. Regarding how this collection opens a space for reception theory, the chapters invite more analyses of the impact and effects of the ontological tactics deployed. This might include, for instance, examining cases in which public and governmental stakeholders take part in a call-and-response engagement, in which readers and scholars can observe *in media res* what changes or developments are the result of community action. Consider the case of the open comment period for the Texas Commission on Environmental Quality's (TCEQ) report on Texas Nonpoint Source Management. The open comment period afforded time to the general public to write comments and concerns about a draft of the report. In the case of climate change and other prospective science, simulations are often cited and used as evidence for making claims about future living conditions and for justifying policies that will impact the economy and quality of life for citizens (Roundtree 2013). Open commenting periods for local, state, and federal governmental reports are required by precedent, if not law, and they usually document both the public's issues and the governmental agency's response.

Simulations present a unique challenge for use in public policy because they involve not only observed data but also predictions about future conditions. In the TCEQ report, experts published plans based on simulated projection of water levels, waste growth, and human behavior. The translation of simulated evidence not only involves interpreting facts between experts and the public, it also enlists translation of physical realities to virtual ones. Simulations derive their meaning from relationships between deliberation and negotiation of facts, from relationships between texts, and from situations in which they are written and distributed (Roundtree 2013). They enlist abduction (an enthymatic form that, from a symbolic logical perspective, cannot achieve tautological status as might induction or deduction), and they yield virtual evidence, which carries the virtue or power of the object it represents without pure verisimilitude. Politics and complexity of arguments underpinning the simulations affect how simulated evidence is perceived and cast by the general public and stakeholders in these high-profile, highly charged, and highly contentious social issues. In the case of open commenting periods, how the public understands those simulated facts and how the experts explain them can make a difference in policy design and implementation.

I conducted a dialogic analysis of the public comments and responses regarding the 2012 and 2017 report, and used text mining software to confirm qualitative findings. Dialogic analysis seeks to uphold principles of what Arthur Frank (2012) calls commitments. Dialogic analysis asks particular questions about utterances. Frank's commitments of dialogic analysis are as follows: (1) any individual voice is actually a dialogue between voices; (2) monologic interpretations are suspicious; and (3) utterances are dialogic because human nature is unfinalized. These were my goals and questions for analyzing the following reports, in order to generate my own model and typography of issues that emerge when translating simulations used to justify a policy.

The nonpoint source water management plan is meant to regulate and minimize nonpoint source pollution, which is caused by rainfall moving over and through the ground. As the runoff moves, it picks up and carries away natural and human-made pollutants, finally depositing them into lakes, rivers, wetlands, coastal waters, and ground waters. Nonpoint source pollution can include excess fertilizers; herbicides and insecticides from agricultural lands and residential areas; oil, grease, and toxic chemicals from urban runoff and energy production; bacteria and nutrients from livestock, pet wastes, and faulty septic systems; and atmospheric deposition and hydromodification (altering the natural flow of water through a landscape).

Ultimately, in the exchanges between the TCEQ and the public, three dialogic modes played an important role—coexisting, competing, and complementary dialogic. First, coexisting dialogic involves interlocutors engaging one another on shared terms, insofar as their comments and response target virtually the same audiences and concerns. For example, when a commenter in the 2012 report complained that "the requirements of the Lake O' the Pines total maximum daily load applicable to point sources have been either ignored or the permits have been renewed without the limits described," the TCEQ replied that it was "working with EPA to put phosphorus load reductions as required by the Water Quality Management Plan in an individual permit. Negotiations between TCEQ and EPA are ongoing, and once this permit is issued, the TMDL I-Plan for Lake O' the Pines can be fully implemented." In this case, the public wanted more transparency from potential nonpoint polluters about their water quality management plans, to which the TCEQ replied that obtaining that information was a work in progress. Here, we see two interlocutors engaging in the same or similar discourses of methods and processes. Both were on topic discussing what should or should not be made more accessible by potential nonpoint polluters. While the TCEQ took no action to make changes based on the public comment, the level of engagement by addressing all points and counterpoints stands in as an action taken to improve comments. In this way, the government acknowledges that they heard the interlocutor, their direct comment, and the dialogue implied by their comments.

Second, competing dialogic exposes how interlocutors talk over one another, insofar as their comments and responses do not meet each other on the same terms. In the 2017 report, a commenter complained that the report should "[acknowledge] the connections between surface and groundwater more specifically and generally [have] more 'teeth' in it for protecting state waters from nonpoint source pollution." The TCEQ replied that "the Nonpoint Source Program does not propose or enforce rules or regulations. The Nonpoint Source Program recognizes the connections between surface and groundwater and the impacts of nonpoint source pollution on groundwater." Therefore "[n]o changes were made to the document as a result of this comment." In this case, the commenter wanted a discussion at the proposal stasis; they wanted more action. The TCEQ, on the other hand, rather than providing recourse or other means for action, maintained a response at the causal stasis; they replied that they only report on causes, not solutions. The parties talk over one another rather than wade into the weeds of specific variables and algorithms embedded with presuppositions about what the future holds for water safety and policy. In this way, both parties engage in monologic interpretations worthy of suspicion—and inevitably counterproductive exchange—because neither accounts for the other's stasis position.

Third, the 2012 report also illustrates complementary dialogic responses, which involve the highest level of mutual exchange, insofar as comments and response advance on the same terms without condition and some yield change. For example, in one exchange, the public commenter requests that the TCEQ focus on clear "environmental results" such as "loading reductions," "percentages of BMPs [best management practices for water conservation] inspected." In response, the TCEQ not only explains that such metrics are out of the purview of their charge, they also define the terms and list the other organizations and reports responsible for keeping the metrics requested. In this case, the individual voice of the TCEQ and the commenter engage at the same level and mode of evidence. Both sides of the dialogue parallel each other by way of their discourse of outcomes and goals as well as their command of the technical variables. In this way, as was the case in dialogical ethos, formal features of comments and responses, such as the use of jargon and citations, create a space for increasing affinity between interlocutors. This example also illustrates how simulated evidence in dialogic exchanges evolve from future conditional toward practice and enforcement. The exchange itself and the framing of questions and concerns therein predict and begin to instantiate future policies that will affect industry, homeowners, and others with stakes in water conservation. In all three dialogical modes, examining the open comments and replies expose how audiences—both the public commenters and the governmental agency—interpreted their engagement and responded to each other. It also reveals the impact of public comments on governmental action—in this case, a water management plan.

The TCEQ case illustrates the value of tracking over time stakeholder reception of and response to rhetorical appeals and activities in water justice. Let us turn to the second trajectory for future work—namely, longitudinal studies. Regarding how this collection opens a space for tracing long-term trends, longitudinal and big data studies might play an important role. Take, for example, a content analysis that I conducted with graduate and undergraduate students investigating visuals of technologies from Twitter postings under a hashtag advocating for clean water in Flint—#flintwatercrisis. Activists took to social media to document the extent of the contaminated water in Flint. In the case of water poisoning and other issues of environmental degradation in the headlines, questions of civic calling and action hinge on questions of technology. Here, calls to action are inextricably tied to attitudes about the severity and risk of the problem. Weighing harms and risks, in turn, requires assessing the technologies that cause and/or mitigate the crisis. Visuals offer their opinions and explanations of technologies and technical terms important to assessing risk and harm, such as lead exposure, iron-rusted piping, and bioaccumulation. Activists share technical and official visuals to enact a counternarrative against governmental sources downplaying or denying the problem. While the images serve to illustrate actor-network theory, the images also seem to reinforce or underline the disparity between material power and the ontological or existential impact that humans and nonhumans wield over one another. It is not a level playing field. And technology has the upper hand.

Overall, my students and I coded 1,174 pages of tweets, 1,594 images—of which 830 were technical images—from Twitter content between 2015 and 2018 using Twitter Advanced Search to find images shared using the popular political hashtag #FlintLivesMatter. Students and I co-coded the images using several categories including sentiment, the function of technology, stasis and appeals of the claim that the image suggests, and the nature of the sign that the image assumes. Of all the images (most of which were images of memes, flyers for events, selfies, protests, and meetings), more than eight hundred included depictions of high and low technologies—everything from knitting products for sale to water filtration and purification technologies to pipes and water towers. Overall, the images of technologies extend commonsense, firsthand eyewitness evidence verified by volume and narrative rather than peer review or citation and source attribution. They included pictures of infographics, amplifiers, bottles, water pallets, towers, faucets, pipes, and filters.

A longitudinal study of rhetorical trends across time revealed important trends that helped understand a broader picture of the movement. For example, we saw that, over time the community response was more negative, even when, after the peak of the crisis in 2016, fervor, frequency of post, and rhetorical activity dissipated (see figure 17.1)

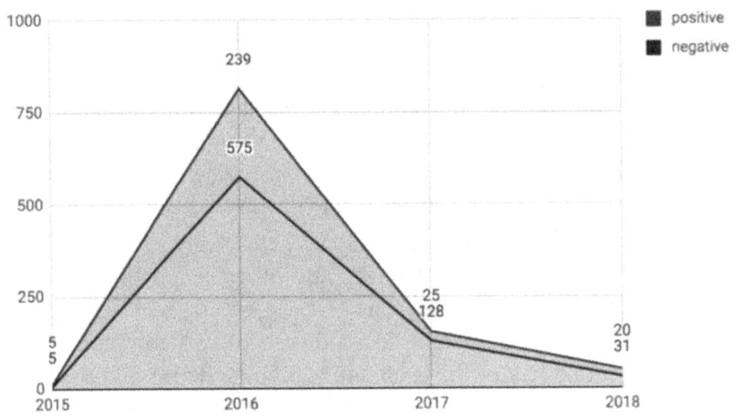

Figure 17.1. Sentiment. Created by author.

A longitudinal study also revealed that, over time, the pattern persisted of mostly sharing information graphics and pictures of water bottles, including palettes of water bottles, but that, proportionally, pictures of megaphones and microphones at demonstrations and press events dominated 2016, compared to the information graphics that were used more often in 2017, indicating a shift in focus and argumentation from epideictic to deliberative rhetoric (see figure 17.2).

The longitudinal study also revealed that the rhetorical appeals used were proportionally consistent over time. Appeals to credibility, emotion, and morality dominated, with very few appeals to logic throughout. Even the information graphics provided more messaging about believability of citizens' experience (or the lack of believability of official communication) than citations of facts and statistics (see figure 17.3).

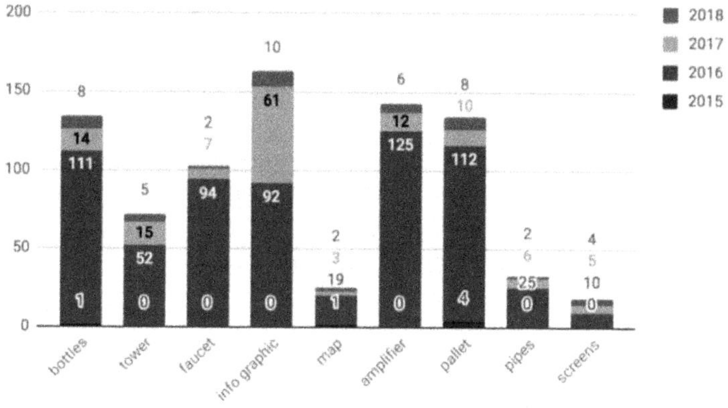

Figure 17.2. Technology Types. Created by author.

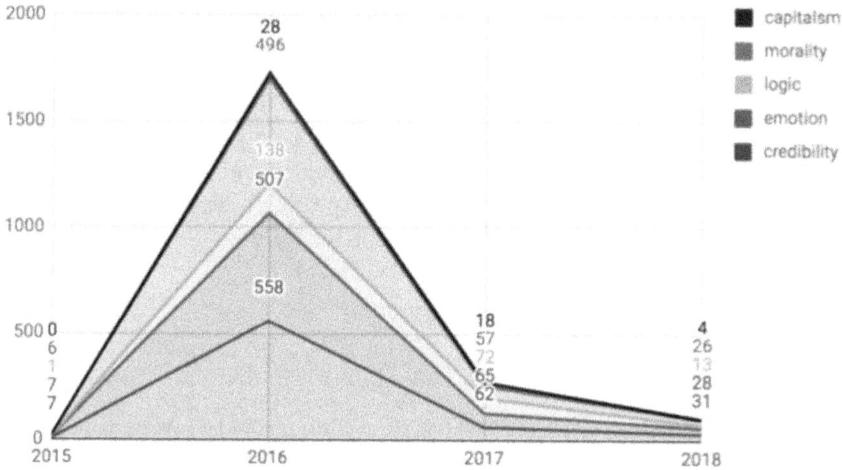

Figure 17.3. Appeals. Created by author.

Overall, repetition was a major rhetorical strategy for producing emphasis, clarity, amplification, or emotional effect in this case. It offered images for consideration and reconsideration, and it opened up a reassessment of meaning and signification in each iteration. In addition to helping emphasize common qualities of the images, the piling up and piling on of these images also created cognitive dissonance during the interpretive process, having to go back and remember, then retell and recount and reassess the meaning. While the images served to demonstrate the veracity of actor-network theory—that human and nonhuman agents act upon one another and have agency in the social milieu and in semiotic meaning—the images seem to reinforce or underline the disparity between material power and the ontological or existential impact that human and nonhumans wield over one another. Ultimately, the visuals triggered a chain of evidence and events that challenge the official narrative and instantiate questions of ethical responsibility and official accountability. The visuals provided tactile, eye-witness arguments for explaining the implications of technical failures. Visual evidence drew attention, provoked reflection, and solicited action from local and national audiences. Visuals created an argument by number to evidence in an almost epidemiological manner the patterns, causes, and effects of local water contamination. In most cases, images of technology serve as indexes of crisis prevalence. However, they also lent credibility and evidentiary weight to an understanding of its severity. A longitudinal study can help show how the rhetorical tactics evolved and changed over time.

In summary, this anthology clearly makes a contribution in its breadth of topics, theoretical orientations, and international cases that hold together in their demonstration of different ways that ideas and action on water justice exist in the world and how the nature of these actions and ideas call into question the fundamental nature of water rights and justice as currently articulated. Furthermore, the collection opens up opportunities for future rhetorical explorations in water justice and rights, including invitations for reception and longitudinal studies. This collection not only presents useful and informative fodder for classes on environmental health, water rights, and social justice, it also advances the field by exploring a broad range of cases involving diverse stakeholders and rhetorical dimensions.

REFERENCES

Asthana, Vandana. 2012. *Water Policy in India: Discourse of Power and Resistance.* New York: Routledge.

Balzotti, Jonathan Mark, and Richard Benjamin Crosby. 2014. "Diocletian's Victory Column: Megethos and the Rhetoric of Spectacular Disruption." *Rhetoric Society Quarterly*, 44, no. 4: 323–42.

Buechler, Stephanie, and Anne-Marie Hanson. 2015. *A Political Ecology of Women, Water, and Global Environmental Change.* New York: Routledge.

Christensen, Clayton M. 2013. *The Innovator's Dilemma: When New Technologies Cause Great Firms to Fail.* Boston: Harvard Business Review Press.

Frandsen, Finn, and Winni Johansen. 2011. "Rhetoric, Climate Change, and Corporate Identity Management." *Management Communication Quarterly* 25, no. 3: 511–30.

Frank, Arthur W. 2012. *Practicing Dialogical Narrative Analysis,* edited by James A. Holstein and Jaber F. Gubrium. Thousand Oaks, CA: Sage.

Getzler, Joshua. 2004. *A History of Water Rights at Common Law.* New York: Oxford University Press.

Grabill, Jeffrey T., and Stacey Pigg. 2012. "Messy Rhetoric: Identity Performance as Rhetorical Agency in Online Public Forums." *Rhetoric Society Quarterly* 42, no. 2: 99–119.

Gross, Alan G. 2010. "Rhetoric, Narrative, and the Lifeworld: The Construction of Collective Identity." *Philosophy & Rhetoric* 43, no. 2: 118–38.

Leff, Michael C. 2012. "Tradition and Agency in Humanistic Rhetoric." *Philosophy & Rhetoric* 45, no. 2: 213–26.

Moriya, Jun, and Yoshihiko Tanno. 2009. "Competition between Endogenous and Exogenous Attention to Nonemotional Stimuli in Social Anxiety." *Emotion* 9, no: 5: 739–43.

Murray, Jeffrey W. 2003. "The Face in Dialogue: Emmanuel Levinas and Rhetorics of Disruption and Supplication." *Southern Journal of Communication* 68, no. 3: 250–66.

Murray, Jeffrey W. 2015. "Critical Thinking Activities and the Enhancement of Ethical Awareness: An Application of a 'Rhetoric of Disruption' to the Undergraduate General Education Classroom." *Open Review of Educational Research* 2, no. 1: 240–58.

Paul, Danette, Davida Charney, Aimee Kendall. 2001. "Moving beyond the Moment: Reception Studies in the Rhetoric of Science." *Journal of Business and Technical Communication* 15, no. 3: 372–99.

Rosenberg, Benjamin D., and Jason T. Siegel. 2016. "The Effect of Inconsistency Appeals on the Influence of Direct-to-Consumer Prescription Drug Advertisements: An Application of Goal Disruption Theory." *Journal of Health Communication* 21, no. 2: 217–27.

Ross, Derek G. 2017. *Topic-Driven Environmental Rhetoric*. New York: Routledge.

Roundtree, Aimee Kendall. 2013. *Computer Simulation, Rhetoric, and the Scientific Imagination: How Virtual Evidence Shapes Science in the Making and in the News*. Lanham, MD: Lexington Books.

Sayes, Edwin. 2014. "Actor–Network Theory and Methodology: Just What Does It Mean to Say That Nonhumans Have Agency?" *Social Studies of Science* 44, no. 1: 134–49.

Index

About the Contributors

Mostafa Aniss (MA, University of the Pacific) is a fourth-year PhD candidate who is passionate about qualitative methods. He leverages that passion to investigate how popular social media users use platforms' features in ways that have implications for romantic relationship development.

Joshua Trey Barnett (PhD, University of Utah; MA, Indiana University) is assistant professor of rhetoric in the Department of Communication Arts and Sciences at the Pennsylvania State University, where he holds a joint appointment in the Huck Institutes of the Life Sciences. He reads, thinks, and writes in the areas of rhetorical theory and criticism, continental philosophy, visual culture, and the sprawling ecological humanities. His scholarship on ecological communication has been published or is forthcoming in the journals *Environmental Communication*, *The Quarterly Journal of Speech*, *Communication, Culture & Critique*, *Culture, Theory & Critique*, *ISLE: Interdisciplinary Studies in Literature and Environment*, and *Ethics & the Environment*. Currently, he is writing a book on the rhetoric of ecological grief.

Elizabeth Brunner (PhD, University of Utah; MA, University of Colorado Denver; MFA, Ohio University) is assistant professor in the Communication, Media, and Persuasion Department at Idaho State University. She has studied at Nankai University in Tianjin and Tsinghua University in Beijing. She is the author of *Environmental Activism, Social Media, and Protest in China: Becoming Activists over Wild Public Networks*, as well as numerous journal articles and essays examining the use of social media in activist efforts.

Catherine J. Bruns (MA, James Madison University) is a communication studies doctoral student at the University of Minnesota–Twin Cities, where she studies water discourse. Her past publications include a 2019 case study in the textbook *Cases in Public Relations: Translating Ethics into Action* and a 2016 journal article in *Carolinas Communication Annual*. She has presented research at state, national, and international conferences, where her work has received top paper recognition in various divisions. She is a 2017 recipient of the Outstanding Graduate Teaching Assistant award from JMU's School of Communication Studies and a 2015 initiate of the Omicron Delta Kappa national leadership honor society.

Theresa R. Castor (PhD, University of Washington; MA, University of California, Davis) is professor and department chair of communication at the University of Wisconsin–Parkside. Her research examines language and social interaction in the communicative constitution of crises and organizational problems. Her recent work examines freshwater policy and water diversion controversies related to the Great Lakes of North America. She is the author of *Climate Risks as Organizational Problems*.

Jordan Christiansen (MA, Kansas State University) is a PhD candidate in the Department of Communication Studies at the University of Kansas. His research program runs along the intersections of indigenous rhetorics, environmental rhetorics, and critical/cultural rhetorics, particularly in colonial and decolonial thought.

Emilie Falc (PhD, MA, Ohio University) studied rhetorical criticism and feminist theory at Ohio University, which perches on a hill in Athens, Ohio, above the beautiful Hocking River. Since 2001, she has been teaching communication studies and women's, gender, and sexuality studies at Winona State University in Winona, Minnesota, a city in the Mississippi River Valley, where she lives with her family. She has research interests in how social activists develop deeply meaningful and powerful communication systems that effectively promote their movements for environmental justice and social change. She has presented her research in Uppsala, Sweden; Boulder, Colorado; Leicester, England; Atlanta, Georgia; and Madison, Wisconsin.

Caroline Gottschalk Druschke (PhD, MA, University of Illinois at Chicago) is associate professor of composition and rhetoric in the Department of English at the University of Wisconsin–Madison. Fusing rhetorical and ecological theories and methodologies, Druschke's work is focused on university-community collaborations to improve Wisconsin's waterways and the human and nonhuman lives that depend on them.

Chris Ingraham (PhD, University of Colorado at Boulder; MA, University of Chicago) is assistant professor of communication at the University of

Utah. His work across the theoretical humanities draws on insights from the study of rhetoric, media, culture, and the environment to think and feel through some of the many challenges and wonders of contemporary life. He is the author of *Gestures of Concern* (forthcoming with Duke University Press), and coeditor of *LEGOfied: Building Blocks as Media* (forthcoming with Bloomsbury).

Alicia Lehrer (MS, University of Rhode Island) joined the Woonasquatucket River Watershed Council (WRWC) as executive director in March 2008. Her work at the WRWC has included spearheading migratory fish passage restoration at the first five dams on the river, working with partners to determine cleanup efforts of the Centredale Manor Superfund site, and leading Woonasquatucket River Greenway expansion, improvement, and maintenance. Alicia joined Clean Water Action's efforts to establish the RI Green Infrastructure Coalition in 2014 and has remained an active member.

Emma Lundberg (MA, Humboldt State University) is a PhD student in the Environment and Resources Program through the Nelson Institute at the University of Wisconsin–Madison. Her dissertation research focuses on identifying and deconstructing settler logics that permeate natural resource management. She uses social and ecological research approaches to understand interactions among beaver and salmonid species, and their management.

James Coleman McGuffey (PhD Indiana University; MA, Ball State University) is lecturer in the Department of Community Development and Applied Economics at the University of Vermont. His research focuses on residual and emergent discursive practices of water culture and rhetorical invention in an age of climate crisis.

Kelsey Mesmer (née Husnick) (MA, MBA, Kent State University) is a PhD candidate in the Department of Communication at Wayne State University in Detroit, Michigan. She studies the intersection of journalism and organizational communication, focusing on resilience, feminist media, and social justice. Kelsey's work can be followed on Twitter at @KelseyMesmer.

Jacob Miller (MA, Kansas State University) graduated in 2019 with a master's in communication studies (emphasis: environmental communication). He has another forthcoming chapter in Crick's (ed.) *The Rhetoric of Fascism* and has been published in *The Land Report*. Upon graduation, he moved to Washington, DC, to direct an environmental campaign and intern for a U.S. sentator. He recently moved back to his home state of Kansas.

Rahul Mitra (PhD, Purdue University; MA, Bowling Green State University) is associate professor of communication whose research is at the intersection of organizational and environmental communication. Prior to obtaining his PhD, he worked in the news broadcast and public relations fields in India for three and a half years. His scholarship focuses on environmental organizing, sustainability and corporate social responsibility (CSR), and meaningful work discourses. He is a critical-interpretive scholar, and uses primarily qualitative methods, such as ethnography, interviews, focus groups, discourse analysis, and arts-based research. His work has appeared in numerous peer-reviewed publications such as *Environmental Communication, Management Communication Quarterly, Human Relations, Communication Theory, International Journal of Business Communication, Public Relations Review*, and the *Journal of Business Ethics*. Rahul is deeply invested in public scholarship and tweets on issues related to environmental organizing @RahulMitra.

Gabi Mocatta (PhD, University of Tasmania; MA, University of Melbourne) is lecturer in journalism, media, and communications at the University of Tasmania. Her research interests are particularly concerned with communicating the environment. Her doctoral research examined environmental communication and protest in Latin America. Gabi is especially interested in the relationship between environmental harms and social inequality, and ways in which this plays out in the media. She also works on media discourse around energy transitions, climate change, and water justice in different national contexts.

Christopher J. Oldenburg (PhD, University of Memphis; MA, University of Dayton) is associate professor of communication and rhetorical studies at Illinois College. He has research interests in the rhetoric of the Catholic Church, Pope Francis, social and environmental justice, and American political discourse. Oldenburg's work has appeared in the *KB Journal* (Kenneth Burke Journal), *Communication Studies, Southern Communication Journal, Presidential Studies Quarterly*, and *Advances in the History of Rhetoric*. He is author of *The Rhetoric of Pope Francis: Critical Mercy and Conversion for the Twenty-First Century*, recipient of the 2019 Religious Communication Association Book of the Year award. He is currently researching papal rhetoric on Catholic social teaching and immigration.

Mark Pedelty (PhD, MA, University of California, Berkeley) is a professor of communication studies and anthropology at the University of Minnesota and fellow at the Institute on the Environment. His most recent books include *Ecomusicology: Rock, Folk and the Environment* (Temple University Press, 2012) and *A Song to Save the Salish Sea: Musical Performance as Environ-*

mental Activism (Indiana University Press, 2016). He has conducted ethnographic field research in El Salvador, Mexico, British Columbia, and Washington State. Pedelty also produces, directs, and composes music for Ecosong.net, a talented team of musicians, dancers, artists, and videographers that make music videos for environmental organizations. Dr. Pedelty teaches courses in environmental communication, research methods, and music.

Aimee Kendall Roundtree (PhD, University of Texas at Austin) is associate dean for research and promotion in the College of Liberal Arts at Texas State University. She teaches courses in usability research, medical and science writing, digital media and technology, web publishing and accessibility, visual design, and technical communication. She has helped with social media campaigns for organizations including the Association of Teachers of Technical Writing and *Connexions* journal. She also serves as a medical writer and qualitative researcher in the Texas Medical Center, where she has participated on research teams at the University of Texas MD Anderson Cancer Center, Baylor College of Medicine, and the Houston Center for Quality of Care and Utilization Studies. She worked as a communications specialist for the Texas Medical Foundation and public relations officer for Air Combat Command at Langley Air Force Base.

Casey R. Schmitt (PhD, University of Wisconsin–Madison; MA, University of Wisconsin–Madison; MA, University of Oregon) is assistant professor of communication studies at Gonzaga University in Spokane, Washington. His research explores environmental narrative, ecological debate, and ethnographic fieldwork, especially in stories of wilderness and natural spaces. His secondary research includes studies on the social import of American folk narrative, representations of indigenous Americans, and popular culture. He is author of more than two dozen peer-reviewed articles and chapters, including essays in the *Western Journal of Communication*, *Environmental Communication*, and *Cultural Analysis*, and is coeditor of *Standing Up, Speaking Out: Stand-Up Comedy and the Rhetoric of Social Change* (Routledge, 2017). He has also served on the editorial staff of multiple academic journals, including *New Directions in Folklore*, the *Oral History Review*, and *Western Folklore*.

Christopher S. Thomas (PhD, University of Iowa; MA, Wake Forest University) is assistant professor in the Department of Communication at the College at Brockport in Brockport, New York. His research explores community-based environmental communication, social change and protests around sustainability, and the role of citizens in the environmental decision-making process. His secondary research focuses on the gendered dynamics of environmental communication, particularly the means by which women organize and lead grassroots campaigns and movements for environmental justice. His research has appeared in the *Southern Communication Journal* and the *Iowa Journal of Communication*, as well as the *Routledge Handbook of the Rhetoric of Social Movements* (forthcoming).

www.ingramcontent.com/pod-product-compliance
Lightning Source LLC
LaVergne TN
LVHW052029200926
848481LV00005B/1626